Algorithms and Combinatorics 4

Bernhard Korte   László Lovász
Rainer Schrader

Greedoids

With 17 Figures

Springer-Verlag Berlin
Heidelberg GmbH

Bernhard Korte
Rainer Schrader
Research Institute for Discrete Mathematics/
Institute of Operations Research
University of Bonn
Nassestraße 2
W-5300 Bonn
Fed. Rep. of Germany

László Lovász
Department of Computer Science
Eötvös Loránd University
Múzeum krt. 6–8
H-1088 Budapest
Hungary

Mathematics Subject Classification (1980):
05 B 35, 05 C 65, 06 A 10, 90 C 10, 90 C 35

ISBN 978-3-642-63499-4   ISBN 978-3-642-58191-5 (eBook)
DOI 10.1007/978-3-642-58191-5

© Springer-Verlag Berlin Heidelberg 1991

Originally published by Springer-Verlag Berlin Heidelberg New York in 1991
Softcover reprint of the hardcover 1st edition 1991

Typesetting: Data conversion by Springer-Verlag and typesetting output by Universitäts-
druckerei H. Stürtz AG, Würzburg
41/3140-543210 – Printed on acid-free paper

Preface

Oh cieca cupidigia, oh ira folle,
Che si ci sproni nella vita corta,
E nell' eterna poi si mal c'immolle!

O blind *greediness* and foolish rage,
That in our fleeting life so goads us on
And plunges us in boiling blood for ever!

Dante, *The Divine Comedy*
Inferno, XII, 17, 49/51.

On an afternoon hike during the second Oberwolfach conference on Mathematical Programming in January 1981, two of the authors of this book discussed a paper by another two of the authors (Korte and Schrader [1981]) on approximation schemes for optimization problems over independence systems and matroids. They had noticed that in many proofs the hereditary property of independence systems and matroids is not needed: it is not required that every subset of a feasible set is again feasible. A much weaker property is sufficient, namely that every feasible set of cardinality k contains (at least) one feasible subset of cardinality $k - 1$. We called this property accessibility, and that was the starting point of our investigations on greedoids.

Very soon we had to find a name for those combinatorial structures which enjoy, besides the classical matroid exchange property, only this accessibility property. We coined the name *greedoid*, as a synthetic blending of the words *greedy* and matr*oid*. We believed that one could live with this artefact, in full knowledge of the fact that the word matroid, derived from the word matrix, has been considered cacophonic by some colleagues. However, later a distinguished colleague and friend from North America stated that only a Hungarian and a German could misuse the English language that much to come up with such a name. We have learned in the meantime that the name we invented evokes the image of extraterrestrial beings rather than that of mathematical objects. However, we hope that this will not affect their further study.

Since greedoids came into life in 1981, many other scholars as well as the authors themselves have developed the subject further. Especially during the academic year 1984/85, while one of the authors was at the Bonn Institute as an Alexander von Humboldt fellow, efforts were devoted to this topic. Already then the idea of writing a book on this theme came up.

This plan benefited greatly from the research project "Algorithmic Principles of Combinatorial Optimization" which has been jointly sponsored by the Hungarian Academy of Sciences (Magyar Tudományos Akadémia) and the German Research Association (Deutsche Forschungsgemeinschaft) for the last nine years. We express our sincere thanks to the Alexander von Humboldt Foundation, the Hungarian Academy of Sciences and the German Research Association.

We are grateful to many colleagues for stimulating discussions on this topic, in particular to Anders Björner, Tom Brylawski, Ulrich Faigle, András Frank and Paul Seymour.

Our special thanks are due to Dirkje Keiper, Claudia Krupp and Beatrix Schaschek for typing and word processing and to Rabe von Randow for editorial help. Last but not least we appreciate the most efficient cooperation of the Springer-Verlag.

This book and the preface was completed, much later than intended, at the 6th Oberwolfach conference on Mathematical Programming in January 1990.

Bernhard Korte
László Lovász
Rainer Schrader

Table of Contents

Chapter I. Introduction

With the advent of computers, algorithmic principles play an ever increasing role in mathematics. Algorithms have to exploit the structure of the underlying mathematical object, and properties exploited by algorithms are often closely tied to classical structural analysis in mathematics. For example, the convexity of a function is essential for the efficiency of an algorithm minimizing it, or the Chinese remainder theorem is fundamental for modular arithmetic.

This connection between algorithms and structure is particularly apparent in discrete mathematics, where proofs are often constructive and can be turned into algorithms more directly.

Matroids are perhaps the best known examples of this paradigm. As a combinatorial structure they were invented by Whitney [1935], who axiomatized the combinatorial properties of linear dependence and circuits in a graph. He used this to describe the combinatorial structure of the duality of planar graphs and related it to the orthogonality of linear subspaces. The same axioms were introduced by van der Waerden [1937] as a common abstraction of linear independence in vector spaces and algebraic independence in fields. Birkhoff [1935] arrived at essentially the same notion from lattice theory, introducing geometric (i.e. atomic and semimodular) lattices. Matroids were also related to projective geometries (MacLane [1936]) and to transversals (Rado [1942]). Since then, many other mathematical structures have been studied with the help of an underlying matroid. Modern matroid theory started with the work of Tutte [1959], who established deep connections between matroid theory, graph theory and matrix algebra.

However, to the best of our knowledge, the first occurrence of a structure equivalent to matroids dates back to a paper of Borůvka [1926]. He arrived at this structure from an algorithmic point of view. He was faced with the real-world problem of constructing an optimal network for electricity in Moravia. He modelled it as a spanning tree problem, solved it by using a simple local selection rule, and proved the optimality of the solution obtained. His algorithmic principle is now called the greedy algorithm. He formulated the general combinatorial exchange properties on which the optimality proof was based. This algorithmic characterization of a structure turned out to be equivalent to the aforementioned non-algorithmic definitions of matroids (Rado [1957], Edmonds [1971]). See Graham and Hell [1985] for more on the history of the greedy algorithm.

Matroids also provide a framework for very general formulations of other, more sophisticated algorithmic principles. For example, the matroid intersection problem is one of the most general situations in which the method of alternating paths can be applied. In turn, the analysis of the matroid intersection algorithm leads to a variety of structural results implying many of the classical min-max theorems in combinatorics.

Min-max results are examples par excellence for the close relationship between algorithmic principles and mathematical structures. They provide direct stopping (optimality) criteria for algorithms. In addition, they give rise to polyhedral characterizations which can then be exploited, e.g. by cutting plane algorithms.

We have seen that matroids can be defined, and were in fact first defined, by the behavior of an algorithm. Whilst this way of defining mathematical structures has not been too common (the only other examples that come to mind are Euclidean rings), we feel that it will become increasingly important in the near future.

Since the concept of matroids has turned out to be so rewarding, many generalizations have been proposed. Amongst these, polymatroids and submodular functions have proved to be the most successful (they are at the same time quite closely related to matroids). Submodularity is a central property of matroid rank functions, which seems to play the same role in discrete mathematics as convex or concave functions do in the continuous case (cf. Lovász [1983]). Other extensions of the concept of matroids, combining the exchange properties with a partial order, were introduced by Dunstan, Ingleton and Welsh [1972] (supermatroids) and by Faigle [1980] (ordered geometries).

A notion reflecting not only linear dependence but also the ordered structure of the underlying field is the notion of oriented matroids, introduced by Las Vergnas [1975], Bland [1977], and Folkman and Lawrence [1978]. This concept has proved particularly successful in optimization, especially by providing a combinatorial description of the simplex algorithm.

Besides linear independence, a similarly important concept in vector spaces is convexity. It is therefore natural to ask for an analogue of matroids providing a combinatorial abstraction of convexity. Such an abstraction indeed exists in the concept of antimatroids. These were introduced by Jamison [1970] and Edelman [1980].

Antimatroids (as their name suggests) are in many respects "antipodal" to matroids. For example, matroids can be defined by a closure operator ("linear hull") which has the Steinitz-MacLane exchange property:

> If x and y are two points and A is a set such that x is not in the linear hull of A but is in the linear hull of $A \cup \{y\}$, then y is in the linear hull of $A \cup \{x\}$.

Antimatroids, on the other hand, can be defined by a closure operator ("convex hull") which obeys an antisymmetric version of the Steinitz-MacLane exchange property:

> If x and y are two points and A is a set such that x is not in the convex hull of A but is in the convex hull of $A \cup \{y\}$, then y is *not* in the convex hull of $A \cup \{x\}$.

One can arrive at antimatroids from different directions. Already in 1940 Dilworth considered a special class of semimodular lattices that are cryptomorphic with antimatroids. In particular, the ideals of a partially ordered set form a distributive lattice, and thereby an antimatroid. Korte and Lovász [1984a and 1984b] introduced an equivalent notion, motivated in part by scheduling problems with alternative precedence constraints and by point-by-point decomposition procedures (shellings). An antimatroid can be described as a scheduling rule: a job can be scheduled if at least one of its alternative precedences has been processed. Boyd and Faigle [1990] characterized antimatroids by the optimality of a greedy scheduling algorithm for a natural class of objective functions.

This latter result also shows that the application of the greedy principle is not restricted to matroids. Since the greedy algorithm selects elements one by one, the natural environment in which it works is not a collection of sets but a collection of ordered sets (i.e. languages).

A greedoid is defined as a collection of ordered sets satisfying an ordered version of the matroid axioms. Thus they generalize matroids, but – somewhat surprisingly – they also generalize antimatroids. They can also be defined as collections of *unordered* sets satisfying the exchange axiom of matroids, but it is not assumed that every subset of a feasible set is feasible. For algorithmic purposes it suffices that every feasible set can be reached from the empty set through feasible sets by single-element augmentations, and this much is indeed retained for greedoids.

Greedoids are not only a common generalization of matroids and antimatroids, but they cover a variety of other combinatorial structures. Many of these are related to algorithms and decomposition procedures. Thus there are natural greedoids associated with Gaussian elimination, matching algorithms, ear-decompositions of graphs, dismantling and retracting of posets, perfect elimination schemes etc. The procedure of searching a graph gives rise to several greedoids; one of these, the branching greedoid, is an ordered analogue of graphic matroids. Of the structures mentioned earlier, the ordered geometries of Faigle [1980] and many of the supermatroids of Dunstan, Ingleton and Welsh [1972] are also greedoids.

Greedoids can also be characterized by the optimality of a greedy procedure for the class of generalized bottleneck functions. Unfortunately this class does not include linear objective functions, and indeed linear objective functions cannot be optimized over greedoids by the greedy algorithm. Optimizing a linear objective function is already NP-hard over very special classes of greedoids.

Since polyhedral descriptions of combinatorial structures reflect the behaviour of linear objective functions, we cannot expect that the powerful polyhedral theory of matroids extends to greedoids. However, algorithms optimizing linear objective functions and polyhedral descriptions can be developed for certain subclasses of greedoids.

For the theory of both matroids and antimatroids, lattice theory was one of the key sources. So it is not surprising that lattices play an important role in the study of wider classes of greedoids. In fact, a class of greedoids called interval greedoids (which play a special role in many other respects as well) turns out to be very closely related to the class of semimodular lattices.

This book is organized as follows. We review some basic concepts of matroid theory in Chapter II. This is done only to the extent that is needed to introduce terminology and motivation for the subsequent chapters. For a detailed treatment of matroids, the reader is referred to Welsh [1976], von Randow [1975] and Recski [1989].

Chapter III gives a comprehensive study of antimatroids. Chapter IV introduces greedoids and gives a large variety of examples.

In Chapters V and VI, we extend the basic toolbox of matroid theory to greedoids. Many notions of matroid theory (rank, closure, minors, flats, connectivity etc.) have meaningful (though sometimes limited) extensions to greedoids. Rank and closure also give cryptomorphic axiomatizations of greedoids. Furthermore, the important subclass of interval greedoids mentioned above is characterized by the nice behavior of these notions.

The next four chapters deal with special classes of greedoids. The first three of these are motivated by the general question: which greedoids can be obtained from matroids and antimatroids by certain construction principles? The simplest such construction is intersection, but more sophisticated procedures yield surprisingly large classes of greedoids. For example, distributive supermatroids arise as intersections of matroids with poset antimatroids, while ordered geometries arise by a trickier operation called "meet".

The class of transposition greedoids treated in Chapter X captures a simple common feature of many decomposition procedures like series-parallel decomposition, dismantling and retracting. Among other things, this yields a proof that these procedures give rise to greedoids, which is often a non-obvious fact by itself.

Chapter XI deals with optimization in greedoids, which was a main motivation for these investigations, as discussed above. In the final Chapter XII we explore some connections between greedoids and topology. The structure in combinatorics called an independence system, is called in topology a simplicial complex. While combinatorialists and topologists are interested in different aspects of one and the same structure, their results often fertilize each other. In particular, the study of topological properties like contractibility of simplicial complexes derived from graphs and matroids has been a powerful tool in the solution of various problems. It turns out that generalizing these constructions to greedoids unifies and also simplifies several of these results. The topological property of shellability makes it possible to generalize (at least partially) the algebraic machinery of Tutte polynomials and the Tutte-Grothendieck decomposition from matroids to greedoids.

For the convenience of the reader, an inclusion chart of subclasses of greedoids treated in this book is shown on the inside back cover. An arrow indicates proper inclusion: A → B denotes A ⊂ B. For a discussion of this chart we refer to Korte and Lovász [1985c, 1985d] and Björner and Ziegler [1987].

1. Set Systems and Languages

A **set system** over a finite ground set E is a pair $(E, \mathcal{F})$, where $\mathcal{F} \subseteq 2^E$ is a collection of subsets. We will always assume that the ground sets we are dealing with are nonempty and finite. Given a set system $(E, \mathcal{F})$, we will call the sets in $\mathcal{F}$ **feasible** and those not in $\mathcal{F}$ **infeasible**.

For a finite ground set E we denote by E^* the set of all sequences $x_1 x_2 \ldots x_k$ of elements $x_i \in E$ for $1 \leq i \leq k$. The set E is sometimes called an **alphabet**, its elements **letters** and the elements of E^* **words** or **strings**. A collection of words $\mathcal{L} \subseteq E^*$ is called a **language** over the alphabet E, its elements $\alpha \in \mathcal{L}$ will be denoted by small Greek letters. The symbol $\emptyset$ will also be used to denote the empty string. The number of letters in a word α is the **length** $|\alpha|$ of α. A subsequence of a given word α is called a **substring** of α.

The **concatenation** $\alpha\beta$ of two words α and β is the string α followed by the string β. α is then called a **prefix** of $\alpha\beta$.

Similarly, αx denotes the concatenation of the word α and the letter x.

The underlying set of letters of a word α is denoted by $\tilde{\alpha}$. $\tilde{\mathcal{L}}$ is the collection of underlying sets of the language $\mathcal{L}$. The notation $x \in \alpha$ stands for $x \in \tilde{\alpha}$.

We will assume throughout that the language $\mathcal{L}$ under consideration is **simple**, i.e. no letter is repeated in any word. A letter that does not occur in any word of $\mathcal{L}$ is called a **loop**. A language is called **normal** if it does not contain loops.

A language $(E, \mathcal{L})$ is **hereditary** if it satisfies the following axioms:

(1.1) $\emptyset \in \mathcal{L}$,
(1.2) $\alpha\beta \in \mathcal{L}$ implies $\alpha \in \mathcal{L}$.

Given a simple hereditary language $(E, \mathcal{L})$ we may associate with it a set system $(E, \mathcal{F}) = (E, \mathcal{F}(\mathcal{L}))$ by setting $\mathcal{F}(\mathcal{L}) = \tilde{\mathcal{L}}$. This system is **accessible**, i.e. has the following properties:

(1.3) $\emptyset \in \mathcal{F}$,
(1.4) for every nonempty set $X \in \mathcal{F}$ there exists some $x \in X$ such that $X \backslash x \in \mathcal{F}$.

Conversely, for any given accessible set system $(E, \mathcal{F})$ we may define a simple hereditary language $(E, \mathcal{L}) = (E, \mathcal{L}(\mathcal{F}))$ by

$$\mathcal{L}(\mathcal{F}) = \{x_1 \ldots x_k : \{x_1, \ldots, x_i\} \in \mathcal{F} \text{ for } 1 \leq i \leq k\} .$$

Hereditary languages are "richer" than accessible set systems in the following sense. If $\mathcal{F} \neq \mathcal{F}'$ are two different accessible set systems then $\mathcal{L}(\mathcal{F}) \neq \mathcal{L}(\mathcal{F}')$. On the other hand, there may be different languages with the same associated set system. To see this, consider $\mathcal{L} = \{x, y, xy, yz, xyz, yzx\}$ and $\mathcal{L}' = \mathcal{L} \cup \{yx, yxz\}$. For these, $\mathcal{F}(\mathcal{L}) = \mathcal{F}(\mathcal{L}')$.

Lemma 1.1. *Let $(E, \mathcal{L})$ be a hereditary language. Then $\mathcal{L} = \mathcal{L}(\mathcal{F}(\mathcal{L}))$ if and only if the following property is satisfied:*

(1.5) *If $\alpha, \beta \in \mathcal{L}$ with $\tilde{\alpha} = \tilde{\beta} \cup \{x\}$ then $\beta x \in \mathcal{L}$.*

Proof. Assume $\mathcal{L} = \mathcal{L}(\mathcal{F}(\mathcal{L}))$ and let $\alpha, \beta \in \mathcal{L}$ with $\tilde{\alpha} = \tilde{\beta} \cup \{x\}$. Then $\beta x \in \mathcal{L}$, since $\beta \in \mathcal{L}$ and $\tilde{\beta} \cup \{x\} = \tilde{\alpha} \in \mathcal{F}(\mathcal{L})$.

Conversely, assume that $(E, \mathcal{L})$ is a hereditary language with property (1.5). Clearly $\mathcal{L} \subseteq \mathcal{L}(\mathcal{F}(\mathcal{L}))$. Consider some $x_1 \ldots x_k \in \mathcal{L}(\mathcal{F}(\mathcal{L}))$ and let i be the largest index such that $\beta = x_1 \ldots x_i \in \mathcal{L}$. If $i < k$, then $x_1 \ldots x_{i+1} \in \mathcal{L}(\mathcal{F}(\mathcal{L}))$ and $\tilde{\alpha} = \{x_1, \ldots x_{i+1}\} \in \mathcal{F}(\mathcal{L})$. Hence $\alpha \in \mathcal{L}$ for some ordering of the elements in $\tilde{\alpha}$. By (1.5) $\beta x_{i+1} = x_1 \ldots x_i x_{i+1} \in \mathcal{L}$, contradicting the choice of i.   □

For a given set system $(E, \mathcal{F})$ and $A \subseteq E$, a subset $C \subseteq A, C \in \mathcal{F}$ is called a **basis** of A if $C \cup x \notin \mathcal{F}$ for all $x \in A \setminus C$. For a language $(E, \mathcal{L})$ and $A \subseteq E$, a word $\alpha \in \mathcal{L}$ with $\tilde{\alpha} \subseteq A$ is called a **basic word** of A if $\alpha x \notin \mathcal{L}$ for all $x \in A$. Bases or basic words of E will be called **bases** of the set system or **basic words** of the language.

A **clutter** is a collection of sets such that no member of the collection contains another. For a given clutter $\mathcal{K}$, the **blocker** $\mathcal{B}(\mathcal{K})$ is defined as the collection of minimal sets that have a nonempty intersection with each element in $\mathcal{K}$. It is well-known (cf., e.g Edmonds and Fulkerson [1970]) that the blocker of a clutter is again a clutter. Moreover, the blocking operation is **involutory**, i.e.

$$\mathcal{B}(\mathcal{B}(\mathcal{K})) = \mathcal{K}.$$

2. Graphs, Partially Ordered Sets and Lattices

In order for the book to be self-contained, we review some notions and definitions from graph theory and partially ordered sets.

A **graph** $G = (V, E)$ consists of a nonempty finite set V of **nodes** or **vertices** and a multiset E of unordered pairs $e = [i, j]$ of elements of V called **edges**. The nodes i and j are the **endnodes** of e and are called **adjacent** or **neighbors**. The set of neighbors of a node v is denoted by $N(v)$ and we set $\overline{N}(v) = N(v) \cup v$. An edge e is **parallel** to an edge e' if e and e' have the same endnodes.

A graph is called **complete** if any two vertices are adjacent. The complete graph on n nodes is denoted by K_n. A subset S of vertices is a **clique** in G if every pair of nodes in S is connected by an edge. A graph $G = (V, E)$ is **bipartite** if its vertex set can be partitioned into two nonempty disjoint subsets V_1 and V_2 such that no edge has both endpoints in the same subset V_i. A special bipartite graph is $K_{n,m}$ where $|V_1| = n, |V_2| = m$ and every vertex in V_1 is connected to every vertex in V_2. A **matching** in a graph is a subset of edges no two of which have an endpoint in common. A **perfect matching** is a matching which covers every node in the graph.

An alternating sequence $v_1 e_1 \ldots v_k e_k v_{k+1}$ of vertices and edges is a **path** if $v_i \neq v_j$ for $i \neq j, e_{i-1} \cap e_i = v_i$ for $2 \leq i \leq k$, v_1 is an endnode of e_1 and v_{k+1} is an endnode of e_k. The number of edges in a path is called the **length** of the path. An alternating sequence of vertices and edges is a **cycle** if $v_1 e_1 \ldots v_k$ is a path, $v_{k+1} = v_1$ and $e_k = [v_1, v_k]$. A **chord** in a cycle is an edge linking two nonconsecutive nodes of the cycle.

A graph is **connected** if any two nodes are joined by a path. A connected graph with at least three nodes is **2-connected** if the removal of one arbitrary node still leaves the graph connected. A **block** is a 2-connected component. A connected graph without cycles is called a **tree**. A **branching rooted at r** is a tree $T = (V, E)$ together with a specified vertex $r \in V$. A **spanning tree** of a graph G is a subgraph of G that is a tree and contains all vertices of G.

A **directed graph** $D = (V, A)$ consists of a finite nonempty set V of **nodes** or **vertices** and of a multiset A of **arcs** which are now ordered pairs of elements of V. An arc $e = (u, v)$ has u as its **initial node (tail)** and v as its **terminal node (head)**. A **directed path** in a graph $G = (V, E)$ is an alternating sequence $v_1 e_1 \ldots v_k e_k v_{k+1}$ of vertices and arcs so that $v_i \neq v_j$ for $i \neq j$, and $e_i = (v_i, v_{i+1})$ for $1 \leq i \leq k$. A directed graph is **strongly connected** if any two nodes are joined by a directed path.

A **directed branching** B in a directed graph is a tree in the underlying undirected graph which has a node, called its **root**, from which every other node can be reached by a directed path.

A **partially ordered set (poset)** $(E, \leq)$ is a finite set E endowed with a reflexive, antisymmetric and transitive relation. If we do not have antisymmetry, then $P = (E, \leq)$ is called a **preordered set**. If $x \leq y$ and $y \nleq x$, then we write $x < y$.

Given a poset $P = (E, \leq)$, an element $x \in E$ is **minimal** if $y \leq x$ implies $y = x$ and **maximal** if $x \leq y$ implies $y = x$. The element x is an **upper neighbor** of an element $y \neq x$ if $y \leq x$ and $y \leq z \leq x$ implies $z = x$ or $z = y$. We then also say that x **covers** y and y is a **lower neighbor** of x. An **ideal** of a poset is a subset $I \subseteq E$ such that $y \in I, x \leq y$ implies $x \in I$. Dually, a **filter (upper ideal)** F is a subset such that $y \in F, y \leq x$ implies $x \in F$. Let $I(x) = \{y \in E : y \leq x\}$. This ideal is called the **principal ideal** generated by x. **Principal filters** are defined analogously. The intersection of a principal ideal $I(y)$ with a principal filter $F(x)$ is an **interval** $[x, y]$, i.e.

$$[x, y] = \{z : x \leq z \leq y\}.$$

Two elements $x, y \in E$ are **comparable** if either $x \leq y$ or $y \leq x$, and **incomparable** otherwise. A subset $C \subseteq E$ such that any two elements of C are comparable is a **chain**. The **length** of the chain is $|C| - 1$. A poset that is a single chain is called **totally ordered**. An **antichain** is a subset $A \subseteq E$ any two elements of which are incomparable. A partial order is called a **forest order** if no element covers more than one other element. The **width** $w(P)$ of a poset is the maximum size of an antichain. The following Theorem of Dilworth [1950] relates width to chain covers.

Theorem 2.1. *The minimum number of chains needed to cover the ground set equals the maximum size of an antichain.*   □

A **linear extension** of a poset is a permutation of the ground set such that $x_i < x_j$ in the poset implies that x_i precedes x_j in the permutation. We also say that the total order defined by this linear extension is **compatible** with the poset. The **intersection** of two posets on the same underlying set is the partial order with

$x_i \leq x_j$ if and only if this holds in both posets. The **order dimension** of a partial order is the minimum number of total orders whose intersection is the poset.

We will visualize posets $(E, \leq)$ with the help of their **Hasse diagrams**. This is a directed graph with node set E and arcs from each element to the elements covering it. In drawing it we will always omit the arrows using the convention that the arcs are oriented upwards. For forest orders, the Hasse diagram is a forest with exactly one minimal element in each component. Finally, the **comparability graph** is the undirected graph with node set E and an edge between comparable elements.

We say that an element u is the **meet** of x and y if for every z we have that $z \leq u$ if and only if $z \leq x$ and $z \leq y$. An element w is the **join** of x and y if for every z we have that $z \geq w$ if and only if $z \geq x$ and $z \geq y$.

A poset is a **lattice** if any two elements $x, y \in E$ have a meet $x \wedge y$ and a join $x \vee y$. If only the meets (joins) exist, the poset is called a **meet- (join)-semilattice**. A finite lattice necessarily has a unique minimal element which we will frequently denote by 0, and a unique maximal element denoted by 1. Elements covering 0 are called **atoms**. The **height** $h(x)$ of an element x is the length of the longest chain from 0 to x.

It is easy to see that a finite meet-semilattice with a unique maximal element is a lattice. To any meet-semilattice we can formally add a 1 and thereby turn it into a lattice.

An element x in a lattice is called **meet-irreducible (join-irreducible)** if $x = y \wedge z$ $(x = y \vee z)$ implies $y = x$ or $z = x$. Two elements $x, y \in L$ are **complements** if $x \vee y = 1$ and $x \wedge y = 0$.

A lattice L is **distributive** if for any $x, y, z \in L : (x \wedge y) \vee (y \wedge z) \vee (x \wedge z) = (x \vee y) \wedge (y \vee z) \wedge (x \vee z)$. The fundamental result of Birkhoff says that every distributive lattice is isomorphic to the lattice of ideals of some poset.

A lattice L is **semimodular** if whenever $x, y \in L$ both cover $x \wedge y$, $x \vee y$ covers x and y. It is well-known that every distributive lattice is also semimodular. Semimodular lattices are in particular **graded** posets, i.e. they have a unique minimal and a unique maximal element and for every element x any two maximal chains in $[0, x]$ have the same length **(Jordan-Dedekind chain condition)**. A lattice L is semimodular if and only if it has the Jordan-Dedekind chain property and a submodular height function, i.e.

$$h(x) + h(y) \geq h(x \wedge y) + h(x \vee y)$$

for all $x, y \in L$. This in particular implies the following version of the defining property: if x covers $x \wedge y$, then $x \vee y$ covers y.

A **geometric lattice** is a semimodular lattice in which every element is a join of atoms. A lattice is called **locally free** if for any $x \in L$ the interval $[x, y]$ is a Boolean algebra, where y is the join of all elements covering x.

Chapter II. Abstract Linear Dependence – Matroids

Matroids were introduced by Whitney [1935] as a combined abstraction of linear independence and the cycle structure of graphs. However, they were already considered by Borůvka [1926] from a different point of view and characterized by a basic algorithmic property. Later, but independently, they were also rediscovered by van der Waerden [1937] as a combination of linear and algebraic independence. Modern matroid theory started with the fundamental work of Tutte [1959].

By today, matroids are among the most intensively studied objects in combinatorics. They play an important role in different fields such as optimization, enumeration, algebraic combinatorics, applications to electrical networks and statics etc. Since their structural and algorithmic behavior also laid the ground for the theory of greedoids, we briefly review their definition, basic properties and some characteristic results in Section 1. In Section 2 we sketch some fundamental results on matroids and optimization, while Section 4 introduces submodularity and some generalizations of matroids. For a detailed treatment of matroids and their applications see, e.g., Welsh [1976], von Randow [1975] and Recski [1989].

1. Matroid Axiomatizations

A **matroid** is a set system $(E, \mathcal{M})$ with the following properties:

(1.1) $\emptyset \in \mathcal{M}$,

(1.2)   if $X \in \mathcal{M}$ and $Y \subseteq X$, then $Y \in \mathcal{M}$,

(1.3)   for $X, Y \in \mathcal{M}$ with $|X| = |Y| + 1$ there exists an $x \in X \setminus Y$ such that $Y \cup x \in \mathcal{M}$.

A set system satisfying only (1.1) and (1.2) is called an **independence system**. (Note that independence systems are accessible.)

It will help to understand the matroid concept and the definitions that follow if we give two basic examples of matroids. First, let E be a finite set of vectors in a linear space and let $\mathcal{M}$ consist of the linearly independent subsets. We call $(E, \mathcal{M})$ a **linear matroid**. Second, let E be the set of edges of a graph G and let $\mathcal{M} = \mathcal{M}(G)$ consist of those subsets that contain no circuit. We call $(E, \mathcal{M})$ a **graphic matroid**.

The **uniform matroid** $U_{k,n}$ contains all subsets of size at most k of an n-element set. $U_{n,n}$ is called the **free matroid**.

As a consequence of (1.3), all bases of any $A \subseteq E$ must have the same cardinality. This cardinality is called the **rank** $r(A)$ of A. Clearly, $X \in \mathcal{M}$ if and only if $r(X) = |X|$. We say that a maximal feasible subset of E is a **basis** and denote by $\mathscr{B}$ the set system of bases.

The **closure operator** is the mapping $\sigma : 2^E \to 2^E$, which assigns to a set X the set of all elements $y \in E$ which do not increase the rank, i.e.

$$\sigma(X) = \{y \in E : r(X \cup y) = r(X)\}.$$

Sets with $\sigma(X) = X$ are called **flats** or **closed sets**. **Circuits** are minimal sets not in $\mathscr{M}$. Clearly this definition originates from graph theory, while the others are more algebraically motivated.

We proceed by giving various axiomatizations of matroids in terms of the previously defined concepts.

Theorem 1.1. *A non-empty collection $\mathscr{B}$ of subsets of E is the set of bases of a matroid $(E, \mathscr{M})$ if and only if it satisfies the following condition:*

(1.4)   *For $B_1, B_2 \in \mathscr{B}$ and $x \in B_1 \setminus B_2$ there exists a $y \in B_2 \setminus B_1$ such that*
$B_1 \cup y \setminus x \in \mathscr{B}$.

Proof. Let $\mathscr{B}$ be the set of bases of some matroid $(E, \mathscr{M})$. Since all bases have the same cardinality and $B_1 \setminus x \in \mathscr{M}$ we can find a $y \in B_2 \setminus B_1$ such that $B_1 \cup y \setminus x \in \mathscr{B}$.

Conversely, let $\mathscr{B}$ be a set system satisfying (1.4). Then all members of $\mathscr{B}$ must have the same cardinality. For if $B_1, B_2 \in \mathscr{B}$ with $|B_1| < |B_2|$, we can successively exchange the elements in $B_1 \setminus B_2$ with elements from $B_2 \setminus B_1$. We arrive at a set $B_3 \in \mathscr{B}$, where B_3 is a proper subset of B_2. Clearly B_3 and B_2 violate condition (1.4).

Now let $\mathscr{M}$ be the collection of all subsets of members of $\mathscr{B}$. Clearly, $(E, \mathscr{M})$ satisfies (1.1) and (1.2). To verify the augmentation property, consider $X \subseteq B_1$ and $Y \subseteq B_2$ with $|X| = |Y| + 1$.

Again we use property (1.4) to successively exchange elements from $(B_2 \setminus B_1) \setminus Y$ with elements from $(B_1 \setminus B_2) \setminus X$. Since, by assumption, $|(B_1 \setminus B_2) \setminus X| < |(B_2 \setminus B_1) \setminus Y|$, we must arrive at a situation where we replace some $b \in (B_2 \setminus B_1) \setminus Y$ by some $x \in X$. Then $Y \cup x$ is contained in a basis, i.e. the augmentation property holds.   □

We also have the following characterization of a matroid:

Theorem 1.2. *A set system $(E, \mathscr{M})$ is a matroid if and only if it satisfies conditions (1.1), (1.2) and has the following property:*

(1.5)   *For $A \subseteq E$ all maximal feasible subsets of A have the same cardinality.*

Proof. If $(E, \mathscr{M})$ is a matroid, then the augmentation property (1.3) clearly implies (1.5). Conversely, if $(E, \mathscr{M})$ satisfies (1.1), (1.2) and (1.5), consider $X, Y \in \mathscr{M}$ with $|X| = |Y| + 1$. Let $A = X \cup Y$. Since all maximal feasible subsets of A have the same cardinality, there exists an element $x \in A$ with $|Y \cup x| = |X|$ and $Y \cup x \in \mathscr{M}$, which proves (1.3).   □

The following two theorems axiomatize matroids in terms of their rank function.

Theorem 1.3. *A function* $r : 2^E \to \mathbb{Z}$ *is the rank function of a matroid if and only if for all sets* $X \subseteq E$ *and elements* $x, y \in E$ *the following holds:*

(1.6) $r(\emptyset) = 0,$

(1.7) $r(X) \leq r(X \cup y) \leq r(X) + 1,$

(1.8) *if* $r(X \cup x) = r(X \cup y) = r(X),$ *then* $r(X \cup x \cup y) = r(X).$

Proof. If r is the rank function of a matroid, the first two conditions are trivially fulfilled. Now suppose $r(X \cup x \cup y) > r(X)$. Then any basis A of X can be augmented to a basis of $X \cup x \cup y$, i.e. $r(X \cup x) > r(X)$ or $r(X \cup y) > r(X)$.

To see the converse, let r satisfy the above three conditions. An inductive argument shows that $r(X) \leq |X|$. Define $\mathcal{M}$ to be the set system given by

$$\mathcal{M} = \{X \subseteq E : r(X) = |X|\}.$$

Then $\emptyset \in \mathcal{M}$. For $X \in \mathcal{M}$ and $Y = X \setminus x$ we have $|X| = r(Y \cup x) \leq r(Y) + 1 \leq |Y| + 1$ and hence $Y \in \mathcal{M}$. Applying this argument inductively, we see that $\mathcal{M}$ is closed under taking subsets. Now suppose that the sets $X, Y \in \mathcal{M}$ with $|X| = |Y| + 1$ violate (1.3). Let $X \setminus Y = \{x_1, \ldots, x_k\}$. Then $r(Y \cup x_i) = r(Y)$ for all $1 \leq i \leq k$. Hence $r(Y \cup x_1 \cup x_i) = r(Y)$. Repeated application of this argument gives $r(Y) = r(Y \cup X \setminus Y) = r(X)$, contradicting the assumption. Thus $(E, \mathcal{M})$ is a matroid and it is easy to see that its rank function is in fact r. $\quad\square$

Condition (1.8) is called "local submodularity" of the rank function. In fact, the rank function is **submodular**, i.e. satifies the inequality

$$r(X \cup Y) + r(X \cap Y) \leq r(X) + r(Y),$$

as the following theorem states.

Theorem 1.4. *A function* $r : 2^E \to \mathbb{Z}$ *is the rank function of a matroid if and only if for all* $X, Y \subseteq E$:

(1.9) $0 \leq r(X) \leq |X|,$

(1.10) $X \subseteq Y$ *implies* $r(X) \leq r(Y),$

(1.11) $r(X \cup Y) + r(X \cap Y) \leq r(X) + r(Y).$

Proof. The first two conditions are obviously fulfilled if r is the rank function of a matroid. Now let A be a basis of $X \cap Y$. By the augmentation property of matroids we can enlarge A to a basis $A \cup B$ of X and to a basis $A \cup B \cup C$ of $X \cup Y$. Then $A \cup C$ is a feasible subset of Y. Hence

$$r(X) + r(Y) \geq |A \cup B| + |A \cup C| = |A \cup B \cup C| + |A| = r(X \cup Y) + r(X \cap Y).$$

Conversely, for a function r satisfying the above conditions, (1.6) and (1.7) clearly hold. Then submodularity implies

$$r(X \cup x \cup y) \leq r(X \cup x) + r(X \cup y) - r(X),$$

i.e. if $r(X \cup x) = r(X \cup y) = r(X)$ we get $r(X \cup x \cup y) = r(X)$.   □

Submodularity seems to be one of the crucial properties of matroids. Before we give a more detailed discussion of submodularity, we continue to axiomatize matroids in terms of their closure and of circuits.

It follows from local submodularity along the lines of the proof of Theorem 1.3 that $r(\sigma(X)) = r(X)$ for every set X. So the closure of X is the unique largest superset of X with the same rank. The next theorem characterizes the closure operator of matroids as a monotone, idempotent hull operator which satisfies the **Steinitz-MacLane exchange property** (1.15) below.

Theorem 1.5. *A function $\sigma : 2^E \to 2^E$ is the closure operator of a matroid if and only if for all $X, Y \subseteq E$ and $x, y \in E$, the following hold:*

(1.12) $X \subseteq \sigma(X),$
(1.13) $X \subseteq Y$ *implies* $\sigma(X) \subseteq \sigma(Y),$
(1.14) $\sigma(X) = \sigma(\sigma(X)),$
(1.15)   *if* $y \notin \sigma(X)$, $y \in \sigma(X \cup x)$, *then* $x \in \sigma(X \cup y).$

Proof. Let σ be the closure operator of a matroid $(E, \mathcal{M})$. Property (1.12) is trivially fulfilled. Let $X \subseteq Y$ and $x \in \sigma(X)$. Then $r(X) + r(Y) = r(X \cup x) + r(Y) \geq r((X \cup x) \cap Y) + r(X \cup x \cup Y) \geq r(X) + r(Y \cup x)$, and hence $x \in \sigma(Y)$. Thus (1.13) holds. Property (1.14) follows from the characterization of closure given above.

To show (1.15), suppose that $y \notin \sigma(X)$, $y \in \sigma(X \cup x)$ but $x \notin \sigma(X \cup y)$. Then $r(X \cup x) = r(X \cup x \cup y) = r(X \cup y) + 1 = r(X) + 2$, contradicting (1.7).

Conversely, let $\sigma : 2^E \to 2^E$ be a function which satisfies (1.12) – (1.15). Define a set system $(E, \mathcal{M})$ by

$$\mathcal{M} = \{X \subseteq E : x \notin \sigma(X \setminus x) \text{ for all } x \in X\}.$$

We claim that $(E, \mathcal{M})$ is a matroid. Clearly, $\emptyset \in \mathcal{M}$. Let $X \in \mathcal{M}$ and suppose $Y \notin \mathcal{M}$ for some $Y \subseteq X$. Then there exists some $y \in Y$ with $y \in \sigma(Y \setminus y)$. Hence by (1.13) $y \in \sigma(X \setminus y)$, a contradiction. Thus $(E, \mathcal{M})$ is an independence system. To prove (1.3), we first prove the following claim:

Claim: If $X \in \mathcal{M}$, $Y \subseteq E$ and $|Y| < |X|$, then $X \not\subseteq \sigma(Y)$.

We prove the claim by induction on $|Y \setminus X|$. Assume $X \subseteq \sigma(Y)$. If $Y \subseteq X$, then this is clearly impossible, so we can choose a $y \in Y \setminus X$. By induction, $X \not\subseteq \sigma(Y \setminus y)$, and so there exists an $x \in X \setminus \sigma(Y \setminus y)$. Then $x \notin \sigma(Y \setminus y)$ but $x \in \sigma((Y \setminus y) \cup y)$, so by (1.15) $y \in \sigma(Y \setminus y \cup x)$. Hence $Y \subseteq \sigma(Y \setminus y \cup x)$, so by (1.13) and (1.14) we get $\sigma(Y) \subseteq \sigma(Y \setminus y \cup x)$. Thus $X \subseteq \sigma(Y \setminus y \cup x)$, which contradicts the induction hypothesis. This proves the claim.

Now assume $X, Y \in \mathcal{M}$ with $|X| = |Y| + 1$. By the claim, there exists an element $x \in X \setminus \sigma(Y)$. We show that $Y \cup x \in \mathcal{M}$. For if not, then there exists a $z \in Y \cup x$ such that $z \in \sigma(Y \cup x \backslash z)$. Clearly $z \neq x$. Since $Y \in \mathcal{M}$, we have $z \notin \sigma(Y \backslash z)$, and hence by (1.15) $x \in \sigma(Y)$, a contradiction.

We leave it to the reader to verify that σ is the closure operator of $(E, \mathcal{M})$. $\quad \square$

Theorem 1.6. *A nonvoid collection $\mathscr{C}$ of nonempty subsets of E is the family of circuits of a matroid on E if and only if the following holds:*

(1.16) $\mathscr{C}$ *is a clutter, i.e. $C_1, C_2 \in \mathscr{C}$ and $C_1 \subseteq C_2$ implies $C_1 = C_2$,*

(1.17)   *For $C_1, C_2 \in \mathscr{C}, C_1 \neq C_2$ and $x \in C_1 \cap C_2$ there exists a $C_3 \in \mathscr{C}$ with*
       $C_3 \subseteq (C_1 \cup C_2) \backslash x.$

Proof. Let $(E, \mathcal{M})$ be a matroid and $\mathscr{C}$ its collection of circuits. Then clearly (1.16) holds. Now let C_1 and C_2 be two circuits with $x \in C_1 \cap C_2$. By the submodularity of the rank function

$$r(C_1 \cup C_2 \backslash x) \leq r(C_1 \cup C_2)$$
$$\leq r(C_1) + r(C_2) - r(C_1 \cap C_2)$$
$$= |C_1| + |C_2| - 2 - |C_1 \cap C_2|$$
$$= |C_1 \cup C_2| - 2 ,$$

i.e. $C_1 \cup C_2 \backslash x$ is not feasible and thus contains a circuit.

Conversely, let $\mathcal{M} = \{X \subseteq E : C \not\subseteq X \text{ for all } C \in \mathscr{C}\}$. Then $(E, \mathcal{M})$ is obviously an independence system. Let $X, Y \in \mathcal{M}$ with $|X| = |Y| + 1$. We show the augmentation property (1.3) by induction on $k = |Y \backslash X|$. The case $k = 0$ is trivial.

Suppose (1.3) does not hold for some sets $X, Y \in \mathcal{M}$ with $|Y \backslash X| \geq 1$. Then for all $x \in X \backslash Y$, $Y \cup x$ contains a set $C_x \in \mathscr{C}$. Since $X \in \mathcal{M}, C_x$ must contain at least one element $y \in Y \backslash X$. Consider $Y' = Y \backslash y \cup x$. Then $Y' \in \mathcal{M}$, for if not, Y' has a subset C_x' containing x with $C_x' \in \mathscr{C}$. Then (1.16) implies that $C_x' \not\subseteq C_x$ and (1.17) yields a set $C \in \mathscr{C}$ with $C \subseteq (C_x \cup C_x') \backslash x \subseteq Y$, contradicting $Y \in \mathcal{M}$.

By induction, Y' can be augmented from X to a set $Z = Y \backslash y \cup \{x, z\} \in \mathcal{M}$. Hence the two sets $C_x, C_z \in \mathscr{C}$ with $C_x \subseteq Y \cup x \notin \mathcal{M}$ and $C_z \subseteq Y \cup z \notin \mathcal{M}$ both contain y. But then we arrive at a contradiction, since (1.17) implies the existence of a set $C \subseteq (C_x \cup C_z) \backslash y \subseteq Z$. $\quad \square$

With a matroid $(E, \mathcal{M})$ we can associate a poset $L(\mathcal{M})$ whose elements are the flats ordered by inclusion. In this poset, a flat X covers a flat Y if and only if $Y \subseteq X$ and $r(X) = r(Y) + 1$. Since $L(\mathcal{M})$ has a unique minimal element $\sigma(\emptyset)$ and a unique maximal element E, the poset rank $r(X)$ of a flat coincides with its matroid rank $r(X)$. Moreover, $L(\mathcal{M})$ satisfies the Jordan-Dedekind chain condition.

Consider two flats X and Y. Then $X \cap Y \subseteq \sigma(X \cap Y) \subseteq \sigma(X) \cap \sigma(Y) = X \cap Y$. Hence flats are closed under intersection. Then $L(\mathcal{M})$ is a lattice with the meet and join operations given by:

$$X \wedge Y = X \cap Y$$
$$X \vee Y = \cap \{A : A \in L(\mathcal{M}), X \cup Y \subseteq A\} = \sigma(X \cup Y).$$

This lattice is semimodular since

$$h(X \vee Y) + h(X \wedge Y) = r(\sigma(X \cup Y)) + r(X \cap Y)$$
$$= r(X \cup Y) + r(X \cap Y)$$
$$\leq r(X) + r(Y)$$
$$= h(X) + h(Y) .$$

We are now ready to prove:

Theorem 1.7. *A lattice L is isomorphic to the lattice of flats of a matroid if and only if it is geometric.*

Proof. For the "only if" direction it remains to prove that the lattice of flats of a matroid is atomic. But this is clear since if Y is a flat, then $Y = \vee\{\sigma(x) : x \in X\}$.

Conversely, let L be a geometric lattice with set E of atoms. Define a set system $(E, \mathcal{M})$ on E by $\mathcal{M} = \{X \subseteq E : h(\vee X) = |X|\}$, where h is the rank function of L. The function $r(X) = h(\vee X)$ is subcardinal and monotone. Moreover, the submodularity of h yields the submodularity of r:

$$r(X) + r(Y) \geq h((\vee X) \vee (\vee Y)) + h((\vee X) \wedge (\vee Y))$$
$$\geq h(\vee(X \cup Y)) + h(\vee(X \cap Y))$$
$$= r(X \cup Y) + r(X \cap Y) .$$

Hence r is a matroid rank function. It is not difficult to see that the lattice of flats of the matroid induced by r is isomorphic to the geometric lattice we started with.   □

2. Matroids and Optimization

Matroids have become well-known in combinatorial optimization because of their close relationship to the greedy algorithm. This algorithm is an abstraction of the algorithm of Kruskal [1956] for finding a minimum weighted spanning tree in a graph.

Let $(E, \mathcal{F})$ be an independence system and $c : E \to \mathbb{R}_+$ a weighting on its ground set. Abbreviating $c(S) = \sum_{e \in S} c(e)$, we want to find a feasible set of maximum weight, i.e.

(2.1) $\max\{c(S) : S \in \mathcal{F}\}$.

The **greedy algorithm** for solving this problem proceeds as follows:

1. Set $X = \emptyset$.
2. Set $T = \{x \in E \setminus X : X \cup x \in \mathcal{F}\}$.
   If $T = \emptyset$, stop.
   If $T \neq \emptyset$, choose $x \in T$ such that $c(x) \geq c(y)$ for all $y \in T$.
3. Set $X = X \cup x$ and go to 2.

Clearly, this algorithm generates a feasible set but for arbitrary independence systems this will not be optimal. For matroids, however, the greedy algorithm always produces an optimal solution.

Theorem 2.1. *Let $(E, \mathcal{M})$ be a matroid and $c : E \rightarrow \mathbb{R}_+$ a weighting. Then the greedy algorithm will generate an optimal solution for the problem (2.1).*

Proof. Suppose not, then let X be the basis generated by the greedy algorithm and choose an optimal basis Y with $c(Y) > c(X)$ and so that $|Y \cap X|$ is maximum. Let $x \in X \setminus Y$ be the first element chosen by the greedy algorithm which is not in Y.

Augment $(X \cap Y) \cup \{x\} \in \mathcal{M}$ from Y to a basis $Y \setminus y \cup x$. Since Y is optimal, $c(y) \geq c(x)$. However, if $c(y) > c(x)$, the greedy algorithm would have chosen y, so $c(y) = c(x)$. Hence, $Y \setminus y \cup x$ is an optimal basis, contradicting the choice of Y. So X must have been optimal. □

The above theorem may be turned into an algorithmic characterization of matroids among independence systems, by also proving that the converse holds.

Theorem 2.2. *An independence system is a matroid if and only if for all weight function $c : E \rightarrow \mathbb{R}_+$ the greedy algorithm generates an optimal solution for problem (2.1).*

Proof. It remains to show that the augmentation property holds. Let $X, Y \in \mathcal{F}$ be feasible sets with $|X| = k$ and $|Y| = k + 1$. Define a weight function c on E by

$$c(e) = \begin{cases} (k+1)/(k+2) & \text{for } e \in X \\ k/(k+1) & \text{for } e \in Y \setminus X \\ 0 & \text{otherwise .} \end{cases}$$

Then the greedy algorithm will build up X first. Since $c(Y) > c(X)$ it must find an element y in $Y \setminus X$ such that $X \cup y \in \mathcal{F}$. □

3. Operations on Matroids

Deletion and contraction of edges of graphs can be generalized to matroids. Let $(E, \mathcal{M})$ be a matroid. For $S \subseteq E$ define the **restriction** of $\mathcal{M}$ to S by

$$\mathcal{M}_S = \{X \in \mathcal{M} : X \subseteq S\} .$$

$(S, \mathcal{M}_S)$ is again a matroid. We say that the restriction of $\mathcal{M}$ to S is obtained by the **deletion** of $T = E \setminus S$, and we also denote it by $(E \setminus T, \mathcal{M} \setminus T)$. The **contraction** of a set $T \subseteq E$ is the set system

$$\mathcal{M}/T = \{X \subseteq E \setminus T : X \cup Y \in \mathcal{M} \text{ for some basis } Y \text{ of } T\} .$$

The contraction $(E \setminus T, \mathcal{M}/T)$ is a matroid.

Given a matroid $(E, \mathcal{M})$ and a matroid $(S, \mathcal{N})$ where $S \subseteq E$, we say that $(S, \mathcal{N})$ is a **minor** of $(E, \mathcal{M})$ if it can be obtained by some combination of restrictions and contractions from $(E, \mathcal{M})$.

Minors play an important role in matroid theory in the characterization of certain classes of matroids. As an illustration, a classical theorem of Whitney states that matroids linear over the two-element field can be characterized as matroids not having $U_{2,4}$ as a minor.

If $\mathscr{B}$ is the set of bases of a matroid $(E, \mathcal{M})$ then it can be shown that $\mathscr{B}^* = \{E \setminus B : B \in \mathscr{B}\}$ also is the set of bases of a matroid. We call this the **dual matroid** of $(E, \mathcal{M})$ and denote it by $(E, \mathcal{M}^*)$. The rank functions r and r^* are related by

$$r^*(E \setminus X) = |E| - r(E) - (|X| - r(X)) .$$

The reader can observe readily that contraction in a matroid results in deletion in the dual and vice versa.

Another operation on matroids is truncation. We define for any positive integer $k \leq r(E)$ the k-**truncation** of a matroid $(E, \mathcal{M})$ as

$$\mathcal{M}_k = \{X \in \mathcal{M} : |X| \leq k\} .$$

It is obvious that $(E, \mathcal{M}_k)$ is again a matroid.

Next we sketch some basic min-max theorems for matroids which form the core of matroid optimization. For two matroids $(E, \mathcal{M}_1)$ and $(E, \mathcal{M}_2)$ their **union** $\mathcal{M}_1 \vee \mathcal{M}_2$ is defined by

$$\mathcal{M}_1 \vee \mathcal{M}_2 = \{X_1 \cup X_2 : X_i \in \mathcal{M}_i, \ i = 1, 2\} .$$

It can be proved that this is again a matroid with rank function

$$r(X) = \min_{Y \subseteq X} \{r_1(Y) + r_2(Y) + |X \setminus Y|\} .$$

We can also define the **intersection** of two matroids as $(E, \mathcal{M}_1 \cap \mathcal{M}_2)$. This is in general not a matroid but we do have a formula for its "rank" (cf. Edmonds [1970]):

Theorem 3.1. *Let $(E, \mathcal{M}_1)$ and $(E, \mathcal{M}_2)$ be two matroids with rank function r_1 and r_2. Then the maximal size of a common feasible set is equal to*

$$\min_{X \subseteq E} \{r_1(X) + r_2(E \setminus X)\} . \qquad \square$$

These two theorems imply many min-max results in graph theory and combinatorics. We illustrate this by the following example. A **transversal** of a finite family $\{A_1, \ldots, A_n\}$ of sets is a set X for which a bijection $\phi : X \to \{1, \ldots, n\}$ exists such that $x \in A_{\phi(x)}$. It can be proved that transversals form the bases of a matroid which is the union of matroids of rank 1. Using this, one can derive from Theorem 3.1 Rado's transversal theorem:

Corollary 3.2. *Let $(E, \mathcal{M})$ be a matroid with rank function r. The set system $\{A_1, \ldots, A_n\}$ has a transversal which is independent in $\mathcal{M}$ if and only if for all $1 \leq i_1 < \ldots < i_k \leq n$,*

$$r(A_{i_1} \cup \ldots \cup A_{i_k}) \geq k . \qquad \square$$

If the matroid is free, this reduces to the classical König-Hall theorem.

4. Submodular Functions and Polymatroids

We have seen that the rank function of a matroid is submodular. Although this will not remain true for the rank function of greedoids, the notion of submodularity will still play an important role in their study. We therefore collect some properties of submodular functions which will be needed in the sequel. For a survey on submodularity see, e.g. Lovász [1983] and Fujishige [1991].

We call a function f **supermodular** if $f(X \cup Y) + f(X \cap Y) \geq f(X) + f(Y)$ holds for all subsets $X, Y \subseteq E$ and **modular** if it is both sub- and supermodular.

Lemma 4.1. *A function $f : 2^E \to \mathbb{R}$ is submodular if and only if for all $e \in E$, $f(X \cup e) - f(X)$ is monotone decreasing.*

Proof. Let f be a submodular function, $X \subseteq Y$ and $e \in E \setminus Y$. Then the submodularity of f implies

$$f(X \cup e) + f(Y) \geq f(Y \cup e) + f(X) .$$

Conversely, let $X, Y \subseteq E$, $Y \setminus X = \{e_1, \ldots, e_k\}$ and set $E_i = \{e_1, \ldots, e_i\}$, $1 \leq i \leq k$ and $E_0 = \emptyset$. Then

$$f((X \cap Y) \cup E_i) - f((X \cap Y) \cup E_{i-1}) \geq f(X \cup E_i) - f(X \cup E_{i-1}) \text{ for } 1 \leq i \leq k .$$

Summing these k inequalities yields

$$f(Y) - f(X \cap Y) \geq f(X \cup Y) - f(X) . \qquad \square$$

We will need two operations preserving submodularity. Given an integer $k \geq 1$, the **k-truncation** of a function f is given by

$$f_k(X) = \min \{f(Y) + k|X \setminus Y| : Y \subseteq X\}.$$

Lemma 4.2. *The k-truncation of a submodular function is submodular.*

Proof. By definition there exist subsets $U \subseteq X$ and $V \subseteq Y$ such that

$$f_k(X) = f(U) + k|X \setminus U| \text{ and } f_k(Y) = f(V) + k|Y \setminus V| .$$

Hence

$$
\begin{aligned}
f_k(X) + f_k(Y) &= f(U) + f(V) + k(|X \setminus U| + |Y \setminus V|) \\
&\geq f(U \cup V) + f(U \cap V) + k(|(X \setminus U)| + |Y \setminus V|) \\
&= f(U \cup V) + f(U \cap V) + k|(X \cup Y) \setminus (U \cup V)| \\
&\quad + k|(X \cap Y) \setminus (U \cap V)| \\
&\geq f_k(X \cup Y) + f_k(X \cup Y) . \qquad \square
\end{aligned}
$$

For a fixed subset $T \subseteq E$, let $e_T \notin E$ be a new element and set $E' = E \cup e_T$. The **extension** f' of f **parallel to** T is defined as

$$f'(X) = f(X)$$

$$f'(X \cup e_T) = f(X \cup T), \quad X \subseteq E.$$

Lemma 4.3. *Let f be a monotone increasing, submodular function. Then its extension f' parallel to some subset T is also monotone increasing and submodular.*

Proof. The proof is straightforward. We only verify the submodularity for sets X, Y with $e_T \in X \setminus Y$.

$$f'(X \cup Y) + f'(X \cap Y) = f(X \cup T \cup Y) + f(X \cap Y)$$
$$\leq f(X \cup T) + f(Y) - f((X \cup T) \cap Y) + f(X \cap Y)$$
$$\leq f'(X) + f'(Y) . \qquad \square$$

A function $f : 2^E \to \mathbb{Z}_+$ is called a **polymatroid rank function** if the following axioms are satisfied:

(4.1) $f(\emptyset) = 0$,
(4.2) $X \subseteq Y \subseteq E$ implies $f(X) \leq f(Y)$,
(4.3) $f(X \cup Y) + f(X \cap Y) \leq f(X) + f(Y)$ for all $X, Y \subseteq E$.

A **polymatroid** is the collection of all vectors $x \in \mathbb{R}_+^{|E|}$ such that $\sum_{e \in S} x_e = x(S) \leq f(S)$ for all $S \subseteq E$. Alternatively, we may think of a polymatroid as a polyhedron P with the following properties:

(4.4) if $x \in P$ and $0 \leq y \leq x$, then $y \in P$,
(4.5) for any $a \in \mathbb{R}_+^E$ and all maximal vectors $x, y \in P$ with $x \leq a$ and $y \leq a$ we have $\sum_{e \in E} x(e) = \sum_{e \in E} y(e)$.

We remark that the k-truncation of a polymatroid rank function is a polymatroid rank function. For $k = 1$, the polymatroid of the 1-truncation f_1 is the **induced matroid**. By Edmonds [1970], this matroid can be described as follows.

Lemma 4.4. *The matroid induced by a submodular function f is given by*

$$\mathcal{M} = \{X \subseteq E : |Y| \leq f(Y) \quad \text{for all} \quad Y \subseteq X\}. \qquad \square$$

Polymatroids can also be characterized by various versions of the greedy algorithm, see Grötschel, Lovász and Schrijver [1988].

Chapter III. Abstract Convexity – Antimatroids

If we abstract the combinatorial properties of convexity in a manner similar to the abstraction of linear dependence in matroid theory, we obtain antimatroids. Interest in these structures has its sources in different fields of mathematics.

The first source was lattice theory. Antimatroids were introduced by Dilworth [1940] as particular examples of semimodular lattices. Since then, several authors have arrived at the same concepts by abstracting various combinatorial situations (Jamison [1970], Hoffman [1979], Graham, Simonovits, Sós [1980], Edelman [1980]). A systematic study of these combinatorial structures was started by Edelman and Jamison emphasizing the combinatorial abstraction of convexity. Jamison [1970] exhibited the relationship between an abstract notion of convexity and the anti-exchange closure. The latter was then shown by Edelman [1980] to be the crucial property of closures induced by what he called meet-distributive lattices – a concept equivalent to antimatroids. Thereby, different notions of convexity in graphs, directed graphs and ordered sets were linked to semimodular lattices (cf. Edelman and Jamison [1985]).

Antimatroids were also studied as special greedoids in Korte and Lovász [1984b], described there as deletion processes determined by alternative precedence constraints. A historical treatment of antimatroids can be found in Monjardet [1985].

The name "antimatroid" finds its justification not only in the anti-exchange property (which is in a certain sense the opposite of the exchange property for matroids) but also in other properties which are in a sense dual or antipodal to matroid properties. We shall discuss several of these properties later.

In Section 1 we introduce the notion of abstract convexity and give several axiomatizations of antimatroids. Section 2 contains a number of examples arising from various "shelling", "searching" and "convexity" concepts. Section 3 gives two further axiomatizations of antimatroids in terms of circuits and paths. In Section 4 we extend classical ideas from convexity theory to antimatroids. Among these are the Caratheodory and Helly numbers. Section 5 contains Ramsey type theorems. The final Section 6 describes two representations of antimatroids in terms of posets and joins.

1. Convex Geometries and Shelling Processes

Consider a family $\mathcal{N}$ of subsets of E with the following properties

(1.1.1) $\emptyset \in \mathcal{N}, E \in \mathcal{N}$,
(1.1.2) $X, Y \in \mathcal{N}$ implies $X \cap Y \in \mathcal{N}$.

The family $\mathcal{N}$ gives rise to the following operator:

(1.2) $\tau(A) = \bigcap \{X : A \subseteq X : X \in \mathcal{N}\}$.

By (1.1.2), $\tau(A)$ is the unique smallest set in $\mathcal{N}$ containing A. It is straightforward to check that τ has the properties of a **closure operator,** namely

(1.3.1) $\tau(\emptyset) = \emptyset$,
(1.3.2) $A \subseteq \tau(A)$,
(1.3.3) $A \subseteq B$ implies $\tau(A) \subseteq \tau(B)$,
(1.3.4) $\tau(\tau(A)) = \tau(A)$.

Conversely, every closure operator τ defines a family $\mathcal{N}$ with property (1.1) as the family of its closed sets.

While both the closure operator of a matroid and the Euclidean convex hull operator conv(X) satisfy the axioms (1.3), they show a different, in a sense opposite, behaviour with respect to exchange properties.

We call a set E endowed with a closure operator τ a **convex geometry** if the following axiom is satisfied:

(1.4)   **Anti-exchange property**
         if $y, z \notin \tau(X)$ and $z \in \tau(X \cup y)$, then $y \notin \tau(X \cup z)$.

Axiom (1.4) is a combinatorial abstraction of a property of the usual convex closure in Euclidean spaces. Namely, for two points y and z not in the convex hull of the set X, if z is in the convex hull of $X \cup y$ then y is outside the convex hull of $X \cup z$ (cf. Figure 1).

Thus every finite subset in a Euclidean space gives rise to a convex geometry. Just as linear matroids are archetypes of matroids, these convex geometries are also archetypal. We shall also use the name convex geometry for the set system $(E, \mathcal{N})$ if it satisfies (1.1) and the corresponding closure operator satisfies (1.4). We call the sets in $\mathcal{N}$ **convex.**

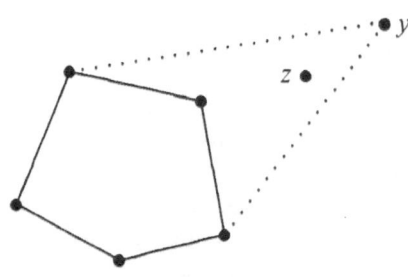

Fig. 1

The name "anti-exchange property" is chosen to indicate the close formal relationship to the Steinitz-MacLane exchange property of the matroid closure (cf. Theorem II.1.5).

In accordance with Euclidean convexity we call an element x of a subset $A \subseteq E$ an **extreme point** of A if $x \notin \tau(A \setminus x)$. For a convex set $A \in \mathcal{N}$ this is equivalent to $A \setminus x \in \mathcal{N}$. The set of extreme points of A is denoted by $ex(A)$. A **minimal spanning set** of a set X is a minimal subset $Y \subseteq X$ such that $\tau(Y) = \tau(X)$. The following characterization shows that convex geometries share some well-known properties of Euclidean convexity, among them an abstract version of the Krein-Milman Theorem.

Theorem 1.1. *Let $\tau : 2^E \to 2^E$ be a closure operator on E. Then the following are equivalent:*

(i)   *(E, τ) is a convex geometry.*
(ii)   *Let $X \neq E$ be a closed set. Then $X \cup y$ is closed for some $y \in E \setminus X$.*
(iii)   *For every closed set $X \subseteq E$, $X = \tau(ex(X))$.*
(iv)   *Every subset $X \subseteq E$ has a unique minimal spanning set.*

Proof. (i) $\Rightarrow$ (ii) Suppose $X \cup y$ is not closed for all $y \in E \setminus X$. Choose a $y \in E \setminus X$ for which $\tau(X \cup y)$ is minimal. Consider an element $z \in \tau(X \cup y) \setminus (X \cup y)$. Then $X \cup z \subseteq \tau(X \cup y)$ implies $\tau(X \cup z) \subseteq \tau(X \cup y)$. Moreover, the anti-exchange property gives $y \notin \tau(X \cup z)$, contradicting the minimality of $\tau(X \cup y)$.

(ii) $\Rightarrow$ (iii)   Clearly, $\tau(ex(X)) \subseteq \tau(X) = X$. If this inclusion is strict, take a largest convex superset Y of $ex(X)$ not containing X. By (ii), Y misses only one element $x \in X$. But then $x \notin \tau(X \setminus x)$ and so $x \in ex(X)$, a contradiction. Hence $\tau(ex(X)) = X$.

(iii) $\Rightarrow$ (iv) Let Y be a spanning set for X. Suppose $x \notin Y$ for some $x \in ex(\tau(X))$. Then $x \in \tau(Y) = \tau(Y \setminus x) \subseteq \tau(\tau(Y) \setminus x) = \tau(\tau(X) \setminus x)$, a contradiction. Hence $ex(\tau(X))$ is contained in every spanning set of X. Since by (iii) $\tau(ex(\tau(X))) = \tau(X)$, it is also minimal.

(iv) $\Rightarrow$ (i)   Suppose the anti-exchange axiom fails to hold. Then there are $y, z \notin \tau(X)$ with $z \in \tau(X \cup y)$ and $y \in \tau(X \cup z)$. Hence $\tau(X \cup y) = \tau(X \cup z)$. Let S be the unique minimal spanning set of $\tau(X \cup y)$. Since both $X \cup y$ and $X \cup z$ are spanning, we must have $S \subseteq (X \cup y) \cap (X \cup z) = X$. Thus $\tau(S) \subseteq \tau(X) \subseteq \tau(X \cup y) \setminus y$ in contradiction to the choice of S. $\square$

For any closure operator τ with properties (1.3) and any τ-closed set A, the relation

$$x \leq_A y \text{ if and only if } x \in \tau(A \cup y)$$

is reflexive and transitive on $E \setminus A$. The anti-exchange property (1.4) then is equivalent to the antisymmetry of $\leq_A$. Thus for a convex geometry we obtain a partial order on $E \setminus A$ for any convex set A.

Another, in a sense complementary approach leading to abstract convexity is an elemination procedure or shelling process. For instance, take the convex geometry of points in the Euclidean space. Starting with the whole set, we may successively shell away extreme points of the remaining set until all the points have been removed and we end up with the empty set. Of course, there is more than one way to do that. Clearly, this elemination process can be defined for every convex geometry: A sequence $x_1 \ldots x_k$ is a shelling sequence if x_i is an extreme point of $E \setminus \{x_1, \ldots, x_{i-1}\}$. This associates a hereditary language with every convex geometry.

The essential property of this shelling is that once a point becomes an extreme point, it remains so. This gives rise to a (seemingly) more general class of selection processes: we select elements from a given set. We have a rule that tells us at any point which elements can be selected. The only assumption about the rule is that once an element becomes eligible, it stays eligible. This can be formalized as follows: for each $x \in E$ let a system $\mathcal{H}_x \subseteq 2^E$ of **alternative precedences** be given. We inductively define feasible words $x_1 \ldots x_k$ by starting with the empty word. A given feasible word $\alpha = x_1 \ldots x_k$ can be enlarged to a feasible word $\alpha x_{k+1} = x_1 \ldots x_k x_{k+1}$ if for some subset $U \in \mathcal{H}_{x_{k+1}}$, $U \subseteq \{x_1, \ldots, x_k\}$. Obviously this system of feasible words is a hereditary language.

We will now give an intrinsic definition of these hereditary languages. We call a language $(E, \mathcal{L})$ an **antimatroid** if it is normal, hereditary and the following augmentation property holds:

(1.5)   for any two words $\alpha, \beta \in \mathcal{L}$ with $\tilde{\alpha} \not\subseteq \tilde{\beta}$ there exists an $x \in \tilde{\alpha}$ with $\beta x \in \mathcal{L}$.

It follows from the definition that any word in an antimatroid can be extended to a basic word containing all elements of E. Thus all basic words are permutations of E.

Associated with the language $(E, \mathcal{L})$ of an antimatroid we have an accessible set system $(E, \mathcal{F}) = (E, \mathcal{F}(\mathcal{L}))$ which we again call an antimatroid.

Lemma 1.2. *For a normal accessible set system $(E, \mathcal{F})$ the following statements are equivalent:*

(i)   *$(E, \mathcal{F})$ is an antimatroid.*
(ii)   *$\mathcal{F}$ is closed under union.*
(iii)   *$A, A \cup x, A \cup y \in \mathcal{F}$ implies $A \cup \{x, y\} \in \mathcal{F}$.*

Proof. Let $(E, \mathcal{F})$ be an antimatroid, $X, Y \in \mathcal{F}$ and let α and β be feasible orderings of X and Y, respectively. Repeated application of (1.5) then yields a word $\alpha \beta' \in \mathcal{F}$ with $\widetilde{\alpha \beta'} = X \cup Y$, i.e. (i) implies (ii).

Since (iii) is a special case of (ii), it remains to show that (iii) implies (i). For this purpose, let $\mathcal{L} = \mathcal{L}(\mathcal{F})$ be the associated simple hereditary language. Obviously the underlying sets of words in $\mathcal{L}$ are exactly the sets in $\mathcal{F}$.

Let $\alpha, \beta \in \mathcal{L}$ be feasible words with $\alpha \not\subseteq \beta$. As before, we may write

$$\alpha = \alpha_1 x \alpha_2 \text{ and } \beta = \alpha_1 y_1 \ldots y_r,$$

where α_1 is the longest common initial substring of α and β. Repeated application of (iii) to $\alpha_1 x$ and $\alpha_1 y_1 y_2 \ldots y_r$ yields $\alpha_1 y_1 \ldots y_r x \in \mathcal{L}$, i.e. $\tilde{\beta} \cup x \in \mathcal{F}$. Hence $(E, \mathcal{F})$ satisfies (1.5) and thus is an antimatroid.   □

As an immediate consequence we note that for a given antimatroid any subset $X \subseteq E$ has a unique basis.

Since the feasible sets of an antimatroid $(E, \mathcal{F})$ are closed under union, the system of complements $\mathcal{N} = \{E \setminus X : X \in \mathcal{F}\}$ is closed under intersection and hence induces a closure operator.

Theorem 1.3. $(E, \mathcal{F})$ *is an antimatroid if and only if* $(E, \mathcal{N})$ *is a convex geometry.*

Proof. Let $(E, \mathcal{F})$ be an antimatroid, $\mathcal{N}$ the system of complements, and τ the associated closure operator. Since the properties (1.3) obviously hold, we only have to prove the anti-exchange property (1.4).

For this purpose, let B be the (unique) basis of $E \setminus X$ and A the basis of $E \setminus (X \cup y)$. Then $A \subseteq B \setminus \{y, z\}$ since, by assumption, we have $y, z \in B$ but $z \notin A$. So we can augment A to some set $A \cup x \in \mathcal{F}$, where $x \in B$. Since A is a basis of $E \setminus (X \cup y)$ we must have $x = y$, i.e. $A \cup y$ is a feasible subset of $E \setminus (X \cup z)$ and thus $y \notin \tau(X \cup z)$.

Conversely, assume that τ is the closure operator of a system $(E, \mathcal{N})$ of subsets satisfying (1.3) and (1.4). Let

$$\mathcal{F} = \{X \subseteq E : \tau(E \setminus X) = E \setminus X\}.$$

Then by (1.3.1), $\emptyset \in \mathcal{F}$. Next we show that $(E, \mathcal{F})$ is accessible. Let $X \in \mathcal{F}$ and $x \in X$ be such that $\tau(E - (X \setminus x))$ is minimal with respect to set theoretic containment. Suppose $X \setminus x \notin \mathcal{F}$. By (1.3.2), there exists an element

$$x' \in \tau(E \setminus (X \setminus x)) \setminus (E \setminus (X \setminus x)) \subseteq X \setminus x.$$

Then

$$E \setminus (X \setminus x') \subseteq \tau(E \setminus (X \setminus x)),$$

hence by (1.3)

$$\tau(E \setminus (X \setminus x')) \subseteq \tau(E \setminus (X \setminus x)),$$

where the inclusion is strict, since, by (1.4), $x \notin \tau (E \setminus (X \setminus x'))$. This, however, contradicts the choice of x, i.e. $(E, \mathcal{F})$ is accessible.

Finally, let $X, Y \in \mathcal{F}$ be feasible sets. Then the conditions (1.3) yield

$$E \setminus (X \cup Y) \subseteq \tau(E \setminus (X \cup Y))$$
$$\subseteq \tau(E \setminus X) \cap \tau(E \setminus Y)$$
$$= (E \setminus X) \cap (E \setminus Y)$$
$$= E \setminus (X \cup Y).$$

Thus $X \cup Y \in \mathcal{F}$ and the result follows from Lemma 1.2. $\square$

Now we are in the position to prove the equivalence of antimatroids with the languages associated with alternative precedences and the shelling of convex sets.

Theorem 1.4. *For a normal language* $(E, \mathscr{L})$ *the following three statements are equivalent:*

(i) $(E, \mathscr{L})$ *is an antimatroid.*
(ii) $(E, \mathscr{L})$ *is the language of shelling sequences of a convex geometry.*
(iii) $(E, \mathscr{L})$ *is the language of feasible words of a system of alternative precedences.*

Proof. (i) $\Rightarrow$ (ii) If we have an antimatroid, we know by Theorem 1.3 that its feasible sets are just the complements of the convex sets of a convex geometry. It is straightforward to verify that the shelling sequences of this convex geometry are just the feasible words in this antimatroid.

(ii) $\Rightarrow$ (iii) For every $x \in E$ let $\mathscr{H}_x$ consist of all sets $U \subseteq E$ such that x is an extreme point of $E \setminus U$.

(iii) $\Rightarrow$ (i) Let $(E, \mathscr{L})$ be the language of feasible words of a system of alternative precedences and let $\alpha, \beta \in \mathscr{L}$ be two feasible words with $\tilde{\alpha} \nsubseteq \tilde{\beta}$. We may write $\alpha = \alpha_1 x \alpha_2$, where x is the first element of α which does not occur in β. Then for some $U \in \mathscr{H}_x$, $U \subseteq \alpha_1 \subseteq \beta$, i.e. $\beta x \in \mathscr{L}$. $\square$

2. Examples of Antimatroids

Various constructions and procedures give rise to antimatroids. In the examples that follow the reader will easily check that one or another of the equivalent systems of axioms is fulfilled. In some cases we give some hints as to this. Further examples can be found in Goecke, Korte and Lovász [1989].

One class of constructions is "shellings", i.e. decompositions of structures by repeated elimination of suitable elements. The first instance is the example we already used in Section 1.

2.1 Shelling of Convex Sets

Let $E \subseteq \mathbb{R}^n$ be a finite set of points and for any $x \in E$, let $\mathscr{H}_x$ consist of all sets $U \subseteq E$ such that x is a vertex of $\mathrm{conv}(E \setminus U)$. The words in this shelling structure correspond to a recursive deletion of extreme points of the convex set spanned by the remaining points. Geometrically, a set U belongs to $\mathscr{H}_x$ if it contains all points in a halfspace defined by a hyperplane through x.

2.2 Lower Convex Shelling

As before let $E \subseteq \mathbb{R}^n$ be a finite set of points and let $C \subseteq \mathbb{R}^n$ be a cone. We define the convex hull of a subset X by $E \cap (\mathrm{conv}(X) + C)$. The name stems from the special case when the cone degenerates to a semi-line pointing "up". Then extreme points of the "lower half" of the convex hull are eligible for elimination.

2.3 Poset Antimatroid

Let $(E, \leq)$ be a poset and $\mathscr{F} = \{X \subseteq E : X \text{ is an ideal}\}$. Clearly $(E, \mathscr{F})$ is an antimatroid, since ideals are closed under union. Basic words are all linear extensions of the poset. The corresponding shelling process consists of repeatedly deleting minimal elements. We can get the poset antimatroid by taking $I(x) \setminus x$ as single alternative precedence for x. Poset antimatroids play an important role in the theory of one-machine scheduling, where the basic words are the feasible schedules of precedence constrained jobs.

The poset antimatroid has the property that also the convex sets are closed under union. This property in fact characterizes poset antimatroids. For, let $(E, \mathscr{F})$ be an antimatroid whose convex sets are closed under union and $X \in \mathcal{N}$. Then by the remark following Theorem 1.1, $(E, \leq_\emptyset)$ is an ordered set. Hence

$$X = \bigcup \{\tau(x) : x \in X\} = \{y : y \leq_\emptyset x \text{ for some } x \in X\},$$

i.e. X is an ideal in $(E, \leq_\emptyset)$. A similar argument shows that conversely any ideal is convex. In other words, the feasible sets correspond to ideals in the poset where the order relation is reversed. Thus poset antimatroids are precisely those antimatroids whose convex sets are closed under union.

Poset antimatroids are fundamental among antimatroids since they have in many respect the simplest structure.

2.4 Double Shelling of a Poset

Let $(E, \leq)$ be a partially ordered set and $\mathscr{F} = \{X \cup Y \subseteq E : X \text{ is an ideal and } Y \text{ is a filter of } E\}$. The corresponding shelling consist of repeatedly deleting minimal or maximal elements from the poset.

2.5 Shelling of a Tree

Let T be a tree. The shelling process is defined on the set of nodes. It consists of repeatedly deleting end-nodes of the tree, and the convex sets are the node-sets of subtrees of T.

2.6 Edge Shelling of a Tree

Again, let T be a tree. The shelling process is defined on $E(T)$ by successively deleting end-edges. Thus the convex sets are the edge-sets of subtrees.

2.7 Simplicial Shelling of a Graph

Let $G = (V, E)$ be a graph. We call a node **simplicial** if its neighbors form a complete subgraph. We now repeatedly eliminate simplicial nodes. We may not be able to delete all elements in this way. It is easy to see that if G contains a cycle of length strictly greater than 3 without chords, then none of the vertices on the cycle will occur in any of the words $\alpha \in \mathscr{L}$, i.e. $(E, \mathscr{L})$ is not normal. However, we do get an antimatroid on the set of nodes occurring in words of $\mathscr{L}$. If G is **triangulated**, i.e. if every cycle of length greater than 3 has a chord, then it can be shown that $(E, \mathscr{L})$ is an antimatroid.

Clearly, we can define antimatroids based on the elimination of nodes with any local property, provided that this property is preserved by the elimination of other nodes. Some of these antimatroids are related to interesting graph properties.

2.8 Simple Elimination

Let $G = (V, E)$ be a graph. A simplicial vertex $v \in V$ is called **simple** if for any two vertices $x, y \in N(v)$

$$\overline{N}(x) \subseteq \overline{N}(y) \text{ or } \overline{N}(y) \subseteq \overline{N}(x) .$$

We can repeatedly eliminate simple vertices to get a language $\mathcal{L}$. This language is an antimatroid on the set of vertices that occur in some word. It is an antimatroid on the whole set V if and only if the graph is **strongly chordal**, i.e. every induced subgraph of G has a simple vertex (for other characterizations of strongly chordal graphs see Farber [1983]).

2.9 Transitivity Antimatroid

Let G be a transitively oriented graph. An edge can be eliminated if the remaining graph is still transitively oriented.

Other types of antimatroids arise from "searching", i.e. building up structures by repeatedly adding suitable elements:

2.10 Line-Search in Graphs

Let $G = (V, E)$ be an undirected graph and $r \in V$ a specified node. Then we can define an antimatroid on E whose feasible sets are edge-sets of connected subgraphs containing r. The basic words are line-searches through all edges of G starting at vertex r.

We can define line-search antimatroids of directed graphs similarly: feasible sets are those sets of arcs in which every element can be reached from the root by a directed path.

2.11 Point-Search in Directed Graphs

Let $G = (V, E)$ be a directed graph with a specified root $r \in V$. We can define an antimatroid on $V \setminus r$ whose feasible sets are those sets $X \subseteq V \setminus r$ such that in the subgraph induced by $X \cup r$ every vertex $x \in X \cup r$ can be reached from r by a directed path. The basic words are the point-searches from r. Point-searches in undirected graphs can be defined analogously.

Note that line-search antimatroids in directed and undirected graphs can be viewed as point-search antimatroids in appropriate line-graphs.

2.12 Point-Line Search in Graphs

Let G be a graph, $E = V(G) \cup E(G)$ and let $\mathcal{F}$ be the collection of all subsets $X \subseteq E$ such that if the edge $[u, v] \in X \cap E(G)$, then either $u \in X$ or $v \in X$. Then $(E, \mathcal{F})$ is an antimatroid.

3. Circuits and Paths

For an antimatroid $(E, \mathcal{F})$ we define the **trace** on $X \subseteq E$ by

$$\mathcal{F} : X = \{A \cap X : A \in \mathcal{F}\}.$$

It is easy to verify that both accessibility and union-closedness are preserved under tracing. Therefore the trace of an antimatroid is again an antimatroid.

We call a set $X \subseteq E$ **free**, if $\mathcal{F} : X = 2^X$. A set $C \subseteq E$ is a **circuit** if it is minimal nonfree, i.e. it is not free but every proper subset of it is free. Note that a set is free if and only if it contains no circuit.

As an example consider the shelling antimatroid of a tree. Every free set consists of the leaves of some subtree. The circuits are triples of nodes lying on a path. As another example, for a poset antimatroid the free sets are precisely the antichains and the circuits are pairs $\{a, b\}$ with $a < b$.

For a set $A \in \mathcal{F}$, let $\Gamma(A) = \{x \in E \setminus A : A \cup x \in \mathcal{F}\}$ be the set of **feasible continuations** of A. In terms of convex geometries, $\Gamma(A)$ is the set of extreme points of the complement, i.e. $\Gamma(A) = \text{ex}(E \setminus A)$. The following lemma exhibits the sets of extreme points of convex sets as precisely the free sets.

Lemma 3.1. *A subset $X \subseteq E$ is free if and only if $X = \Gamma(A)$ for the basis A of $E \setminus X$.*

Proof. Let $A \in \mathcal{F}$ be a feasible set and $X = \Gamma(A)$. Then $\mathcal{F} : X = 2^X$ since $\mathcal{F}$ is closed under union.

Conversely, let X be free and $A \in \mathcal{F}$ be the basis of $E \setminus X$. Clearly $\Gamma(A) \subseteq X$. For any $x \in X$ there exists some $B \in \mathcal{F}$ with $B \cap X = \{x\}$. Then $B \nsubseteq A$ and by (1.5) there exists a $y \in B$ such that $A \cup y \in \mathcal{F}$. Since A is the basis of $E \setminus X$, we have $y = x$ and so $x \in \Gamma(A)$. □

Lemma 3.2. *Let C be a circuit of the antimatroid $(E, \mathcal{F})$ and A be the basis of $E \setminus C$. Then $\Gamma(A) \subseteq C$ and $|C \setminus \Gamma(A)| = 1$.*

Proof. Clearly $\Gamma(A) \subseteq C$ and the inclusion is strict, since otherwise C would be free. So there exists some $r \in C \setminus \Gamma(A)$. Then A is the basis of $E \setminus (C \setminus r)$ as well and thus, since $C \setminus r$ is free, we have $C \setminus r = \Gamma(A)$. □

Consider the antimatroid of the poset in Figure 2 and set $C = \{a, b, x\}$. Then $\Gamma(\emptyset) \subseteq C$ and $C - \Gamma(\emptyset) = x$, however C is not a circuit. This shows that Lemma 3.2 gives only a necessary but not sufficient condition for circuits.

We call the unique element r of a circuit C with $\{r\} = C \setminus \Gamma(A)$ the **root** of the circuit and $C \setminus r$ the **stem**. To make this distinction clear, we sometimes write (C, r) for a circuit C with root r.

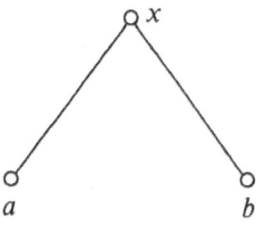

Fig. 2

Lemma 3.3. *Let $(E, \mathscr{F})$ be an antimatroid, $C \subseteq E$ and $r \in C$. Then C is a circuit with root r if and only if $\mathscr{F} : C = 2^C \setminus \{r\}$.*

Proof. If $\mathscr{F} : C = 2^C \setminus \{r\}$, then C is minimal nonfree, i.e. a circuit, and clearly r is its root.

Now let C be a circuit with root r. Then $C \setminus r$ is free and thus $2^{C \setminus r} \subseteq \mathscr{F} : C$. Suppose $\{r\} \in \mathscr{F} : C$, i.e. $B \cap C = \{r\}$ for some $B \in \mathscr{F}$. As in the proof of Lemma 3.1 we may then conclude that $A \cup \{r\} \in \mathscr{F}$, where A is the basis of $E \setminus C$. But this contradicts $r \in C \setminus \Gamma(A)$.

Suppose now that $\{r, x\} \notin \mathscr{F} : C$ for some $x \in C \setminus r$. Then $r \notin \mathscr{F} : (C \setminus x)$, and C is not minimal nonfree. Since $\mathscr{F} : C$ is closed under union, we obtain $\mathscr{F} : C = 2^C \setminus \{r\}$. □

As an immediate consequence we derive:

Corollary 3.4. *Let (C, r) be a circuit of the antimatroid $(E, \mathscr{F})$. Then $r \in \tau(C \setminus r)$.*

Proof. Suppose $r \notin \tau(C \setminus r)$. Then $[E \setminus \tau(C \setminus r)] \cap C = \{r\}$, i.e. $r \in \mathscr{F} : C$, in contradiction to Lemma 3.3. □

Next we describe how the set of circuits of an antimatroid determines other basic objects.

Lemma 3.5. *Let $(E, \mathscr{F})$ be an antimatroid and $X \subseteq E$. Then $r \in \tau(X) \setminus X$ if and only if $X \cup r$ contains a circuit rooted at r.*

Proof. The "if" part follows from the previous corollary. Conversely, let $C \subseteq X$ be a minimal subset with $r \in \tau(C)$. We claim that $(C \cup r, r)$ is a circuit. Let A be the basis of $E \setminus C$, then $C \subseteq E \setminus (A \cup r)$. Hence $r \in \tau(C) \subseteq \tau(E \setminus (A \cup r))$, i.e. $r \notin \text{ex}(E \setminus A)$ and so $r \notin \Gamma(A)$.

Let $x \in C$, then $r \notin \tau(C \setminus x)$ by the choice of C. Moreover, $x \notin \tau(C \setminus x)$ also by the choice of C. Hence $E \setminus \tau(C \setminus x) \in \mathscr{F}$ and

$$(C \cup r) \cap (E - \tau(C \setminus x)) = \{x, r\} \in \mathscr{F} : (C \cup r).$$

Since the trace of an antimatroid is closed under union we get $\mathscr{F} : (C \cup r) = 2^{C \cup r} \setminus \{r\}$, i.e. $(C \cup r, r)$ is a circuit by Lemma 3.3. $\qquad\square$

Lemma 3.6. *Let $(E, \mathscr{F})$ be an antimatroid and $X \subseteq E$. Then $X \in \mathscr{F}$ if and only if for each circuit (C, r), $X \cap C \neq \{r\}$.*

Proof. By the definition of circuits, if X is feasible then it has to satisfy this condition. Conversely, suppose that $X \subseteq E$ and $X \notin \mathscr{F}$. Consider the basis B of X, and let $x \in X \setminus B$. Then $\Gamma(B) \cup x$ is nonfree by Lemma 3.1 and hence it contains a circuit C. Trivially $x \in C$, and in fact x is the root of C. Also trivially, $\Gamma(B) \cap X = \emptyset$. Hence $X \cap C = \{x\}$, i.e. X does not satisfy the condition formulated in the lemma. $\qquad\square$

The structure of circuits in antimatroids is (contrary to first appearances) simpler than in matroids. This is due to the fact that any collection of rooted sets gives rise to a shelling structure. More precisely, a **rooted set** is a pair (C, a) where $|C| \geq 2$ and $a \in C$. Let $\mathscr{K}$ be any collection of rooted subsets of E. We denote by $\mathscr{L}_{\mathscr{K}}$ the set of words $x_1 \ldots x_k$ such that for every $(C, a) \in \mathscr{K}$ if $a = x_i$ then $C \cap \{x_1, \ldots, x_{i-1}\} \neq \emptyset$. We call $\mathscr{L}_{\mathscr{K}}$ the **language determined by** $\mathscr{K}$.

Lemma 3.7. *If $(E, \mathscr{L}_{\mathscr{K}})$ is normal then it is an antimatroid.*

Proof. Simply consider the nonempty subsets of $C \setminus a$ as alternative precedences for a. $\qquad\square$

The next result states a circuit exchange property similar to the one for matroids.

Lemma 3.8. *Let $(E, \mathscr{F})$ be an antimatroid and $(C_1, r_1), (C_2, r_2)$ two circuits of $(E, \mathscr{F})$ with $r_1 \in C_2 \setminus r_2$. Then there exists a rooted circuit (C_3, r_2) with $C_3 \subseteq (C_1 \cup C_2) \setminus r_1$.*

Proof. Set $X = (C_1 \cup C_2) \setminus \{r_1, r_2\}$. By Corollary 3.4, $r_2 \in \tau(C_2 \setminus r_2) \subseteq \tau(X \cup r_1)$. Similarly, $r_1 \in \tau(X \cup r_2)$. Then the anti-exchange property (1.4) implies either $r_1 \in \tau(X)$ or $r_2 \in \tau(X)$. If $r_1 \in \tau(X)$, then by (3.1) $r_2 \in \tau(C_2 \setminus r_2) \subseteq \tau(X \cup r_1) = \tau(X)$. Hence in both cases $r_2 \in \tau(X) \setminus X$ and the result follows from Lemma 3.5. $\qquad\square$

We are now ready to characterize antimatroids in terms of circuits, a result which is due to Dietrich [1987].

Theorem 3.9. *Let $\mathscr{K}$ be a set of rooted subsets of E. Then $\mathscr{K}$ is the set of rooted circuits of an antimatroid $(E, \mathscr{F})$ if and only if $\mathscr{K}$ satisfies the following conditions:*

(3.1.1) $(C_1, r), (C_2, r) \in \mathscr{K}$ and $C_1 \subseteq C_2$ implies $C_1 = C_2$,

(3.1.2)   for $(C_1, r_1), (C_2, r_2) \in \mathscr{K}$ with $r_1 \in C_2 \setminus r_2$ there exists a $(C_3, r_2) \in \mathscr{K}$ with $C_3 \subseteq C_1 \cup C_2 \setminus r_1$.

Proof. The necessity of the conditions follows from Lemmas 3.2 and 3.8.

To see the other direction, define a closure operator $\tau : 2^E \to 2^E$ by

$$\tau(X) = X \cup \{r \in E \setminus X : C \subseteq X \cup r \text{ for some } (C, r) \in \mathscr{K}\}.$$

We claim that τ is the closure operator of a convex geometry. The conditions (1.3.1)–(1.3.3) are obviously satisfied by τ.

Suppose τ is not idempotent, i.e. there exists an element $r_2 \in E \setminus \tau(X)$ and rooted sets $(C, r_2) \in \mathscr{K}$ with $C \subseteq \tau(X) \cup r_2$. Among these, choose $(C_2, r_2) \in \mathscr{K}$ such that $|C_2 \setminus X|$ is minimal. By assumption $C_2 \setminus r_2 \not\subseteq X$, hence there exists an element $r_1 \in (C_2 \setminus r_2) \setminus X \subseteq \tau(X) \setminus X$. Thus r_1 is the root of some set $(C_1, r_1) \subseteq X \cup r_1$. Condition (3.1.2) then yields some rooted set (C_3, r_2) with $C_3 \subseteq C_1 \cup C_2 \setminus r_1$. This contradicts the choice of C_2, since

$$|C_3 \setminus X| = |(C_2 \setminus r_1) \setminus X| < |C_2 \setminus X| \ .$$

Suppose now that the anti-exchange property does not hold. Then there are sets $(C_1, z), (C_2, y) \in \mathscr{K}$ with $C_1 \subseteq X \cup y \cup z$, $C_1 \not\subseteq X \cup z$ and $C_2 \subseteq X \cup y \cup z$, $C_2 \not\subseteq X \cup y$. Hence $y \in C_1$ and $z \in C_2$. So there exists a rooted set $(C_3, y) \in \mathscr{K}$ with $C_3 \subseteq C_1 \cup C_2 \setminus z \subseteq X \cup y$, in contradiction to $y \notin \tau(X)$.

So τ is the closure operator of an antimatroid $(E, \mathscr{F})$, where

$$\mathscr{F} = \{X \subseteq E : \tau(E \setminus X) = E \setminus X\}.$$

It remains to show that $\mathscr{K}$ is the collection of circuits of $\mathscr{F}$. For this purpose, let $(C_1, r) \in \mathscr{K}$ be some rooted set. Then $r \in \tau(C_1 \setminus r)$ and hence, using Lemma 3.5, there exists a circuit (C_2, r) with $C_2 \subseteq C_1$. In view of Corollary 3.4 and the definition of τ, we must have a rooted set (C_3, r) with $C_3 \subseteq (C_2 \setminus r) \cup r = C_2 \subseteq C_1$. Condition (3.1.1) now assures that equality must hold, i.e. (C_1, r) is a circuit.

Conversely, let (C_1, r) be a circuit. As before, $r \in \tau(C_1 \setminus r)$ and hence there is some $(C_2, r) \in \mathscr{K}$ with $C_2 \subseteq (C_1 \setminus r) \cup r = C_1$. But since (C_2, r) is also a circuit, $C_2 = C_1$ and we are done.   □

Corollary 3.10. *An antimatroid is a poset antimatroid if and only if all of its circuits have cardinality 2.*

Proof. We have remarked before that the circuits of the poset antimatroid are of the form $(\{a, b\}, b)$ with $a < b$.

Conversely, set $a < b$ if $(\{a, b\}, b)$ is a circuit. Then Theorem 3.9 implies that this relation is acyclic and transitive and thus defines a partial order on E. Moreover, the circuits of the corresponding poset antimatroid are precisely the circuits we started with.   □

Theorem 3.9 describes how to reconstruct an antimatroid from its set of circuits. In the case of a poset antimatroid it is clear that instead of considering all comparabilities, the subset of covering relations would suffice to determine the antimatroid uniquely. We shall now show that also for arbitrary antimatroids such a unique minimal system exists.

Let $(E, \mathscr{F})$ be an antimatroid, (C, a) a rooted set, and A the basis of $E \setminus C$. Then (C, a) is a **critical circuit** if $A \cup a \notin \mathscr{F}$ but $A \cup c \cup a \in \mathscr{F}$ for all $c \in C \setminus a$.

It follows from Lemma 3.3 that every critical circuit is a circuit. The critical circuits of the poset antimatroid are the covering pairs $a < b$. The critical circuits of the tree shelling antimatroid are the paths of length 3. The critical circuits of a point-search antimatroid are those rooted sets (C, x), where $C \setminus x$ consists of those nodes y for which $yx \in E$ and y can be reached from the root by a directed path avoiding x. In particular, every node is the root of at most one critical circuit.

Theorem 3.11. *Let $(E, \mathscr{L})$ be an antimatroid. Then*
(i)   *$(E, \mathscr{L})$ is determined by its critical circuits, and*
(ii) *every system of circuits determining $(E, \mathscr{L})$ contains all critical circuits.*

Proof. (i) Let $\mathscr{K}_0$ be the set of critical circuits and let $\mathscr{L}(\mathscr{K}_0)$ be the set of all words $x_1 x_2 \ldots x_k$ such that if $x_i = a$ for some critical circuit (C, a), then $C \cap \{x_1, \ldots x_{i-1}\} \neq \emptyset$. Clearly, $\mathscr{L} \subseteq \mathscr{L}(\mathscr{K}_0)$. So suppose $\mathscr{L}(\mathscr{K}_0) \neq \mathscr{L}$, then there exists a shortest word $\alpha a = x_1 x_2 \ldots x_{k-1} a \in \mathscr{L}(\mathscr{K}_0) \setminus \mathscr{L}$. Let β be a longest continuation of α with $\beta a \notin \mathscr{L}$. If we set $C = \Gamma(\beta) \cup a$, then β is a basic word of $E \setminus C$. By definition, $\beta c a \in \mathscr{L}$ for any $c \in \Gamma(\beta)$ and $\beta a \notin \mathscr{L}$. Hence (C, a) is a critical circuit, but $C \cap \tilde{\alpha} = \emptyset$ in contradiction to the definition of $\mathscr{L}(\mathscr{K}_0)$.

(ii) Let $\mathscr{K}$ be a system of circuits determining $(E, \mathscr{L})$, let (C, a) be a critical circuit and $A \in \mathscr{F}$ a basis of $E \setminus C$. Then in view of Lemma 3.5, $A \cup a \notin \mathscr{F}$ implies that there exists some circuit $(D, a) \in \mathscr{K}$ with $A \cap D = \emptyset$. Consider $c \in C \setminus a$. Then $A \cup c \cup a \in \mathscr{F}$ by the definition of a critical circuit. Since $\mathscr{F} : D = 2^D \setminus \{a\}$, we must have $(A \cup c \cup a) \cap D \neq \{a\}$ and thus $C \subseteq D$. Since both (C, a) and (D, a) are circuits, (3.1.1) yields $C = D$, and so $(C, a) \in \mathscr{K}$.   □

We now turn our attention to the concept of paths which turns out to be dual to that of circuits. Let $(E, \mathscr{F})$ be an antimatroid and X a feasible set. An **endpoint** of X is an element $e \in X$ such that $X \setminus e \in \mathscr{F}$. A feasible set is a **path** if it has a single endpoint. We sometimes consider a path with endpoint e as a rooted set (P, e). Equivalently, P is a minimal feasible set containing e.

For poset antimatroids, a path (P, e) is a principal ideal with maximal element e. Let $(T, \mathscr{F})$ be the tree shelling antimatroid of a tree (T, E). A path (P, e) can be obtained by deleting one component of $T \setminus e$. The paths (P, e) of the double shelling antimatroid of an ordered set $(E, \leq)$ are principal ideals $I(x)$, where x is not maximal in E, and principal filters $F(x)$, where x is not a minimal element in E.

Lemma 3.12. *Let $(E, \mathscr{F})$ be an antimatroid and X a feasible set with k endpoints. Then X is the union of k paths.*

Proof. For each endpoint e let (P, e) be a minimal feasible subset of X containing e. Then (P, e) is a path and $\bigcup(P, e) = X$.   □

In particular, a set is feasible if and only if it is a union of paths. This leads to the following characterization of antimatroids in terms of paths (cf. Goecke [1986]):

Theorem 3.13. *A finite collection* $\mathscr{P} = \{(P_i, e_i) : i \in I\}$ *of rooted sets is the collection of paths of an antimatroid* $(E, \mathscr{F})$ *if and only if the following conditions hold:*

(3.2.1) $\bigcup_{i \in I} P_i = E$,

(3.2.2)   *for all* $(P, e) \in \mathscr{P}$ *and* $f \in P \setminus e$,

   (a) there exists a $(P', f) \in \mathscr{P}$ *with* $P' \subseteq P \setminus e$, *and*

   (b) there does not exist a $(P', f) \in \mathscr{P}$ *with* $e \in P' \subseteq P$.

Proof. The "only if" part is trivial. Conversely, the conditions (3.2) imply that

$$\mathscr{F} = \left\{ \bigcup \{P_j : j \in J\} : J \subseteq I \right\}$$

is a normal accessible set system which is closed under union. Hence $(E, \mathscr{F})$ is an antimatroid.

It remains to verify that $\mathscr{P}$ is the collection of paths of $(E, \mathscr{F})$. First let (P, e) be a path in $\mathscr{F}$. Then $P = \bigcup P_j, j \in J$. Let $e \in P_j$. Then $P_j \in \mathscr{F}$. Since P is a minimal set containing e, we must have $P = P_j$. It is then clear that $e = e_j$. Conversely, P_j is feasible and it is trivial from (3.2) that its only endpoint is e_j, so (P_j, e_j) is a path.   □

For more details on paths, cf. Goecke [1986]. In the remaining part of this section we are going to demonstrate that circuits and paths are in blocking relation to each other.

For this purpose let $\mathscr{P}_x$ be the collection of paths with endpoint x. More precisely,

$$\mathscr{P}_x = \{P \setminus x \in \mathscr{F} : P \cup x \in \mathscr{P}\}.$$

Similarly, let

$$\mathscr{K}_x = \{C \setminus x : (C, x) \in \mathscr{K}\}$$

be the stems of circuits rooted at x. Both $\mathscr{K}_x$ and $\mathscr{P}_x$ form a clutter.

Theorem 3.14. *Let* $(E, \mathscr{F})$ *be an antimatroid. Then* $\mathscr{P}_x$ *is the blocker of* $\mathscr{K}_x$ *and vice versa, i.e.*

$$\mathscr{B}(\mathscr{P}_x) = \mathscr{K}_x \text{ and } \mathscr{B}(\mathscr{K}_x) = \mathscr{P}_x \text{ for all } x \in E.$$

Proof. Since the blocking operation is in general involutory, it suffices to prove $\mathscr{B}(\mathscr{P}_x) = \mathscr{K}_x$. This relation is trivially fulfilled if $\mathscr{P}_x = \{\emptyset\}$ or, equivalently, if $x \in \mathscr{F}$.

Assume now that $x \notin \mathscr{F}$ and let $C \in \mathscr{K}_x$ be the stem of some circuit rooted at x. Then $C \cap X \neq \emptyset$ for all $X \in \mathscr{F}$ containing x, in particular $C \cap P \neq \emptyset$ for all $P \in \mathscr{P}_x$.

Consider any proper subset $C' \subset C$ and some $y \in C \setminus C'$. Since $C \cup x \in \mathcal{X}_x$, there exists a set $X \in \mathcal{F}$ such that

$$X \cap (C \cup x) = \{x, y\}.$$

The set X contains some path $P \in \mathcal{P}_x$ for which $P \cap C' = \emptyset$, i.e. no proper subset of C intersects all paths with endpoint x. Hence $\mathcal{X}_x \subseteq \mathcal{B}(\mathcal{P}_x)$.

Conversely, choose any $C \in \mathcal{B}(\mathcal{P}_x)$. Then the same argument as before proves $x \notin \mathcal{F} : (C \cup x)$. Now let A be the basis of $E \setminus (C \cup x)$ and consider some $y \in C$. Since C is minimal with respect to set inclusion, there exists some path $P \in \mathcal{P}_x$ such that

$$P \cap (C \setminus y) = \emptyset.$$

Then $y \in P \in \mathcal{F}$ and, by the accessibility of $\mathcal{F}$, we obtain a subset $B \subseteq P \setminus y \subseteq A$ with $B \cup y \in \mathcal{F}$. Hence $A \cup y = A \cup (B \cup y) \in \mathcal{F}$.

Thus $\mathcal{F} : (C \cup x) = 2^{C \cup x} \setminus x$ and $(C \cup x, x) \in \mathcal{X}_x$ by Lemma 3.3. $\qquad \square$

4. Helly's Theorem and Relatives

A number of geometric ideas involving convexity generalize to antimatroids. For each antimatroid $(E, \mathcal{F})$ we can define the **Caratheodory number** $c(\mathcal{F})$ as the least integer d for which the following property holds: for all $X \subseteq E$ and all $a \in \tau(X)$ there exists a subset $X' \subseteq X$ such that $|X'| \leq d$ and $a \in \tau(X')$.

Caratheodory's Theorem asserts that if $(E, \mathcal{F})$ is the convexity antimatroid in $\mathbb{R}^d$ then its Caratheodory number is at most $d + 1$.

Proposition 4.1. *For every antimatroid* $(E, \mathcal{F})$,

$$c(\mathcal{F}) = \max\{|C| : C \in \mathcal{C}\} - 1.$$

Proof. Let $(C, r) \in \mathcal{C}$ be a circuit. Then $r \in \tau(C \setminus r)$ by Lemma 3.4. For any proper subset D of $C \setminus r$, however, we have $C \setminus D \in \mathcal{F} : C$, i.e. $r \notin \tau(D)$. Hence $c(\mathcal{F}) \geq |C| - 1$.

Conversely, consider an element $a \in \tau(X)$. If $a \in X$, then we may clearly set $X' = \{a\}$. Otherwise, by Lemma 3.5, we have a circuit (C, a) such that $C \subseteq X \cup a$. Hence $c(\mathcal{F}) \leq \max\{|C| : C \in \mathcal{C}\} - 1$. $\qquad \square$

We define the **Helly number** $h(\mathcal{F})$ of an antimatroid as the least integer k for which the following property holds: for any family $\mathcal{H}$ of convex sets such that $\bigcap \mathcal{H} = \bigcap\{X : X \in \mathcal{H}\} = \emptyset$ there exists a subfamily $\mathcal{H}' \subseteq \mathcal{H}$ such that $|\mathcal{H}'| \leq k$ and $\bigcap \mathcal{H}' = \emptyset$.

Proposition 4.2. *The Helly number is the least integer k such that for all $X \subseteq E$ with $|X| \geq k + 1$, $\bigcap\{\tau(X \setminus y) : y \in X\} \neq \emptyset$.*

Proof. Let k be the Helly number of $(E, \mathcal{F})$ and $X \subseteq E$ a subset with $|X| \geq k+1$. Set $\mathcal{H} = \{\tau(X \setminus y) : y \in X\}$. Then for any $\mathcal{H}' \subseteq \mathcal{H}$ with $|\mathcal{H}'| = k$ there exists an element $z \in X$ such that $\tau(X \setminus z) \notin \mathcal{H}'$. Now $z \in X \setminus y \subseteq \tau(X \setminus y)$ for all $y \in X \setminus z$, i.e. $\bigcap \mathcal{H}' \neq \emptyset$. Thus $\bigcap \mathcal{H} = \bigcap \{\tau(X \setminus y) : y \in X\} \neq \emptyset$.

Conversely, let $\mathcal{H} = \{X_1, \ldots, X_{k+m}\}$ be a family of convex sets with $|\mathcal{H}| \geq k + 1$. By induction we may assume that $\bigcap(\mathcal{H} \setminus \{X_i\}) = \bigcap\{X : X \in \mathcal{H}, X \neq X_i\} \neq \emptyset$ for any set $X_i \in \mathcal{H}$. Hence we may choose a $y_i \in \bigcap(\mathcal{H} \setminus \{X_i\})$ and set $Y = \{y_1, \ldots, y_{k+m}\}$. If $|Y| < k + m$, there must exist a $y_i \in X_i$, i.e. $\bigcap \mathcal{H} \neq \emptyset$. So we may assume $|Y| = k + m$. Since $Y \setminus y_i \subseteq X_i$, we have $\tau(Y \setminus y_i) \subseteq X_i$ and thus $\emptyset \neq \bigcap\{\tau(Y \setminus y_i) : y_i \in Y\} \subseteq \bigcap \mathcal{H}$.   □

With this proposition we can prove a nice characterization of the Helly number due to Hoffman [1979] and Jamison [1981] :

Theorem 4.3. *The Helly number of an antimatroid is equal to the maximum size of a free convex set.*

Proof. For any free convex set X, $\bigcap\{\tau(X \setminus y) : y \in X\} = \emptyset$. Hence the Helly number is at least as great as the maximum size of a free convex set.

Conversely, we know from Proposition 4.2 that there exists a set $X \subseteq E$ with $|X| = k$ such that $\bigcap\{\tau(X \setminus x) : x \in X\} = \emptyset$. Choose such an X with $\tau(X)$ minimal. It is clear that X is free. We show that it is also convex. Suppose not, then there exists $p \in \tau(X) \setminus X$. By the definition of X there must exist an element $x \in X$ such that $p \notin \tau(X \setminus x)$. Choose p so that the number of such x is minimum and let $q \in X$ be such that $p \notin \tau(X \setminus q)$.

Consider the set $Y = X \setminus q \cup p$. Clearly, $|Y| = k$ and $\tau(Y) \subset \tau(X)$. So by the minimality of X, $\bigcap\{\tau(Y \setminus y) : y \in Y\} \neq \emptyset$. Select an element $z \in \bigcap\{\tau(Y \setminus y) : y \in Y\}$. Then, in particular, $z \in \tau(Y \setminus p) = \tau(X \setminus q)$ and so by the choice of p, there must exist an element $w \in X$ such that $z \notin \tau(X \setminus w)$ but $p \in \tau(X \setminus w)$. Then $Y \setminus w \subseteq \tau(X \setminus w)$, hence $\tau(Y \setminus w) \subseteq \tau(X \setminus w)$. This implies that $z \notin \tau(Y \setminus w)$, which is a contradiction.   □

As an illustration we give some examples of the Helly number in antimatroids. First, consider the node shelling antimatroid of a tree. Free convex sets are subtrees with at most two nodes and hence its Helly number is at most two. This implies the well-known fact that, if any two members of a family of subtrees intersect, then all subtrees have a common node.

Secondly, in the edge shelling antimatroid of a tree, a free convex set is a star. Hence its Helly number is the maximum degree d of the tree. So we get from Theorem 4.3 that if any d members of a family of subtrees have a common edge then all have a common edge. This is a result of Lehel [1982].

Furthermore, Doignon [1973], Bell [1977] and Scarf [1977] independently proved that a system of linear inequalities $Ax \leq b$ in d variables has an integer solution if any subsystem $A'x \leq b'$ with up to 2^d rows has an integer solution. This again is implied by Theorem 4.3 if we consider the convexity antimatroid restricted to integer points. If X is a convex free set in this antimatroid and $|X| > 2^d$, then there are at least two points $x, y \in X$ whose coordinates are congruent modulo 2. Hence $z = (x + y)/2$ is an integer point and $z \in X$, because

X is convex. Then, however, X is not free. So the size of a convex free set is at most 2^d and the unit cube shows that this is best possible.

The **Radon number** $\mathrm{rd}(\mathcal{F})$ of an antimatroid $(E, \mathcal{F})$ is the smallest integer such that for any subset $A \subseteq E$ of size $|A| = \mathrm{rd}(\mathcal{F})$ there exists a nonempty proper subset $B \subseteq A$ with $\tau(B) \cap \tau(A \setminus B) \neq \emptyset$.

The following results (due to Levi [1951] and Kay and Womble [1971]) describe some relationships between the Caratheodory, Helly and Radon numbers of antimatroids.

Lemma 4.4. *For any antimatroid* $(E, \mathcal{F})$

$$\mathrm{h}(\mathcal{F}) \leq \mathrm{rd}(\mathcal{F}) - 1.$$

Proof. For any subset $A \subseteq E$ with $|A| = \mathrm{rd}(\mathcal{F})$ there exists a proper nonempty subset B such that $\tau(B) \cap \tau(A \setminus B) \neq \emptyset$. Since $\tau(B) \subseteq \bigcap\{\tau(A \setminus a) : a \in A \setminus B\}$ and $\tau(A \setminus B) \subseteq \bigcap\{\tau(A \setminus a) : a \in B\}$, we get

$$\emptyset \neq \tau(B) \cap \tau(A \setminus B) \subseteq \bigcap\{\tau(A \setminus a) : a \in A\},$$

i.e. $\mathrm{h}(\mathcal{F}) \leq \mathrm{rd}(\mathcal{F}) - 1$. □

Lemma 4.5. *For any antimatroid* $(E, \mathcal{F})$

$$\mathrm{rd}(\mathcal{F}) \leq \mathrm{c}(\mathcal{F}) \cdot \mathrm{h}(\mathcal{F}) + 1.$$

Proof. Let $S \subseteq E$ be any subset with $|S| = \mathrm{c}(\mathcal{F}) \cdot \mathrm{h}(\mathcal{F}) + 1$. Define a family of subsets by $\mathcal{H}' = \{X \subseteq S : |X| \geq \mathrm{c}(\mathcal{F}) \cdot \mathrm{h}(\mathcal{F}) + 1 - \mathrm{c}(\mathcal{F})\}$. Set $\mathcal{H} = \{\tau(X) : X \in \mathcal{H}'\}$. Then each collection of $\mathrm{h}(\mathcal{F})$ members of $\mathcal{H}$ has nonempty intersection. Hence, there exists an $x \in \bigcap \mathcal{H}$. Since $S \in \mathcal{H}'$ and thus $\tau(S) \in \mathcal{H}$, we have $x \in \tau(S)$. As $|S| > \mathrm{c}(\mathcal{F})$, there exists a proper subset $T \subseteq S$ with $x \in \tau(T)$ and $|T| \leq \mathrm{c}(\mathcal{F})$. Now $|S \setminus T| \geq \mathrm{c}(\mathcal{F})\mathrm{h}(\mathcal{F}) + 1 - \mathrm{c}(\mathcal{F})$ implies $\tau(S \setminus T) \in \mathcal{H}$, and so $x \in \tau(S \setminus T)$. Thus $(T, S \setminus T)$ is a Radon partition, which shows $\mathrm{rd}(\mathcal{F}) \leq \mathrm{c}(\mathcal{F})\mathrm{h}(\mathcal{F}) + 1$. □

Furthermore, we associate with every antimatroid the **Erdös-Szekeres number** $\mathrm{es}(\mathcal{F})$ which is defined as the maximum size of a free set (by Theorem 4.3 this is the maximum Helly number of traces of the antimatroid). In the convex shelling antimatroid the Erdös-Szekeres number is the maximum number of points that are the vertices of a convex polytope. The name is motivated by the following classical result of Erdös and Szekeres [1935] :

Theorem 4.6. *Every set of* $\binom{2k-2}{k} + 1$ *points in the plane in general position contains the vertices of a convex k-gon.* □

It is conjectured that the threshold $\binom{2k-2}{k-1} + 1$ can be replaced by $2^{k-2} + 1$. Clearly, $\mathrm{h}(\mathcal{F}) \leq \mathrm{es}(\mathcal{F})$ and $\mathrm{c}(\mathcal{F}) \leq \mathrm{es}(\mathcal{F})$, since if $a \in \tau(X)$ and we choose a minimal subset $Y \subseteq X$ such that $a \in \tau(Y)$, then Y must be free.

Lemma 4.7. *For any antimatroid* $(E, \mathcal{F})$,

$$\mathrm{rd}(\mathcal{F}) \leq \mathrm{es}(\mathcal{F}) + 1.$$

Proof. For a given subset A with $|A| = \mathrm{es}(A) + 1$, the set $B = \mathrm{ex}(\tau(A))$ is free and hence $|B| \leq \mathrm{es}(\mathcal{F})$. Then $\tau(A \setminus B) \subseteq \tau(A) = \tau(B)$, i.e. $(B, A \setminus B)$ is a Radon partition. $\square$

The Erdös-Szekeres number has a characterization analogous to the Helly number.

Theorem 4.8. *The Erdös-Szekeres number is the smallest integer k such that every family $\mathcal{H}$ of convex sets contains a subfamily $\mathcal{H}'$, $|\mathcal{H}'| \leq k$ with $\bigcap \mathcal{H} = \bigcap \mathcal{H}'$.*

Proof. Let $\mathcal{H}' = \{X_1, \ldots, X_k\} \subseteq \mathcal{H}$ be a smallest subfamily such that $\bigcap \mathcal{H}' = \bigcap \mathcal{H}$. Then for every $1 \leq i \leq k$ there exists an $a_i \in \bigcap \mathcal{H}' \setminus X_i$, $a_i \notin X_i$. Set $A = \{a_1, \ldots, a_k\}$ and observe that $|A| = k$. Then $A \setminus a_i \subseteq X_i$ implies $\tau(A \setminus a_i) \subseteq X_i$, and hence $a_i \notin \tau(A \setminus a_i)$. Thus A is free, and hence $|\mathcal{H}'| \leq \mathrm{es}(\mathcal{F})$.

Conversely, for a free set A let $\mathcal{H} = \{\tau(A \setminus a) : a \in A\}$. Since $a \notin \tau(A \setminus a)$, we have $\bigcap \mathcal{H}' \neq \bigcap \mathcal{H}$ for every proper subfamily $\mathcal{H}'$ of $\mathcal{H}$, which completes the proof. $\square$

This generalizes a result of Golumbic and Jamison [1985].

5. Ramsey-Type Results

The Erdös-Szekeres Theorem 4.6 is one of the early and classical Ramsey-type results. We will show that this result can be generalized to antimatroids. In order to prove this, we have to introduce a class of very homogeneous antimatroids.

Let $E = \{1, \ldots, n\}$ and $0 \leq a, b \leq n$. We define a language $\mathcal{L}$ on E such that $x_1 \ldots x_k \in \mathcal{L}$ if for $1 \leq i \leq k$, x_i is at most the ath element from below or the bth element from above in $E \setminus \{x_1, \ldots, x_{i-1}\}$, i.e.

(5.1)
$$|(E \setminus \{x_1, \ldots, x_{i-1}\}) \cap \{1, 2, \ldots, x_i\}| \leq a \quad \text{or}$$
$$|(E \setminus \{x_1, \ldots, x_{i-1}\}) \cap \{x_i, x_i + 1, \ldots, n\}| \leq b.$$

We call $(E, \mathcal{L})$ the (a, b)-**path shelling antimatroid** and denote it by $G(a, b, n)$.

Lemma 5.1. $G(a, b, n)$ *is an antimatroid.*

Proof. Since $(E, \mathcal{L})$ is a normal hereditary language, we verify property (1.5) to prove the lemma. Let $\alpha = x_1 x_2 \ldots x_k$, $\beta \in \mathcal{L}$ and let x_i be the first element in α which is not in β. Then

$$|(E \setminus \beta) \cap \{1, \ldots, x_i\}| \leq |(E \setminus \{x_1, \ldots, x_{i-1}\}) \cap \{1, \ldots, x_i\}| \quad \text{and}$$

$$|(E \setminus \beta) \cap \{x_i, x_i + 1, \ldots, n\}| \leq |(E \setminus \{x_1, \ldots, x_{i-1}\}) \cap \{x_i, x_i + 1, \ldots, n\}|,$$

so that $\beta x_i \in \mathscr{L}$. $\square$

Let us mention some special cases. For $a = b = 0$, $\mathscr{L}$ consists only of the empty word. For $a = 1$ and $b = 0$, $(E, \mathscr{L})$ is the poset antimatroid of E with the natural order. For $a = b = 1$, $(E, \mathscr{L})$ is the double shelling antimatroid of E with the natural order. For $a = 2$ and $b = 0$, $\mathscr{L}$ consists of all words for which $i \in E$ does not occur in position $i - 2$ or earlier. For $a + b = n - 1$, E is the only circuit and its root is $a + 1$. For $a + b \geq n$, we get the free antimatroid.

As these special cases suggest, the circuits of an (a, b)-path shelling antimatroid have a very regular structure. A characterization of this structure is contained in the following

Lemma 5.2. *The circuits of $G(a, b, n)$ are the sets $X \subseteq E$ with $|X| = a + b + 1$. The root of such a circuit X is the $(a + 1)$th element from below.*

Proof. We first show that every set $X \subseteq E$ with $|X| = a + b$ is free. Let $x_1 < \ldots < x_a < \ldots < x_{a+b}$. Then $X' = E \setminus (X \cup \{x_a + 1, \ldots, x_{a+1} - 1\}) \in \mathscr{F}$ and $\Gamma(X') = X$. Hence X is free. Similarly one verifies that for a set $X = \{x_1, \ldots, x_{a+b+1}\}$, $\mathscr{F} : X = 2^X \setminus \{x_{a+1}\}$, i.e. (X, x_{a+1}) is a circuit with root x_{a+1}. $\square$

Since circuits with their roots are preserved under tracing, we obtain the following

Corollary 5.3. *The trace of $G(a, b, n)$ on an m-element subset is a path-shelling antimatroid $G(a, b, m)$.* $\square$

We now state the first Ramsey type theorem (Korte and Lovász [1984b]).

Theorem 5.4. *There exists a function $f : \mathbb{N} \to \mathbb{N}$ with the following property: for every antimatroid $(E, \mathscr{F})$ with $|E| \geq f(n)$ there exists a set $T \subseteq E$ with $|T| = n$ such that $(T, \mathscr{F} : T)$ is isomorphic to a $G(a, b, n)$ for some $a, b \geq 0$.*

Proof. Denote by $R_t^k(n)$ the least integer such that if all k-subsets of a set E, $|E| = R_t^k(n)$, are colored with t colors , then there exists some n-element subset all of whose k-element subsets have the same color.

Let

$$g(n) = \max\{R_{k+1}^{k+1}(n), \; 1 \leq k \leq n - 1\}$$

and

$$f(n) = R_2^1(R_2^2(\ldots R_2^n(g(n))\ldots)) \, .$$

Consider an antimatroid $(E, \mathscr{F})$ with $|E| = f(n)$. Label the elements of E with $1, 2, \ldots, f(n)$. Color the free subsets of E red and the nonfree subsets blue. Then by Ramsey's Theorem there exists some set S with $|S| = g(n)$ such that for every k with $1 \leq k \leq n$, all k-subsets of S have the same color.

If the n-element subsets of S are red, i.e. free, then choose any such n-set $T \subseteq S$. Then $(T, \mathcal{F} : T)$ is isomorphic to $G(n, n, n)$.

If for some $1 \leq k \leq n$ the k-sets are red but the $(k + 1)$-sets are blue, i.e. circuits, then we color a circuit with color a if its $(a + 1)$th element from below is its root, $0 \leq a \leq k$. Using Ramsey's Theorem again, we can find a subset T of S with $|T| = n$ such that all $(k + 1)$-subsets of T have the same color, say a. But then $(T, \mathcal{F} : T)$ is isomorphic to $G(a, k - a, n)$.   □

Hence every large antimatroid has a large substructure which is homogeneous.

We now turn to our second result which holds only for the subclass of antimatroids satisfying the following property.

(5.2)   If $(C \cup x, x)$ and $(C \cup y, y)$ are circuits, then there exists a unique subset $C' \subseteq C$ such that $(C' \cup x \cup y, y)$ is a circuit.

The convex shelling in $\mathbb{R}^n$ has property (5.2). For let C span a simplex A with interior points x and y, then the ray from x through y intersects the boundary of A in some face. The vertices of this face are the desired subset C'.

Note that not all antimatroids satisfy (5.2) as is readily seen in the case of a poset antimatroid.

Lemma 5.4. *Let* $(E, \mathcal{F})$ *be an antimatroid with property (5.2) such that every circuit* $C \in \mathcal{K}$ *has at least four elements. Then for every* $X \subseteq E$ *there is a point* $y \in X$ *such that* $X \setminus y$ *is not a circuit.*

Proof. Suppose $X \setminus y \in \mathcal{K}$ for every $y \in X$. Then $X \neq \Gamma(A)$, where A is the basis of $E \setminus X$.

Consider some $x_0 \in X \setminus \Gamma(A)$. Then A is also the basis of $E \setminus (X \setminus x_0)$ and $X \setminus x_0 \in \mathcal{K}$. So $\Gamma(A) = X \setminus \{x_0, x_1\}$ for some $x_1 \in X \setminus x_0$, i.e. x_1 is the root of $X \setminus x_0$. Again, $X \setminus x_1 \in \mathcal{K}$, A is the basis of $E \setminus (X \setminus x_1)$ and x_0 is the root of $X \setminus x_1$.

By (5.2) there is a unique subset $Y \subseteq X \setminus \{x_0, x_1\}$ such that $Y \cup \{x_0, x_1\}$ is a circuit rooted at x_0. But then $|Y \cup \{x_0, x_1\}| = |X| - 1$ and so $Y = X \setminus \{x_0, x_1, y_0\}$ for some $y_0 \in X \setminus \{x_0, x_1\}$. Hence $(X \setminus y_0, x_0) \in \mathcal{K}$ and, by symmetry, $(X \setminus y_1, x_1) \in \mathcal{K}$ for some $y_1 \in X \setminus \{x_0, x_1\}$.

By assumption, every circuit has at least four elements, so we have $|X| \geq 5$. Consider $z \in X \setminus \{x_0, x_1, y_0, y_1\}$. Since $\Gamma(A) = X \setminus \{x_0, x_1\}$, we get for any $y \in X \setminus \{x_0, x_1, z\}$

$$(A \cup y) \cap (X \setminus z) = \{y\}.$$

Hence, by Lemma 3.3, the root of the circuit $X \setminus z$ is either x_0 or x_1. In view of the uniqueness, this contradicts the choice of y_0.   □

The above lemma is the basic tool in the proof of our second Ramsey theorem due to Korte and Lovász [1986c]. It extends the Erdös-Szekeres Theorem mentioned before from convex shelling antimatroids to a larger class.

Theorem 5.5. *There exists a function $f : \mathbf{N} \to \mathbf{N}$ such that every antimatroid contains a free set of size n provided $|E| \geq f(n)$, $(E, \mathscr{F})$ satisfies (5.2) and every circuit has at least four elements.*

Proof. As before, we color the free sets red and the nonfree sets blue. By Ramsey's Theorem, if $|E| \geq f(n) = R_2^1(R_2^2 \ldots (R_2^n(n+1)) \ldots)$, there exists a subset $X \subseteq E$ with $|X| = n+1$ such that all k-element subsets, $1 \leq k \leq n$, have the same color.

Again, either the n-sets are free or there exists a k such that all subsets of X with cardinality k are nonfree and all smaller subsets are free. Thus the k-sets are circuits but, by Lemma 5.4, this is only possible for $k = n+1$. Hence in all cases the n-element subsets of X are free. $\qquad\square$

6. Representations of Antimatroids

We shall now describe two constructions that produce an antimatroid from poset antimatroids. The first of these constructions originated from work by Korte and Lovász [1986c] and was extended by Edelman and Saks [1988].

Let $\varphi : E \to E'$ be a surjective mapping and define the **homomorphic image** $(E', \mathscr{F}')$ of $(E, \mathscr{F})$ under φ by

$$\mathscr{F}' = \{\varphi(X) : X \in \mathscr{F}\}.$$

It is well-known from matroid theory that the homomorphic image of a matroid is again a matroid. Observe that the homomorphic image of an accessible set system is still accessible. Moreover, closedness under unions is also preserved under homomorphisms. Thus we have the following

Lemma 6.1. *The homomorphic image of an antimatroid is again an antimatroid.* $\qquad\square$

For a given antimatroid $(E, \mathscr{F})$ denote by $(\mathscr{P}, \mathscr{I})$ its **path poset antimatroid,** i.e. the collection of ideals in the set of paths ordered by inclusion. Define a surjection $h : \mathscr{P} \to E$ by letting $h(P, e) = e$, and set

(6.1) $h(X) = \{e : (P, e) \subseteq X \text{ is a path}\}.$

As a direct consequence of Theorem 3.13 we obtain the following

Corollary 6.2. *Every antimatroid is the homomorphic image of its path poset antimatroid $(\mathscr{P}, \mathscr{I})$ under the map h.* $\qquad\square$

Hence there is a canonical representation of an antimatroid as a homomorphic image of a poset antimatroid. Our next result shows that this representation is in a certain sense also minimal.

Lemma 6.3. *Let $(E, \mathscr{F})$ be an antimatroid with path poset antimatroid $(\mathscr{P}, \mathscr{I})$. If $(Q, \mathscr{I})$ is another poset antimatroid such that $\varphi(Q) = \mathscr{F}$ for some homomorphism φ, then $|\mathscr{P}| \leq |Q|$.*

Proof. Let $(P,e) \in \mathscr{P}$ be a path with endpoint e. By definition there exists an ideal X in Q with $\varphi(X) = P$. Let $x \in X$ be such that $\varphi(x) = e$, and $Y \subseteq X$ be the principal ideal generated by x. Then $\varphi(Y) \in \mathscr{F}$, and clearly $e \in \varphi(Y) \subseteq P$. Since (P,e) is a path, we must have $\varphi(Y) = \varphi(X)$. Thus $|\mathscr{P}| \leq |Q|$.   □

Note, however, that even in the case where $(\mathscr{P},\mathscr{I})$ and $(Q,\mathscr{I})$ define the same antimatroid and $|\mathscr{P}| = |Q|$, the two corresponding orders need not be isomorphic as the following example shows.

Example 6.4. Consider the two ordered sets P_1 and P_2 in Figure 3.

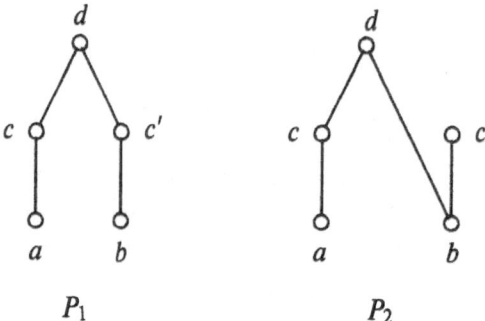

Fig. 3 Nonisomorphic orders with the same poset antimatroid

In both cases let $E = \{a, b, c, d\}$ and

$$\varphi(x) = \begin{cases} c & \text{if } x = c' \\ x & \text{otherwise} . \end{cases}$$

Then P_1 and P_2 yield the same antimatroid whose path poset is P_1, but clearly P_1 and P_2 are not isomorphic.

Next we give a partial answer to the following two questions: Under what conditions is the canonical minimal representation of an antimatroid by its path poset also unique, and secondly, if the minimal representation is not unique, how does it relate to the canonical one?

We start with some simple observations. Let $Q = (E, \leq)$ be an ordered set and φ a homomorphism mapping the ideals of Q to some antimatroid $(E, \mathscr{F})$.

Lemma 6.5. *If $\varphi(x) = \varphi(y)$ for some $x < y$, then the restriction φ' of φ to $Q \setminus y$ applied to the ideals of $Q \setminus y$ yields the same antimatroid.*

Proof. Obviously, the only ideals we have to check are those ideals in Q which have y as a maximal element. More specifically, since ideals are closed under union, we only have to check the principal ideal $I(y)$ in Q. But clearly, if $x_1, \ldots, x_k$ are the elements covered by y in Q, then

$$\varphi(I(y)) = \bigcup_{i=1}^{k} \varphi(I(x_i))$$

$$= \bigcup_{i=1}^{k} \varphi'(I(x_i)) . \qquad \square$$

So from now on, it suffices to consider only homomorphisms φ for which the preimage $\varphi^{-1}(x)$ is an antichain for every $x \in E$, i.e.

(6.2) $x < y$ implies $\varphi(x) \neq \varphi(y)$.

We have seen in Lemma 6.3 that paths in the antimatroid correspond to principal ideals in the defining order. Our next result characterizes those principal ideals which are mapped onto paths.

Lemma 6.6. *The homomorphic image $\varphi(I(x))$ of a principal ideal in Q is a path in the antimatroid if and only if for every $y \in Q$, $\varphi(I(y)) \subset \varphi(I(x))$ implies $\varphi(x) \neq \varphi(y)$.*

Proof. Assume $\varphi(I(y))$ is a proper subset of $\varphi(I(x))$ for some $x, y \in E$ with $\varphi(x) = \varphi(y)$. Let $x_1, \ldots, x_k$ be the elements covered by x, then

$$\varphi(I(x)) = \bigcup_{i=1}^{k} \varphi(I(x_i)) \cup \varphi(x)$$

$$= \bigcup_{i=1}^{k} \varphi(I(x_i)) \cup \varphi(I(y)) .$$

Hence $\varphi(I(x))$ is the union of some of its proper feasible subsets. Thus $\varphi(I(x)) \in \mathcal{F}$ cannot have a unique endpoint, i.e. is not a path.

Conversely, if $\varphi(I(x))$ is not a path, then there exists some $a \in E$ such that $\varphi(I(x)) \setminus \varphi(a) \in \mathcal{F}$ and $\varphi(a) \neq \varphi(x)$. Hence $\varphi(I(x)) \setminus \varphi(a) = \bigcup_{i=1}^{k} \varphi(I(z_i))$ for some appropriate ideals $I(z_1), \ldots, I(z_k)$. So one of these ideals contains an element y with $\varphi(y) = \varphi(x)$ such that $\varphi(I(y))$ is a proper subset of $\varphi(I(x))$. $\square$

We are now ready to answer the second question.

Theorem 6.7. *Let $Q = (E, \leq)$ be an ordered set and $(E, \mathcal{F})$ the antimatroid obtained as the homomorphic image of the poset antimatroid of Q under the homomorphism φ. Then $|Q| = |\mathcal{P}|$ if and only if $\varphi(x) = \varphi(y)$ implies $\varphi(I(x)) \not\subseteq \varphi(I(y))$.*

Proof. By Lemmas 6.3 and 6.6, $|Q| = |\mathcal{P}|$ if and only if for all $x, y \in E$, $\varphi(I(x)) \neq \varphi(I(y))$ and $\varphi(I(x)) \subseteq \varphi(I(y))$ implies $\varphi(x) \neq \varphi(y)$. So if there are two elements with $\varphi(x) = \varphi(y)$ and $\varphi(I(x)) \subseteq \varphi(I(y))$, then one of these conditions fails to hold.

Conversely, assume that $\varphi(x) = \varphi(y)$ implies that $\varphi(I(x)) \not\subseteq \varphi(I(y))$. Then every principal ideal is a path in $(E, \mathcal{F})$. It remains to show that $\varphi(I(x)) \neq \varphi(I(y))$ for any two $x, y \in E$. Suppose $\varphi(I(x)) = \varphi(I(y))$. Then $\varphi(z) = \varphi(x)$ for some $z < y$. Hence $I(z)$ and $I(y)$ violate the assumption. $\square$

In order to give a necessary condition for our first question, we consider only homomorphisms which satisfy condition (6.2) and

(6.3) $\varphi(I(x)) \subseteq \varphi(I(y))$ if and only if $x \leq y$.

It is easily seen that the mapping h defined in (6.1) satisfies (6.2) and (6.3).

Corollary 6.8. *Let* $Q = (E, \leq)$ *be an ordered set and* $(E, \mathcal{F})$ *the homomorphic image of the poset antimatroid for* Q *under a homomorphism* φ *satisfying* (6.2) *and* (6.3). *Then* Q *is the path poset and* $\varphi = h$.

Proof. If (6.2) and (6.3) hold, then the conditions of Theorem 6.8 are satisfied. Hence $|Q| = |\mathcal{P}|$ and the path poset is equal to the homomorphic image $\varphi(I(x))$ of principal ideals in Q ordered by inclusion. By (6.3), this is equivalent to the order of the path.

Furthermore, let $(P, e) = \varphi(I(e))$ be a path with endpoint e. Then, since $I(e) \setminus e \in \mathcal{F}$, we may assume inductively that

$$\varphi(I(e) \setminus e) = P \setminus e = h(P \setminus e)$$

Hence $\varphi(e) = h(e)$ for all $e \in E$. $\square$

Thus, up to isomorphism, every antimatroid has a unique representation as the homomorphic image of a poset antimatroid under homomorphisms satisfying (6.2) and (6.3).

Another operation in matroid theory which can be carried over to antimatroids is "union". Let $(E, \mathcal{F}_1)$ and $(E, \mathcal{F}_2)$ be antimatroids on the same ground set. We define their **union** $(E, \mathcal{F}_1 \vee \mathcal{F}_2)$ by

$$\mathcal{F}_1 \vee \mathcal{F}_2 = \{X_1 \cup X_2 : X_i \in \mathcal{F}_i\}.$$

It is straightforward to verify that this is again an antimatroid. A natural problem is to try to represent antimatroids as the union of simpler antimatroids. The following results were obtained by Edelman and Jamison [1985]; see also Edelman and Saks [1988]and Hexel [1988].

Given a basic word $\alpha = x_1 x_2 \dots x_n$ of $(E, \mathcal{F})$, we can consider the poset antimatroid $(E, \mathcal{F}_\alpha)$ on the chain $x_1 < x_2 < \dots < x_n$. Clearly

$$\mathcal{F} = \bigvee \{\mathcal{F}_\alpha : \alpha \ \text{a basic word of} \ (E, \mathcal{F})\}.$$

In the special case where $(E, \mathcal{F})$ is a poset antimatroid, such a representation corresponds to a representation of the partial order as an intersection of total orders. In analogy to the Dushnik-Miller dimension of posets, we say that the **convex dimension** cdim$(\mathcal{F})$ of $(E, \mathcal{F})$ is the minimum number of basic words whose union gives the antimatroid.

Theorem 6.9. *The convex dimension of an antimatroid is equal to the width of its path poset.*

Proof. Let $w(\mathscr{P})$ be the width of the path poset of $(E, \mathscr{F})$. By Dilworth's Theorem, we can find $w(\mathscr{P})$ basic words α_i such that every path is an initial subsequence of at least one of the α_i's. By Lemma 3.12, $\mathscr{F} = \bigvee\{\mathscr{F}_{\alpha_i} : 1 \leq i \leq w(\mathscr{P})\}$. Hence $\mathrm{cdim}(\mathscr{F}) \leq w(\mathscr{P})$.

If we take fewer than $w(\mathscr{P})$ basic words, then at least one path does not occur as the initial word of any of these. Hence their union will miss this path.   □

Denote by $\dim(\mathscr{F})$ the order dimension of the lattice of feasible sets of an antimatroid $(E, \mathscr{F})$.

Corollary 6.10. *For any antimatroid* $(E, \mathscr{F})$,

$$\mathrm{es}(\mathscr{F}) \leq \dim(\mathscr{F}) \leq \mathrm{cdim}(\mathscr{F}).$$

Proof. For a basic word $\alpha = x_1 x_2 \ldots x_n$ define a linear extension $\ell(\alpha)$ of $(\mathscr{F}, \subseteq)$ by letting $F_1 < F_2$ if the element with the highest index (relative to the numbering in α) of the symmetric difference $(F_1 \setminus F_2) \cup (F_2 \setminus F_1)$ occurs in F_2.

If $\alpha_1, \ldots, \alpha_k$ are basic words such that $\mathscr{F} = \bigvee \mathscr{F}_{\alpha_i}$, then clearly $F_1 < F_2$ in all associated extensions $\ell(\alpha_i)$ whenever $F_1 \subseteq F_2$. If F_1 and F_2 are incomparable in $(\mathscr{F}, \subseteq)$, consider $x \in F_1 \setminus F_2$. Then there exists a path P_x in F_1 which is not contained in F_2. Hence there exists a basic word α_i for some $1 \leq i \leq k$, such that x precedes any $y \in F_2 \setminus F_1$. Similarly, there exists a basic word α_j such that some $y \in F_2 \setminus F_1$ precedes any $x \in F_1 \setminus F_2$. Hence $F_1 < F_2$ in $\ell(\alpha_i)$ and $F_2 < F_1$ in $\ell(\alpha_j)$. This proves

$$\mathrm{cdim}(\mathscr{F}) \geq \dim(\mathscr{F}).$$

Since the dimension is monotone with respect to taking subposets, and a free set of size k induces a Boolean sublattice of rank k and order dimension k, we get

$$\dim(\mathscr{F}) \geq \mathrm{es}(\mathscr{F}).$$ □

Finally we study the question: given a poset P, we would like to represent it by convex sets in an antimatroid so that $x < y$ in P if and only if $K_x \supseteq K_y$, where K_x is the convex set representing the element x.

Lemma 6.11. *Every poset of dimension d can be represented by convex sets in an antimatroid with convex dimension d.*

Proof. Let $\alpha_1, \ldots, \alpha_d$ be basic words of the poset antimatroid such that P is the intersection of the corresponding total orders. Let $\mathscr{F} = \bigvee \mathscr{F}_{\alpha_i}$, then $\mathrm{cdim}(\mathscr{F}) \leq d$.

For every $x \in P$ and $1 \leq i \leq d$, let $A_i(x)$ be the set of elements following x in α_i including x. Then $K_x = \bigcap_{i=1}^{d} A_i(x)$ is a convex set in $(E, \mathscr{F})$. It is straightforward to check that this is a representation as desired.   □

Chapter IV. General Exchange Structures – Greedoids

Now we introduce greedoids. They can be described as hereditary languages satisfying an ordered version of the Steinitz exchange property of matroids, or as set systems, which are not necessarily independence systems but satisfy the Steinitz exchange property.

We introduce the interval property, an important property shared by very many, but not all, greedoids. We observe that matroids and antimatroids are precisely those greedoids for which appropriate stronger versions of this property hold. Most of these basic results on greedoids appeared in Korte and Lovász [1983] and [1984b].

Section 2 contains a collection of examples and constructions of greedoids.

1. Basic Facts

The first definition of greedoids may be viewed as an ordered analogue of matroids. Matroids can be considered as independence systems with an additional exchange property. We define greedoids as hereditary languages with an analogous exchange property.

A simple language $(E, \mathscr{L})$ is a **greedoid** if

(1.1) $\emptyset \in \mathscr{L}$,
(1.2) $\alpha\beta \in \mathscr{L}$ implies $\alpha \in \mathscr{L}$,
(1.3) If $\alpha, \beta \in \mathscr{L}$ with $|\alpha| > |\beta|$, then there exists an $x \in \tilde{\alpha}$ such that $\beta x \in \mathscr{L}$.

Note that since we assume $\mathscr{L}$ to be simple, (1.3) in particular implies $x \notin \tilde{\beta}$.

Obviously the associated accessible set system $(E, \mathscr{F}) = (E, \mathscr{F}(\mathscr{L}))$ of a greedoid satisfies

(1.4) $\emptyset \in \mathscr{F}$,
(1.5) if $X, Y \in \mathscr{F}$ with $|X| > |Y|$, then there exists an $x \in X \setminus Y$ such that $Y \cup x \in \mathscr{F}$.

Observe that if $\mathscr{F}$ is closed under taking subsets, (1.4) and (1.5) are equivalent to the matroid axioms. The following converse of the above statement is true.

Theorem 1.1. *A hereditary language $(E, \mathscr{L})$ is a greedoid if and only if $\mathscr{L} = \mathscr{L}(\mathscr{F}(\mathscr{L}))$ and $(E, \mathscr{F}(\mathscr{L}))$ satisfies (1.5).*

Proof. If $(E, \mathcal{L})$ is a greedoid, then, by Lemma I.1.1, $\mathcal{L} = \mathcal{L}(\mathcal{F}(\mathcal{L}))$ and $\mathcal{F}(\mathcal{L})$ satisfies (1.5).

Conversely, assume that $\mathcal{L} = \mathcal{L}(\mathcal{F}(\mathcal{L}))$ and that $(E, \mathcal{F}(\mathcal{L}))$ satisfies (1.5). We show that $\mathcal{L}$ satisfies condition (1.3). Let $\alpha = x_1 \ldots x_k \in \mathcal{L}$ and $\beta = y_1 \ldots y_i \in \mathcal{L}$ with $k > i$. Then $\{x_1 \ldots x_{i+1}\} \in \mathcal{F}(\mathcal{L})$. Hence, using (1.5), there exists an element $x \in \{x_1, \ldots, x_{i+1}\}$ such that $\beta \cup \{x\} \in \mathcal{F}(\mathcal{L})$, which then implies $\beta x \in \mathcal{L}(\mathcal{F}(\mathcal{L})) = \mathcal{L}$. □

The following Theorem shows that greedoids can also be characterized as set systems.

Theorem 1.2. *Let $(E, \mathcal{F})$ be a set system satisfying (1.4) and (1.5). Then there exists a unique greedoid $(E, \mathcal{L})$ with $\mathcal{F}(\mathcal{L}) = \mathcal{F}$.*

Proof. Suppose (1.4) and (1.5) hold for $(E, \mathcal{F})$. Then obviously $(E, \mathcal{F})$ is an accessible set system. By Theorem 1.1, the hereditary language $(E, \mathcal{L})$ with

$$\mathcal{L} = \mathcal{L}(\mathcal{F}) = \{x_1 \ldots x_k : \{x_1, \ldots, x_i\} \in \mathcal{F} \text{ for } 1 \leq i \leq k\}$$

is a greedoid and $\mathcal{F}(\mathcal{L}) = \mathcal{F}(\mathcal{L}(\mathcal{F})) = \mathcal{F}$, since $\mathcal{F}$ is accessible.

It remains to show that $(E, \mathcal{L})$ is the unique greedoid with $\mathcal{F}(\mathcal{L}) = \mathcal{F}$. Suppose there is another greedoid $(E, \mathcal{L}')$ with $\mathcal{F}(\mathcal{L}') = \mathcal{F}$. By Lemma I.1.1 we get

$$\mathcal{L} = \mathcal{L}(\mathcal{F}(\mathcal{L})) = \mathcal{L}(\mathcal{F}(\mathcal{L}')) = \mathcal{L}'.$$ □

We have observed in the previous proof that (1.4) and (1.5) imply the accessibility of $(E, \mathcal{F})$. On the other hand, assuming accessibility it is an easy exercise to verify that (1.5) is equivalent to

(1.6)   if $X, Y \in \mathcal{F}$ and $|X| = |Y| + 1$, then there exists an $x \in X \backslash Y$ such that $Y \cup x \in \mathcal{F}$.

Therefore greedoids can be considered as relaxations of matroids where subclusiveness is replaced by accessibility. It may be a little surprising that antimatroids, whose exchange property is opposite to that of matroids, are also a special case of greedoids. In fact, exchange axiom (III.1.5) is a strengthening of exchange axiom 1.3.

The above theorem allows us to treat greedoids either as languages or as certain set systems. In the sequel we will make use both of the ordered and of the unordered version depending on which is more convenient.

A seemingly even weaker version of the exchange property is the following:

(1.7)   If $A \subseteq E$ and $x, y, z \in E \backslash A$ such that $A \cup x \in \mathcal{F}$, $A \cup y \in \mathcal{F}$, $A \cup x \cup z \in \mathcal{F}$ and $A \cup x \cup y \notin \mathcal{F}$, then $A \cup y \cup z \in \mathcal{F}$.

Note that (1.7) is an immediate consequence of (1.6) by augmenting $A \cup y$ from $A \cup x \cup z$. The following result shows the converse of this observation.

Theorem 1.3. *Let $(E, \mathcal{F})$ be an accessible set system satisfying (1.7). Then $(E, \mathcal{F})$ is a greedoid.*

Proof. Since $(E, \mathcal{F})$ is accessible, it suffices to prove (1.6). Let X, Y be in $\mathcal{F}$ with $|X| = |Y| + 1$. Let C be a maximal subset of $X \cap Y$ in $\mathcal{F}$. We will verify (1.6) by induction on $|X \cup Y| + |Y \backslash C|$ and can assume $C \neq Y$.

If we apply the induction hypothesis to Y and C, we obtain an element $a \in Y \backslash C$ such that $C \cup a \in \mathcal{F}$. Then clearly $a \notin X$. Using induction again, we can augment $C \cup a$ from X to a set $Y' \in \mathcal{F}$ such that $C \cup a \subseteq Y' \subseteq X \cup a$ and $|Y'| = |Y|$.

We now claim that Y' can be augmented from X. This follows from the induction hypothesis if $|X \cup Y'| < |X \cup Y|$. Otherwise we have $Y \backslash X = \{a\}$. We can augment C from X to a set $C \cup x \in \mathcal{F}$ for some $x \in X \backslash C$. If $x \in C$, then $X \cap Y'$ contains a larger feasible set than $X \cap Y$ and we can augment Y' again by the induction hypothesis. So we may assume that $x \in X \backslash Y'$. As before, augment $C \cup x$ from Y' to get a set $X' \in \mathcal{F}$ such that $|X'| = |Y'|$ and $C \cup x \subseteq X' \subseteq Y' \cup x$. We distinguish two cases.

If $a \notin X'$, then $X' = Y' \cup x \backslash a$. Let $A = X \cap Y' = Y' \backslash a$, $y = a$ and $z = X \backslash X'$. Then $A \cup x = X' \in \mathcal{F}$, $A \cup y = Y' \in \mathcal{F}$, $A \cup x \cup z = X \in \mathcal{F}$, hence either $Y' \cup x \in \mathcal{F}$ or $Y' \cup z \in \mathcal{F}$.

If $a \in X'$, then we can use the induction hypothesis to augment X' from X by some $u \in X \backslash X'$. But now $C \cup a \subseteq Y' \cap (X' \cup u)$, so again by the induction hypothesis, we can augment Y' from $(X' \cup u) \backslash Y' \subseteq X \backslash Y'$.

In any case, there exists an element $b \in X \backslash Y'$ such that $Y' \cup b \in \mathcal{F}$. By assumption, $C \cup a \subseteq Y \cap (Y' \cup b)$, so we can augment Y from $(Y' \cup b) \backslash Y \subseteq X \backslash Y$. $\square$

One might suspect that it suffices to restrict condition (1.7) to feasible sets $A \in \mathcal{F}$. The following example, however, demonstrates that this is not the case. Let $E = \{a, x, y, z\}$ and $\mathcal{F} = \{\emptyset, x, y, \{a, x\}, \{a, y\}, \{a, x, z\}\}$. Then the restricted version (1.7) holds, whereas (1.7) fails for $A = \{a\}$.

An alternative way of stating condition (1.5) is given by the following

Lemma 1.4. *For a given set system $(E, \mathcal{F})$ property (1.5) holds if and only if for any $A \subseteq E$ all bases of A have the same cardinality.*

Proof. Suppose that for some $A \subseteq E$ two bases of A have different cardinalities, then (1.5) implies that the smaller one can be augmented, which leads to a contradiction. Conversely, let $X, Y \in \mathcal{F}$ with $|X| > |Y|$ and assume that Y cannot be augmented from X. Then Y is a basis of $X \cup Y$ and X is contained in some basis of $X \cup Y$ which is strictly larger than Y, contradicting the assumption. $\square$

We say that a greedoid is **full** if $E \in \mathcal{F}$. Antimatroids are always full greedoids.

In the next chapter we will extend the notions of the rank function and the closure operator from matroids to greedoids. It will turn out that, as in the case of matroids, both concepts lead to axiomatizations of greedoids.

A fundamental property which is shared by both matroids and antimatroids and other substantial classes of greedoids but not by all greedoids is the interval property. A greedoid $(E, \mathcal{F})$ is said to have the **interval property** (or to be an **interval greedoid**) if for all $A, B, C \in \mathcal{F}$ with $A \subseteq B \subseteq C$ and all $x \in E \backslash C$ the following holds :

$$A \cup \{x\} \in \mathcal{F} \text{ and } C \cup \{x\} \in \mathcal{F} \text{ implies } B \cup \{x\} \in \mathcal{F}.$$

In particular, we say that $(E, \mathcal{F})$ is an **interval greedoid without lower (resp. upper) bounds** if for all $A, B \in \mathcal{F}$ with $A \subseteq B$ (resp. $B \subseteq A$) and all $x \in E \backslash (A \cup B)$

$$B \cup \{x\} \in \mathcal{F} \text{ implies } A \cup \{x\} \in \mathcal{F}.$$

Lemma 1.5. *A normal interval greedoid $(E, \mathcal{F})$ is a matroid if and only if $x \in \mathcal{F}$ for all $x \in E$.*

Proof. Since normal matroids obviously have this property, it remains to show the reverse direction. Let $X \in \mathcal{F}$ be a feasible set and suppose $Y \subseteq X$ is a minimal subset with $Y \notin \mathcal{F}$.

Let $x \in Y$ be an arbitrary element. Then $Y \backslash x \in \mathcal{F}$ and it can be augmented from $X \backslash (Y \backslash x)$. Hence there exists a subset $Z \in \mathcal{F}$ such that $Y \backslash x \subseteq Z$, $x \notin Z$ and $Z \cup x \in \mathcal{F}$. So x can be added to $\emptyset$ and to Z but not to $Y \backslash x$, contradicting the interval property. $\qquad\square$

Corollary 1.6. *A greedoid is a matroid if and only if it is an interval greedoid without lower bounds.* $\qquad\square$

The following lemma stresses the fact that antimatroids are in a sense dual to matroids.

Lemma 1.7. *A greedoid is an antimatroid if and only if it has the interval property without upper bounds.*

Proof. Since by Lemma III.1.2 the feasible sets of an antimatroid are closed under union, it has the interval property without upper bounds.

Now let $(E, \mathcal{F})$ be a greedoid with interval property without upper bounds. Let $\alpha, \beta \in \mathcal{L}$ be feasible words with $\tilde{\alpha} \nsubseteq \tilde{\beta}$. We may write $\alpha = \alpha_1 x \alpha_2$ with $x \notin \tilde{\beta}$ and $\tilde{\alpha}_1 \subseteq \tilde{\beta}$. Since $\tilde{\alpha}_1 \cup x \in \mathcal{F}$, the interval property without upper bounds then implies $\tilde{\beta} \cup x \in \mathcal{F}$, i.e. $\beta x \in \mathcal{L}$. Hence $(E, \mathcal{L})$ is an antimatroid. $\qquad\square$

Several extensions of the notion of greedoids have been proposed. It is natural to relax the condition that no repeated letters are allowed. Greedoids with repetition were first studied by Björner [1985]. He showed that such generalized greedoids arise from Coxeter groups. In particular, antimatroids with repetition can be defined; these were introduced by Björner, Lovász and Shor [1988], who applied this notion in the study of a diffusion process on graphs.

Another interesting connection between greedoids and Coxeter groups occurs in the work of Gelfand and Serganova [1987]. In studying a stratification of the Grassmannian, they introduce a generalized matroid associated with Coxeter groups. Both matroids and greedoids arise by appropriate special choices of the Coxeter group.

2. Examples of Greedoids

Many procedures and structures in combinatorics and some other fields of mathematics give rise to greedoids. In this section we give an illustration of the rich variety of these structures. For some of them we will prove that they are greedoids, while for others we will refer to later chapters or leave it as an exercise. Further examples can be found in Goecke, Korte and Lovász [1989].

The first example of course is the motivating source of the theory of greedoids, namely

2.1 Matroids

We have just seen that matroids are precisely the interval greedoids without lower bound.

2.2 Antimatroids

Antimatroids are precisely the interval greedoids without upper bound.

The principles of shelling and searching also give rise to greedoids which are not necessarily antimatroids, as the following examples show.

2.3 Gaussian Elimination for Matrices

Let $A = (a_{ij})$ be an (m, n)-matrix with full row rank. The Gaussian elimination algorithm selects for row 1 a column index j_1 such that $a_{1j_1} \neq 0$ and performs a pivot operation. Continuing thus for rows $2, \ldots, m$ yields pivot sequences $j_1 \ldots j_m$ which form the basic words of a greedoid $(E, \mathscr{F})$, where $E = \{1, \ldots, n\}$. In other words $\mathscr{F}$ consists of all those subsets $\{j_1, \ldots, j_k\}$ such that the submatrix $(a_{ij_i})_{i=1}^k$ is nonsingular.

The relation between Gaussian elimination and greedoids was first studied by Goecke [1986]. We will verify the greedoid properties for these sequences in Chapter IX on Gaussian elimination greedoids.

Recall that tree shelling and repeated elimination of simplicial nodes define antimatroids. We will give similar elimination rules which lead to greedoids but not to antimatroids.

2.4 Bisimplicial Elimination

Bisimplicial elimination, also called perfect elimination (cf. Golumbic [1980]), is a pivot rule for Gaussian elimination that maintains sparsity of the matrix. We formulate it in combinatorial terms.

Let (S, T, E) be a bipartite graph. An edge $(s, t) \in E$ is called **bisimplicial** if the vertices adjacent to s or to t induce a complete bipartite subgraph. A sequence of edges $(s_1, t_1) \ldots (s_k, t_k)$ is called an **elimination sequence** if for any $1 \leq i \leq k$, (s_i, t_i) is bisimplicial in $G \setminus \{s_1, t_1, \ldots, s_{i-1}, t_{i-1}\}$.

Define a hereditary language $(S, \mathscr{L})$ over S, where $s_1 s_2 \ldots s_k \in \mathscr{L}$ if for all $1 \leq i \leq k$ there exist $t_1, \ldots, t_i \in V$ such that $(s_1, t_1) \ldots (s_k, t_i)$ is an elimination sequence. The proof that $(S, \mathscr{L})$ is a greedoid will be given in Chapter X (cf. Lemma 2.6).

In the theory of posets and in combinatorial topology, retraction and dismantling are two important decomposition methods. For example, they were used in Duffus and Rival [1976] to analyse the fixed point property in posets.

The following two examples formulate these procedures in graph-theoretic terms.

2.5 Retract Elimination

Let x, y be two vertices of a directed graph $G = (V, A)$, where we assume that a loop is attached to every vertex in G. We say that x is **retractable** to y if $(x, z) \in A$ implies $(y, z) \in A$ and $(z, x) \in A$ implies $(z, y) \in A$; in other words, the map $\varphi_{xy} : V \to V \setminus \{x\}$ defined by

$$\varphi_{xy}(z) = \begin{cases} y & \text{if } z = x \\ z & \text{if } z \neq x \end{cases}$$

is arc-preserving. The **retract greedoid** $(V, \mathscr{L})$ is defined by

$$\mathscr{L} = \{x_1 \ldots x_k : x_i \text{ is retractable in } V \setminus \{x_1, \ldots, x_{i-1}\} \text{ for all } 1 \leq i \leq k\}$$

(cf. Chapter X).

2.6 Dismantling in Directed Graphs

Again, let $G = (V, A)$ be a directed graph. A vertex x is **dismantlable** if there exists another vertex $y \neq x$ such that x and y are adjacent in G and x is retractable to y. The feasible words of the **dismantling greedoid** are sequences $x_1 \ldots x_k$ such that x_i is dismantable in $G \setminus \{x_1, \ldots, x_{i-1}\}$ (cf. Chapter X) .

A fundamental procedure in graph theory is the following.

2.7 Series-Parallel Reduction in Graphs

Let $G = (V, E)$ be a graph without loops. An edge $e \in E$ is **reducible** if it is a pending edge or parallel with another edge or in series with another edge. If e is a reducible edge, let G/e be the graph obtained from G by either deleting e if e has a parallel edge, or by contracting e if it is pending or in series with another edge.

Call a sequence $e_1 \ldots e_k$ of edges a **series-parallel reduction sequence** if e_i is reducible in $G/e_1/e_2 \ldots /e_{i-1}$. Then the system $\mathscr{L}$ of all series-parallel reduction sequences forms a greedoid $(E, \mathscr{L})$ (cf. Chapter X).

2.8 Simplicial Clique Elimination

Let $G = (V, E)$ be a graph. Define a language $(E, \mathscr{L})$ with $E = V(G)$ consisting of all words $x_1 \ldots x_k$ such that x_i is a simplicial vertex in $G \setminus (\overline{N}(x_1) \cup \ldots \cup \overline{N}(x_{i-1}))$. It can easily be proved that $(E, \mathscr{L})$ is a greedoid, e.g. by verifying property (1.7).

Every feasible set of this greedoid is stable in the graph. If G is triangulated, every basis is a maximum size stable set (cf. Gavril [1972] and Section IX.3).

The principle of stepwise searching of structures yields the following classes of greedoids:

2.9 Undirected Branching

Let $G = (V, E)$ be a graph and $r \in V$ a specified root. Let $\mathscr{F}$ be the family of edge-sets of subtrees of G containing r.

We show that this is indeed a greedoid. By definition, $(E, \mathscr{F})$ is a hereditary language. Let $X, Y \in \mathscr{F}$ with $|X| > |Y|$. Then there exists a vertex in G which is covered by the edges in X but not by the edges in Y. Let e be an edge in X such that one endpoint of it is covered by Y and the other is not. Then $Y \cup x$ is in $\mathscr{F}$.

2.10 Directed Branching

Let $D = (V, A)$ be a directed graph and $r \in V$. The system $\mathscr{F}$ consists of arc-sets of arborescences in G rooted at r. Then $(A, \mathscr{F})$ is a greedoid, which can be verified by similar arguments as above. The bases of $(A, \mathscr{F})$ are maximal branchings in G.

2.11 Ear Decomposition in 2-Connected Graphs

It is well-known that every 2-connected graph G has an ear decomposition

$$G = e_0 \cup P_1 \cup \ldots \cup P_r \, ,$$

where e_0 is a specified edge and P_i is an "ear", i.e. a path having exactly its two endpoints in common with $e_0 \cup P_1 \cup \ldots \cup P_{i-1}$. (Such a decomposition is closely related to depth-first search).

Ear decompositions of G give rise to a greedoid in the following way. Let $G = (V, E')$ be a 2-connected graph, $e_0 \in E'$ and $E = E' \setminus \{e_0\}$. Let $\mathscr{F}$ be the collection of all subsets $X \subseteq E$ such that

(i) $X \cup \{e_0\}$ is a connected subgraph.
(ii) every block of $X \cup \{e_0\}$ not containing e_0 is a single edge.

2.12 Ear Decomposition in Strongly Connected Digraphs

A similar construction also applies to strongly connected digraphs. Let $G = (V, A)$ be a strongly connected digraph and $r \in V$. Let $\mathscr{F}$ consist of all arc-sets $X \subseteq A$ of the form

$$X = X_0 \cup \{X_v : v \in V(X)\} ,$$

where X_0 is a strongly connected subdigraph containing r and X_v is an arborescence rooted at v with

$$V(X_v) \cap V(X_u) = \emptyset \qquad u, v \in V(X_0),\ u \neq v,$$
$$V(X_v) \cap V(X_0) = \{v\} \qquad v \in V(X_t).$$

Similarly to the previous construction it can be shown that this defines a greedoid. Morever, every basic word of this greedoid induces an ear decomposition of G into directed paths.

2.13 Blossom Greedoid

Let $G = (V, E)$ be a graph and M a matching. We define a greedoid $(E', \mathscr{F})$ on $E' = E \setminus M$ so that a feasible sequence of edges corresponds to an edge labeling in the matching algorithm. Let us sketch the Edmonds matching algorithm in a form which is convenient for us.

A graph is **matching-critical** if deleting any of its nodes results in a graph that has a perfect matching. Obviously such a graph has an odd number of nodes. A **blossom forest** B (with respect to M) is a subgraph of G that has a set A of nodes (inner nodes) with the following properties. Every connected component of $G \setminus A$ is matching-critical and almost perfectly matched up by M, every node of A is a cut-point, has degree two and is adjacent to exactly one edge of $M \cap G$.

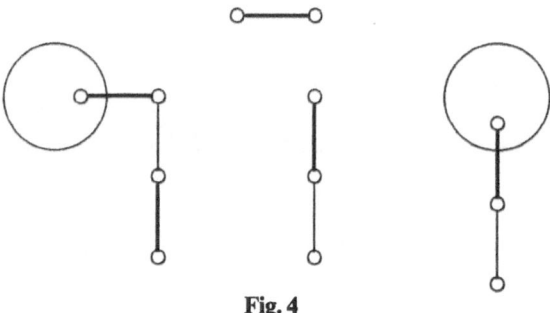

Fig. 4

We briefly mention some properties of the blossom forest. The connected components of $G \setminus A$ are called the **pseudonodes** or **blossoms** of the blossom forest, and if we contract every pseudonode we get a forest. Every connected component of B contains exactly one node that is not covered by M.

The Edmonds algorithm consists of building up a blossom forest one edge at a time (not counting the edges in M). If it finds an edge connecting two pseudonodes in different components of the blossom forest, then it is easy to construct a larger matching and the algorithm starts again. If it finds an edge connecting a pseudonode to a node v outside the blossom forest, then it adds this edge and the edge of M incident with v to the blossom forest. If it finds an edge that connects two pseudonodes in the same connected component or two nodes in the same pseudonode, then it adds this to the blossom forest. Finally, if none of these steps can be carried out, the algorithm stops and concludes that the matching M is maximum.

If M is a maximum matching, the edge set of the blossom forests restricted to $E(G) \setminus M$ with respect to M forms a greedoid. To prove this, we use the fact from matching theory that every maximum matching forest has the same set A of inner nodes and the same pseudonodes. Hence they all have the same number of edges, i.e. all bases have the same cardinality. Since this also holds for all subgraphs containing M, we conclude by Lemma 1.4 that we have a greedoid. A run of the last phase of the Edmonds algorithm corresponds to a basic word of this greedoid.

The last group of examples arises by borrowing ideas from standard matroid constructions.

2.14 Bipartite Matching Greedoid (Medieval Marriage Greedoid)

The following construction is similar to transversal matroids. Let (S, T, E) be a bipartite graph and $s_1, \ldots, s_m$ an ordering on the nodes of S. Let $\mathscr{F}$ consist of all subsets X of T such that the subgraph induced by X and the first $|X|$ nodes of S has a perfect matching. (Cf. the rule of medieval marriage: older daughters must be married off first.) The proof of property (1.3) is postponed to Chapter IX. We remark that these greedoids are special Gaussian elimination greedoids. This is an analogue of the fact that transversal matroids are representable.

2.15 Linking Greedoid

Let $G = (V, E)$ be a directed graph, $W \subseteq V$ some subset of nodes, and let $v_1, v_2, \ldots, v_m$ be an ordering of the nodes in W. A subset $U \subseteq V$ is feasible in the linking greedoid if there exist node-disjoint directed paths linking the nodes in U to the first U elements of W. These are generalizations of bipartite matching greedoids and analogues of gammoids.

2.16 Polymatroid Greedoid

Recall that the rank function f of a polymatroid is normalized, monotone and submodular, i.e. satisfies

$$f(\emptyset) = 0,$$
$$X \subseteq Y \text{ implies } f(X) \leq f(Y),$$
$$f(X \cup Y) + f(X \cap Y) \leq f(X) + f(Y).$$

Given such a function $f : 2^E \to \mathbb{N}$, define a hereditary language $(E, \mathscr{L})$ by

$$\mathscr{L} = \{x_1 x_2 \ldots x_k : f(x_1 \ldots x_i) = i \text{ for all } 1 \leq i \leq k\}.$$

We will show in Chapter VII that $(E, \mathscr{L})$ is a greedoid. This construction is similar to the standard construction associating matroids with polymatroids.

2.17 Paving Greedoids

Let $\mathscr{P} \subseteq 2^E$ be a family of subsets not containing the empty set such that for all $A, B \in \mathscr{P}$ with $|A| = |B|$ and $|A \triangle B| = 2$, the intersection $A \cap B$ is a member of $\mathscr{P}$. Then $(E, 2^E \setminus \mathscr{P})$ is a **paving greedoid** . These greedoids generalize **paving matroids** (cf. Welsh [1976]). As in the matroid case, paving greedoids tend to be pathological and therefore often serve as counterexamples.

To see that $(E, 2^E \setminus \mathscr{P})$ is a greedoid, observe that $\emptyset \in 2^E \setminus \mathscr{P}$. Now let $X, Y \notin \mathscr{P}$ with $|X| > |Y|$. Suppose the augmentation property (1.3) does not hold, then

$$Y \cup \{x\} \in \mathscr{P} \text{ for all } x \in X \setminus Y .$$

If $|X \setminus Y| = 1$, then $Y \cup \{x\} = X \notin \mathscr{P}$. If $|X \setminus Y| > 1$, then there exist $x_1, x_2 \in X \setminus Y$ with $x_1 \neq x_2$. Let $A = Y \cup \{x_1\}$ and $B = Y \cup \{x_2\}$. Since $A, B \in \mathscr{P}$, we have

$$Y = A \cap B \in \mathscr{P},$$

contradicting the assumptions.

A particular example of a paving greedoid is given by any family $\mathscr{P}$ of nonempty subsets closed under intersection.

2.18 Twisted Matroid

Twisted matroids as examples of greedoids were introduced by Björner [1983]. Let $(E, \mathscr{M})$ be a matroid and $A \in \mathscr{M}$. Define

$$\mathscr{F} = \{X \triangle A : X \in \mathscr{M}\}.$$

Clearly $\emptyset \in \mathscr{F}$. Suppose $X \triangle A, Y \triangle A \in \mathscr{F}$ with $|X \triangle A| > |Y \triangle A|$, then we have the following Venn diagram (cf. Figure 5).

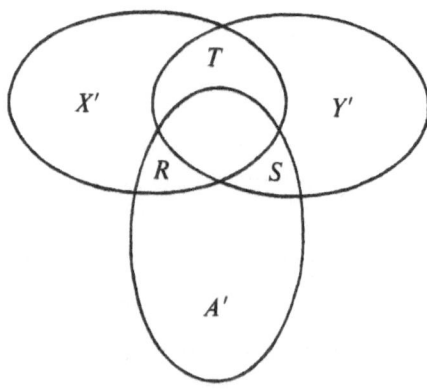

Fig. 5

Hence

$$0 < |X \triangle A| - |Y \triangle A|$$
$$= (|X'| + |T| + |A'| + |S|) - (|Y'| + |T| + |A'| + |R|)$$
$$\leq |X'| - |Y'| + |S|.$$

If $|S| > 0$, then for any $x \in S$ we have $x \in X \triangle A$ and

$$(Y \setminus x) \triangle A = Y \setminus A \cup A \setminus Y \cup x = (Y \triangle A) \cup x.$$

If $|S| = 0$, then $|X'| > |Y'|$ and hence $|X \setminus R| > |Y|$. By the augmentation property for matroids there exists an element $x \in X'$ such that $Y \cup x \in \mathcal{M}$, i.e. $(Y \cup x) \triangle A = (Y \triangle A) \cup x$, which proves property (1.3).

Recall that repeatedly deleting maximal elements from a poset yields a basic antimatroid. In some applications one has restrictions on which maximal elements can be deleted. The next example shows that often such restrictions lead to greedoids which are not antimatroids.

Upper bipartite elimination was investigated in Faigle, Gierz and Schrader [1985] in connection with the setup number of posets.

2.19 Upper Bipartite Elimination in Posets

Let $(P, \leq)$ be a poset. A maximal element $x \in P$ with lower neighbors A is **upper bipartite** in P if either all elements of A are maximal in $P \setminus x$ or the nonmaximal elements are only covered by maximal elements whose lower neighbors are in A and the maximal elements in A are isolated. The feasible words of the **upper bipartite elimination greedoid** are sequences $x_1 \ldots x_k$ such that x_i is upper bipartite in $P \setminus \{x_1, \ldots, x_{i-1}\}$.

Chapter V. Structural Properties

We would like to develop the basic machinery of greedoid theory. Some basic notions like rank, closure and (to a certain extent) minors can be introduced by extending the appropriate notions from matroids, while some others will be more specific for greedoids.

The rank function of a greedoid associates with every set the size of a maximal feasible subset. In Section 1 we observe that a greedoid rank function is monotone, subcardinal and locally submodular. In contrast to the rank function of a matroid, it is not unit-increasing and not submodular. We prove that the above-mentioned characteristics may serve as an equivalent axiomatization.

Section 2 analyses three different generalizations of the matroid closure operator. The most direct extension from matroids is the rank closure; the kernel closure is a variant that plays an important role in associating lattices with greedoids. These two operators, however, are not closure operators in the usual sense since they are not monotone. Nevertheless, they give rise to cryptomorphic descriptions of greedoids. We also introduce a third version of the closure which is monotone, but does not capture enough structure to determine the greedoid uniquely.

The motivation for the studies in Section 3 is the observation that the rank function and the closure operator are even more similar to their matroid analogues if we restrict them to certain subsets. This leads to the notions of rank and closure feasibility. These concepts will play a role in the analysis of interval greedoids given in Section 5, as well as in the theory of optimization in greedoids (Chapter XI).

Section 4 collects some basic facts about minor producing operations and reciprocal constructions leading to extensions of greedoids to larger ground sets.

In the final section we show that for interval greedoids the rank and closure are in a sense better behaved. For example, rank feasibility and closure feasibility coincide, and interval greedoids are characterized by the monotonicity of the kernel closure and by a relaxed monotonicity of the rank closure.

1. Rank Function

We define the **rank** $r(X)$ of a subset $X \subseteq E$ in a greedoid $(E, \mathscr{F})$ as

$$r(X) = \max\{|A| : A \subseteq X, \ A \in \mathscr{F}\}.$$

Recall from Lemma IV.1.4 that all bases of X have the same cardinality $r(X)$.

The purpose of this section is to show that the properties of the rank function give rise to an axiomatization of greedoids.

Theorem 1.1. *A function $r : 2^E \to \mathbb{Z}_+$ is the rank function of a greedoid if and only if for all $X, Y \subseteq E$ and all $x, y \in E$ the following conditions hold:*

(1.1) $r(\emptyset) = 0$,
(1.2) $r(X) \leq |X|$,
(1.3) $X \subseteq Y$ *implies* $r(X) \leq r(Y)$,
(1.4) $r(X) = r(X \cup x) = r(X \cup y)$ *implies* $r(X) = r(X \cup x \cup y)$
        (local submodularity).

Moreover, the greedoid is uniquely determined by its rank function.

Proof. Suppose r is the rank function of a greedoid $(E, \mathcal{F})$, then the conditions (1.1)–(1.3) are obviously fulfilled. If

$$r(X) < r(X \cup x \cup y),$$

then $|A| < |B|$ for any basis A of X and any basis B of $X \cup x \cup y$. By property (IV.1.3), we can augment A by some element from $B \setminus A$. Hence either $A \cup x \in \mathcal{F}$ or $A \cup y \in \mathcal{F}$, i.e. either

$$r(X) < r(X \cup x) \text{ or } r(X) < r(X \cup y),$$

which asserts that (1.4) is necessary.

Conversely, let the function r satisfy (1.1)–(1.4). Let $(E, \mathcal{F})$ be the set system over the ground set E defined by

$$\mathcal{F} = \{X \subseteq E : r(X) = |X|\}.$$

Clearly, $\emptyset \in \mathcal{F}$. Suppose (IV.1.5) is violated. Then there exist two sets $X, Y \in \mathcal{F}$ with $|X| > |Y|$ such that $r(Y \cup x) = r(Y)$ for all $x \in X \setminus Y$. By repeatedly applying (1.4) we obtain $|Y| = r(Y) = r(Y \cup X \setminus Y) = r(X) = |X|$, in contradiction to the assumption. Hence $(E, \mathcal{F})$ is a greedoid.

A similar argument shows that the rank function of $(E, \mathcal{F})$ is indeed r. Moreover, since X is feasible if and only if $r(X) = |X|$, any greedoid is uniquely determined by its rank function. □

The proof of (1.4) in fact yields the following stronger version of local submodularity:

Proposition 1.2. *If $X, Y \subseteq E$ such that $r(X) = r(Y) = r(X \cap Y)$, then $r(X) = r(Y) = r(X \cup Y)$.* □

On the other hand we do not have submodularity, as any of the non-matroid examples shows.

Since antimatroids are special greedoids, they have a more specific rank function characterization. The following result by Dietrich [1989] is an immediate consequence of Theorem 1.1 and Lemma III.1.2.

Corollary 1.3. *A greedoid rank function r is the rank function of an antimatroid if and only if for all $X \subseteq E$ and all $x, y \in E, x \neq y$ the following holds:*

(1.5) $r(X \cup x) > r(X)$ *and* $r(X \cup y) > r(X)$ *implies* $r(X \cup x \cup y) > r(X) + 1.$ □

2. Closure Operators

Given a greedoid $(E, \mathscr{F})$, let the **(rank) closure operator** $\sigma : 2^E \to 2^E$ be defined by

$$\sigma(X) = \{x \in E : r(X \cup x) = r(X)\}.$$

Example. Let $P = (E, \leq)$ be the poset in Figure 6 and $(E, \mathscr{F})$ the poset antimatroid on P. Then $\sigma(\{x\}) = \{x, z\}$ and $\sigma(\{x, y\}) = \{x, y\}$. This example shows that in contrast to the usual understanding of a closure operator, σ is not necessarily monotone.

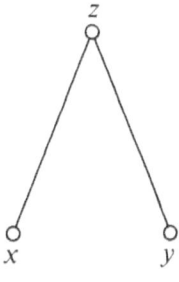

Fig. 6

As in the case of antimatroids, we also define for $A \in \mathscr{F}$ the set of **feasible continuations** as $\Gamma(A) = \{x \in E \setminus A : A \cup x \in \mathscr{F}\}$. Obviously $\sigma(A) = E \setminus \Gamma(A)$. The following lemmas give some relations between the rank function of a greedoid and its closure operator.

Lemma 2.1. *Let $(E, \mathscr{F})$ be a greedoid and $X \subseteq E$. Then*

(i) $X \subseteq \sigma(X)$,
(ii) $r(\sigma(X)) = r(X)$,
(iii) X *is feasible if and only if* $x \notin \sigma(X \setminus x)$ *for all* $x \in X$.

Property (ii) implies that $\sigma(X)$ is the unique largest superset of X with the same rank.

Proof. (i) is obvious. (ii) follows from Propositon 1.2.
    (iii) If X is feasible, then for any $x \in X$, $r(X \setminus x) \leq |X| - 1 < r(X)$, which implies $x \notin \sigma(X \setminus x)$. Conversely, if X is not feasible, consider any basis B of X. Then for any $x \in X \setminus B, r(X \setminus x) = r(X)$ and hence $x \in \sigma(X \setminus x)$.   □

As mentioned above, the closure operator is not monotone; however, a slightly weaker property holds.

Lemma 2.2. *Let $(E, \mathcal{F})$ be a greedoid. Then*

(i) $X \subseteq \sigma(Y)$ *and* $Y \subseteq \sigma(X)$ *implies* $\sigma(X) = \sigma(Y)$,
(ii) $\sigma(\sigma(X)) = \sigma(X)$.

Proof. (i)   From Lemma 2.1 we get $r(X) = r(Y) = r(\sigma(X)) = r(\sigma(Y))$. Since $X \subseteq \sigma(Y)$ and $r(X) = r(\sigma(Y))$, Lemma 2.1 implies $\sigma(Y) \subseteq \sigma(X)$. By symmetry $\sigma(X) \subseteq \sigma(Y)$.
   (ii)   follows from (i) by letting $Y = \sigma(X)$.   □

Recall that the Steinitz-MacLane exchange property of matroids generalizes the following fact from linear algebra: if some vector x is not contained in the subspace generated by a set A of vectors but is contained in the span of $A \cup y$, then y lies in the subspace of $A \cup x$. The Steinitz-MacLane exchange property does not carry over to general greedoids. For example, take $E = \{x, y, z\}$, $\mathcal{F} = \{\emptyset, x, y, \{x, y\}, \{y, z\}\}$. Then $x \notin \sigma(z)$, $x \in \sigma(\{y, z\})$, but $y \notin \sigma(\{x, z\})$.
   However, a slightly weaker version of the exchange property holds, as the next lemma shows:

Lemma 2.3. *Let* $X \subseteq E$ *and* $x, y \in E \setminus X$, *and suppose* $X \cup x \in \mathcal{F}$. *Then* $x \in \sigma(X \cup y)$ *implies* $y \in \sigma(X \cup x)$.

Proof. By assumption we have $r(X \cup x \cup y) = r(X \cup y) \leq |X| + 1 = r(X \cup x)$, hence $y \in \sigma(X \cup x)$.   □

We will return later to show that the properties we formulated in the last three lemmas characterize the closure operator of greedoids. Before that, we investigate a related concept.
   Two sets $X, Y \subseteq E$ are called **cospanning** if $\sigma(X) = \sigma(Y)$. Note that the cospanning relation (as an equivalence relation on 2^E) determines the greedoid uniquely. In fact, X is feasible if and only if it is not cospanning with any proper subset of X. The following theorem characterizes the cospanning relation of greedoids.

Theorem 2.4. *Let E be a finite set and $R \subseteq 2^E \times 2^E$ an equivalence relation on 2^E. Then R is the cospanning relation of a greedoid if and only if the following conditions hold whenever $X, Y, Z \subseteq E$ and $x, y \in E \setminus X$:*

(2.1)   *if* $(X, Y) \in R$, *then* $(X, X \cup Y) \in R$,
(2.2)   *if* $X \subseteq Y \subseteq Z$ *and* $(X, Z) \in R$, *then* $(X, Y) \in R$,
(2.3)   *if* $(X \cup y, X \cup x \cup y) \in R$ *but* $(X \cup x, X \cup x \cup y) \notin R$, *then there exists an element* $z \in X \cup x$ *such that* $(X \cup x \setminus z, X \cup x) \in R$.

Proof. Necessity: (2.1) and (2.2) follow immediately from Lemma 2.2(i). If the conclusion in (2.3) does not hold, then $X \cup x \in \mathcal{F}$ and we get a contradiction to Lemma 2.3.

Sufficiency: Let $\mathscr{F} = \{X : (X, X') \notin R$ for all $X' \subset X\}$.

Claim 1. If $A \in \mathscr{F}$ and $a \in E \setminus A$ such that $(A, A \cup a) \notin R$, then $A \cup a \in \mathscr{F}$.

Suppose not. Then there exists an element $b \in A \cup a$ such that $(A \cup a \setminus b, A \cup a) \in R$. Clearly, $b \neq a$. Set $X = A \setminus b, x = b, y = a$. Then $(X \cup y, X \cup x \cup y) = (A \cup a \setminus b, A \cup a) \in R$ but $(X \cup x, X \cup x \cup y) = (A, A \cup a) \notin R$. Hence, by (2.3), there exists a $z \in A$ such that $(A \setminus z, A) \in R$, in contradiction to $A \in \mathscr{F}$.

Claim 2. Let $X \subseteq E$ and $A \in \mathscr{F}, A \subseteq X$. Then there exists an element $x \in X \setminus A$ such that $A \cup x \in \mathscr{F}$ if and only if $(A, X) \notin R$.

In fact, if $(A, X) \in R$, then $(A, A \cup x) \in R$ for each $x \in X \setminus A$ by (2.2), and hence $A \cup x \notin \mathscr{F}$. Conversely, if $A \cup x \notin \mathscr{F}$ for all $x \in X \setminus A$, then $(A, A \cup x) \in R$ for each $x \in X \setminus A$ by Claim 1. Hence by (2.1), $(A, X) \in R$.

Note that Claim 2 implies that $\mathscr{F}$ is accessible. We now prove that it has the augmentation property. Let $A, B \in \mathscr{F}$ with $|A| < |B|$. Assume that A cannot be augmented from B. Let $C \subseteq A \cap B, C \in \mathscr{F}$, and choose A, B, C such that $|C|$ is maximal. By Claim 2, $(A, A \cup B) \in R$. This, in particular, yields that $A \cup B \notin \mathscr{F}$ and hence $A \not\subseteq B$.

Again by Claim 2, there exists a $c \in B \setminus C$ such that $C \cup c \in \mathscr{F}$. By the maximality of $C, c \notin A$. Let D be a maximal set such that $C \cup c \subseteq D \subseteq A \cup c$ and $D \in \mathscr{F}$. Since $A \cup c \notin \mathscr{F}$ by assumption, we have $D \subset A \cup c$ and hence $|D| \leq |A|$. Since $A \subseteq A \cup D \subseteq A \cup B$, we have by (2.2) $(A \cup D, A \cup B) \in R$. Moreover, by the maximality of D and Claim 2 we have $(D, D \cup A) \in R$. Hence $(D, A \cup B) \in R$. So by Claim 2 again, D cannot be augmented from B. But since $C \cup c \subseteq D \cap B$, this contradicts the choice of A, B and C.

Hence $(E, \mathscr{F})$ is a greedoid and it only remains to show that its cospanning relation is R. To this end, let us remark first that for every $X \subseteq E, (X, \sigma(X)) \in R$. In fact, if A is a maximal feasible set contained in X then, by Claim 2, $(A, X) \in R$ and also $(A, \sigma(X)) \in R$, hence $(X, \sigma(X)) \in R$.

So if $\sigma(X) = \sigma(Y)$, then $(X, \sigma(X)) \in R$ and $(Y, \sigma(X)) \in R$ and so $(X, Y) \in R$. Conversely, let $(X, Y) \in R$. Let A be a maximal feasible subset of X. Then $(A, X) \in R$ by Claim 2 and $(X, X \cup Y) \in R$ by (2.1). So $(A, X \cup Y) \in R$ and, by Claim 2, A is a maximal feasible subset of $X \cup Y$. Hence $X \cup Y \subseteq \sigma(A) = \sigma(X)$. So $Y \subseteq \sigma(X)$ and, similarly, $X \subseteq \sigma(Y)$, which by Lemma 2.2(i) implies that $\sigma(X) = \sigma(Y)$. $\qquad\square$

Now we are ready to prove a characterization of the closure operator of greedoids.

Theorem 2.5. *Let E be a finite set and $\tau : 2^E \to 2^E$ a mapping. Then τ is the closure operator of a greedoid if and only if the following conditions hold whenever $X, Y, Z \subseteq E$ and $x, y \in E \setminus X$:*

(2.4) $X \subseteq \tau(X)$,
(2.5) $\tau(\tau(X)) = \tau(X)$,

(2.6) *if $\tau(X) = \tau(Y)$, then $\tau(X) = \tau(X \cup Y)$,*

(2.7) *if $X \subseteq Y \subseteq Z$ and $\tau(X) = \tau(Z)$, then $\tau(X) = \tau(Y)$,*

(2.8) *if $\tau(X \cup y) = \tau(X \cup x \cup y)$ but $\tau(X \cup x) \neq \tau(X \cup x \cup y)$, then there exists a $z \in X \cup x$ such that $\tau(X \cup x \setminus z) = \tau(X \cup x)$.*

Proof. We already know that the conditions are necessary. Now assume that τ satisfies (2.4)–(2.8). Define an equivalence relation R on 2^E by $(X, Y) \in R$ if and only if $\tau(X) = \tau(Y)$. Then (2.6)–(2.8) imply that R is the cospanning relation of a greedoid $(E, \mathscr{F})$. Let σ be the closure operator of this greedoid. By (2.4), $\sigma(X) \subseteq \tau(\sigma(X))$. Since X and $\sigma(X)$ are cospanning, we have $\tau(\sigma(X)) = \tau(X)$ and hence $\sigma(X) \subseteq \tau(X)$. Conversely, by (2.5) and the definition of R, X and $\tau(X)$ are cospanning. Hence $\tau(X) \subseteq \sigma(X)$.   □

We have seen that the closure operator fails to be monotone in general. Crapo [1982] and Schmidt [1985a] introduce another notion of closure which – at least for the important subclass of interval greedoids – turns out to be monotone but forfeits the idea of a hull operation, since for this new closure we cannot guarantee that every set is contained in its hull.

For this purpose, let $(E, \mathscr{F})$ be a hereditary language. Following Crapo [1982], we call a subset of E a **partial alphabet** if it is the union of feasible sets of $(E, \mathscr{F})$. The collection of all partial alphabets is denoted by $\mathscr{A}$. (Clearly $(E, \mathscr{A})$ is an antimatroid).

We now associate with every subset $A \subseteq E$ a partial alphabet $\lambda(A)$ via

$$\lambda(A) = \bigcup \{X \in \mathscr{F} : r(A \cup X) = r(A)\} = \bigcup \{X \in \mathscr{F} : X \subseteq \sigma(A)\}$$

and call $\lambda : 2^E \to 2^E$ the **kernel closure operator** of the greedoid $(E, \mathscr{F})$.

For matroids without loops we obviously have $\lambda = \sigma$. For general greedoids, however, the two closure operators differ.

The kernel closure is not monotone in general. For example, take $E = \{x, y, z\}$ and $\mathscr{F} = 2^E \setminus \{y, \{x, z\}\}$. Then $\lambda(x) = \{x, z\}$ and $\lambda(\{x, y\}) = \{x, y\}$.

Observe, however, that this example does not have the interval property. Indeed, we will see in Section 5 that for interval greedoids λ does have the monotonicity property, and these are the only greedoids with a monotone kernel closure.

Define the **kernel** of a set $X \subseteq E$ by

$$v(X) = \bigcup \{A \subseteq X : A \in \mathscr{F}\}.$$

Trivially, λ can be expressed in terms of the closure and the kernel as follows:

$$\lambda(X) = v(\sigma(X)).$$

Let us remark also that

$$\sigma(v(X)) = \sigma(X),$$

since every basis of X is also a basis of $v(X)$. Hence it also follows that $\sigma(\lambda(X)) = \sigma(v(\sigma(X))) = \sigma(\sigma(X)) = \sigma(X)$ and $\lambda(\sigma(X)) = v(\sigma(\sigma(X))) = v(\sigma(X)) = \lambda(X)$.

Lemma 2.6. *For every* $X, Y \subseteq E$, $\lambda(X) = \lambda(Y)$ *if and only if* $\sigma(X) = \sigma(Y)$.

Proof. Trivial by the above identities.   □

This lemma shows that we could also define the cospanning relation in terms of λ. In particular, the operator λ determines the greedoid.

Lemma 2.7. *Let* $(E, \mathcal{F})$ *be a greedoid with partial alphabets* $\mathcal{A}$ *and kernel closure* λ. *For any two sets* $X, Y \subseteq E$ *the following holds:*

(i) $X \subseteq \lambda(X)$ *if* $X \in \mathcal{A}$,
(ii) $r(\lambda(X)) = r(X)$,
(iii) $X \subseteq \lambda(Y)$ *and* $Y \subseteq \lambda(X)$ *implies* $\lambda(X) = \lambda(Y)$,
(iv) $\lambda(\lambda(X)) = \lambda(X)$.

Proof. (i) is immediate from the definition of λ.

(ii) Consider a basis Y of X. Clearly we have $r(Y) = r(\lambda(Y))$. Moreover, (i) yields $Y \subseteq \lambda(Y) = \lambda(X)$. Thus $r(X) = r(\lambda(X))$.

(iii) Let $X \subseteq \lambda(Y)$ and $Y \subseteq \lambda(X)$. Then $X \subseteq \sigma(Y)$ and $Y \subseteq \sigma(X)$. So $\sigma(X) = \sigma(Y)$ by Lemma 2.2, and hence $\lambda(X) = \lambda(Y)$ by Lemma 2.6.

(iv) For $X \in \mathcal{F}$, the result follows from (i) and (iii) by letting $Y = \lambda(X)$. For an arbitrary subset $X \subseteq E$ consider a basis $Y \in \mathcal{F}$. Then by the above, $\lambda(X) = \lambda(Y) = \lambda(\lambda(Y)) = \lambda(\lambda(X))$.   □

For a greedoid $(E, \mathcal{F})$, we have $\lambda = \sigma$ if and only if every singleton $\{x\}$ is feasible. To see this, observe that the feasible kernel operator v is the identity mapping if $\{x\} \in \mathcal{F}$ for any $x \in E$. Conversely if $\lambda = \sigma$, then in particular $\sigma(\emptyset) = \{x : r(\{x\}) = 0\} = \lambda(\emptyset) = \emptyset$, i.e. $\{x\} \in \mathcal{F}$ for all $x \in E$.

One can characterize the kernel closure of greedoids in a similar way as for the closure. The conditions become more technical and we refer the interested reader to Schmidt [1985a].

In order to overcome the lack of monotonicity in both closure operators we define the **monotone closure operator** as follows

$$\mu(X) = \bigcap \{Y : X \subseteq Y, \ \sigma(Y) = Y\}.$$

It is not difficult to see that the monotone closure satisfies the following conditions:

(2.10) $A \subseteq \mu(A)$,
(2.11) $\mu(A) = \mu(\mu(A))$,
(2.12) $A \subseteq B$ implies $\mu(A) \subseteq \mu(B)$.

However, μ does not give rise to an axiomatization of greedoids, since a greedoid is not uniquely determined by its monotone closure operator. In particular, for every full greedoid we have $\mu(X) = X$. To see this, suppose for some full greedoid $(E, \mathcal{F})$ and $X \subseteq E$ there exists an $a \in \mu(X) \setminus X$. Then $E \setminus a$ cannot be σ-closed, i.e. $r(E \setminus a) = r(E) = |E|$, contradicting condition (1.2).

3. Rank and Closure Feasibility

In general, the intersection of a set with some basis may have a larger cardinality than the rank of this set. This gives rise to the following definition. Let $(E, \mathscr{F})$ be an accessible set system. For $X \subseteq E$ define

(3.1) $\beta(X) = \max\{|X \cap B| : B \in \mathscr{F}\}$

to be the **basis rank** of X. Clearly,

(3.2) $r(X) \leq \beta(X)$,

and for matroids obviously $\beta = r$. We call a subset **rank feasible** if

$$\beta(X) = r(X)$$

and denote by $\mathscr{R} = \mathscr{R}(E, \mathscr{F})$ the family of all rank feasible sets.

It is clear that $\mathscr{F} \subseteq \mathscr{R}$, and $\mathscr{F} = \mathscr{R}$ if $(E, \mathscr{F})$ is a full greedoid. It is also easy to see that $\mathscr{R} = 2^E$ if and only if the greedoid is a matroid. Thus a greedoid is not uniquely determined by its rank feasible sets. We will show later that the set system $(E, \mathscr{R})$ is always accessible; however, it is in general not a greedoid, as the following example shows:

Example. $E = \{a, b, c, d\}$, $\mathscr{F} = \{\emptyset, a, b, c, \{a, d\}, \{b, d\}, \{c, d\}, \{a, b, d\}, \{a, c, d\}\}$. Then $\mathscr{R} = \mathscr{F} \cup \{\{b, c\}, \{b, c, d\}\{a, b, c, d\}\}$. In $(E, \mathscr{R})$ the singleton $\{a\}$ cannot be augmented from $\{b, c\}$, i.e. $(E, \mathscr{R})$ is not a greedoid.

This example also shows that $\mathscr{R}$ is not closed under union.

We may always assume that the basis rank of a given set X is obtained as the intersection of X with suitable supersets of any of its bases:

Lemma 3.1. For $X \subseteq E$ and $A \subseteq X$ with $A \in \mathscr{F}$, there exists a set $B \in \mathscr{F}$ with $A \subseteq B$ such that $\beta(X) = |B \cap X|$.

Proof. Let $C \in \mathscr{F}$ be such that $\beta(X) = |C \cap X|$. If $|A| \geq |C|$, we are done. Otherwise we can augment A from C to a set $B \subseteq A \cup C$ with $B \in \mathscr{F}$ and $|B| = |C|$. Then $|B \cap X| \geq |B| - |(A \cup C) \setminus X| = |C| - |C \setminus X| = \beta(X)$.   □

The next theorem gives two characterizations of rank feasible sets in terms of the rank function.

Theorem 3.2. For any set $A \subseteq E$ the following statements are equivalent:

(i) A is rank feasible.
(ii) $r(A \cup X) \leq r(A) + |X|$ for all $X \subseteq E \setminus A$.
(iii) $r(A \cup X) = r(B \cup X)$ for any basis B of A and any $X \subseteq E \setminus A$.

Proof. (i) $\Rightarrow$ (iii). Let B be a basis of A, then clearly $r(B \cup X) \leq r(A \cup X)$. To show the converse, augment B to a basis B' of $A \cup X$. Since $A \in \mathscr{R}$, we have

$|B| \leq |B' \cap A| \leq r(A) = |B|$, and hence $B = B' \cap A$. But then $B' \subseteq B \cup X$, which yields $r(B \cup X) \geq |B'| = r(A \cup X)$.

· (iii) $\Rightarrow$ (ii) is trivial, since $r(A \cup X) = r(B \cup X) \leq |B| + |X| = r(A) + |X|$.

(ii) $\Rightarrow$ (i). If $A \notin \mathcal{R}$, there exists a feasible set $Y \in \mathcal{F}$ such that $|A \cap Y| > r(A)$. Take $X = Y \setminus A$, then $r(A \cup X) \geq |Y| = |A \cap Y| + |X| > r(A) + |X|$. □

Note that (ii) in particular implies the unit increase property for rank feasible sets. A further basic property of rank feasible sets is that the rank function is submodular on $\mathcal{R}$:

Theorem 3.3.

$$\beta(A \cup B) + r(A \cap B) \leq \beta(A) + \beta(B) \text{ for all } A, B, \subseteq E.$$

Consequently $r(A \cup B) + r(A \cap B) \leq r(A) + r(B)$ *for all* $A, B \in \mathcal{R}$.

Proof. Let X_1 be a basis of $A \cap B$ and $U \in \mathcal{F}$ such that

$$|U \cap (A \cup B)| = \beta(A \cup B).$$

Extend X_1 to a basis X_2 of $A \cup B \cup U$. By definition of the basis rank function β, we get $|X_2 \cap A| \leq \beta(A)$ and $|X_2 \cap B| \leq \beta(B)$.

Hence

$$\begin{aligned}
\beta(A) + \beta(B) &\geq |X_2 \cap A| + |X_2 \cap B| = |X_2 \cap (A \cap B)| + |X_2 \cap (A \cup B)| \\
&= |X_1| + |X_2 \cap (A \cup B)| \geq |X_1| + |X_2| - |X_2 \setminus (A \cup B)| \\
&\geq |X_1| + |U| - |U \setminus (A \cup B)| = |X_1| + |U \cap (A \cup B)| \\
&= r(A \cap B) + \beta(A \cup B).
\end{aligned}$$

□

The following result relates rank feasible sets to the monotone closure.

Theorem 3.4. *A set* $U \subseteq E$ *is rank feasible if and only if* $X \subseteq U \subseteq \mu(X)$ *for any basis* X *of* U.

Proof. Let $U \in \mathcal{R}$ and let X be a basis of U. We claim that $U \subseteq \mu(X)$, i.e. $U \subseteq \sigma(Y)$ for all Y containing X. Suppose $U \nsubseteq \sigma(Y)$ for some Y with $X \subseteq Y$. Augment X to a basis B of Y. Then there exists an element $u \in U \setminus \sigma(Y) = U \setminus \sigma(B)$, i.e. $B \cup u \in \mathcal{F}$. But then

$$|U \cap (B \cup u)| \geq |X \cup u| > |X| = r(U),$$

contradicting $U \in \mathcal{R}$.

Conversely, suppose $X \subseteq U \subseteq \mu(X)$ for some basis X of U. Consider any basis $B \in \mathcal{F}$. Augment X from B until a set $A \supseteq X$ is obtained with $A \in \mathcal{F}$ and $|A| = |B|$. If at some step an element $u \in U$ is added to some $Y \supseteq X$, $Y \in \mathcal{F}$, then $u \notin \sigma(Y)$, i.e. $U \nsubseteq \mu(X)$. Hence $A \cap U = X$ and clearly $A \setminus X = A \setminus U \subseteq B \setminus U$. Therefore $|B \cap U| \leq |A \cap U| = |X| = r(U)$. □

As a consequence of this we get that $U \in \mathcal{R}$ if and only if $\mu(U) = \mu(X)$ for any basis X of U.

The λ-closure of rank feasible sets satisfies a weak version of the Steinitz-MacLane exchange property.

Lemma 3.5. *Let* $(E, \mathcal{F})$ *be a greedoid,* $X \in \mathcal{R}$, *and* $y, z \in E \setminus X$. *Then* $\lambda(X) \neq \lambda(X \cup y)$, $y \in \lambda(X \cup z)$ *implies* $z \in \lambda(X \cup y)$.

Proof. Since $\lambda(X) \neq \lambda(X \cup y)$ and $X \in \mathcal{R}$,

$$r(X) < r(X \cup y) \leq r(X) + 1 \,,$$

hence $r(X \cup y) = r(X) + 1$. Similarly, $y \notin \lambda(X), y \in \lambda(X \cup z)$ implies $\lambda(X) \neq \lambda(X \cup z)$ and so $r(X \cup z) = r(X) + 1$, as before. Furthermore, $y \in \lambda(X \cup z)$ implies that $r(X \cup \{y, z\}) = r(X) + 1$. Hence $r(X \cup \{y, z\}) = r(X \cup y) = r(X \cup z)$, which implies $\lambda(X \cup y) = \lambda(X \cup \{y, z\}) = \lambda(X \cup z)$. Thus $z \in \lambda(X \cup y)$.   □

Although, as we have shown, $(E, \mathcal{R})$ is not a greedoid in general, it defines an accessible set system.

Lemma 3.6. *For every* $U \in \mathcal{R}$ *there exists an element* $u \in U$ *such that* $U \setminus u \in \mathcal{R}$. *Moreover,* $r(U \setminus u) = r(U) - 1$ *if* $U \in \mathcal{F}$ *and* $r(U \setminus u) = r(U)$ *otherwise.*

Proof. Since $\mathcal{F}$ is accessible, we have to prove the claim only for $U \in \mathcal{R} \setminus \mathcal{F}$. By Theorem 3.5, we have for any basis X of $U, X \subseteq U \subseteq \mu(X)$ and for any $u \in U \setminus X$ we have $X \subseteq U \setminus u \subseteq \mu(X)$. Hence $U \setminus u \in \mathcal{R}$ and $r(U \setminus u) = r(U)$.   □

The above characterization of rank feasible sets with respect to the monotone closure gives rise to a stronger property. We call a set $X \subseteq E$ **closure feasible** if

(3.3) $X \subseteq \sigma(A)$ implies $X \subseteq \sigma(B)$ for all $A \subseteq B \subseteq E$.

In terms of the monotone closure this definition is equivalent to:

(3.4) $X \subseteq \sigma(A)$ implies $X \subseteq \mu(A)$ for all $A \subseteq E$.

The class of all closure feasible sets is denoted by $\mathcal{C} = \mathcal{C}(E, \mathcal{F})$. The following observations are immediate:

Lemma 3.7. $\mathcal{C} \subseteq \mathcal{R}$.

Proof. Let $X \in \mathcal{C}$ and let A be a basis of X. Then $X \subseteq \sigma(A)$ and hence $X \subseteq \mu(A)$. Theorem 3.5 now shows that $X \in \mathcal{R}$.   □

Lemma 3.8. $\mathcal{C}$ *is closed under union.*

Proof. Suppose X and Y are closure feasible and $X \cup Y \subseteq \sigma(A)$. Then $X \subseteq \sigma(A)$ implies $X \subseteq \mu(A)$. Similarly $Y \subseteq \mu(A)$, hence $X \cup Y \subseteq \mu(A)$.   □

In particular, this lemma implies that in general $\mathscr{C} \neq \mathscr{R}$, since $\mathscr{R}$ is not necessarily closed under union. The closure feasible sets do not form a greedoid in general, as the following example shows.

Example. Let $E = \{a, b, c, d\}$ and $\mathscr{F} = 2^E \setminus \{\{a, c\}, \{b, d\}\}$. Then $\mathscr{C} = 2^E \setminus \{a, b, c, d, \{a, c\}, \{b, d\}\}$.

However, let $\mathscr{C}' = \{\{x_1, \ldots, x_k\} : \{x_1, \ldots, x_i\} \in \mathscr{C}$ for $1 \leq i \leq k\}$ be the accessible kernel of $\mathscr{C}$, and $E' = \cup \mathscr{C}'$. Then Lemma 3.8 implies:

Corollary 3.9. $(E', \mathscr{C}')$ is an antimatroid. $\square$

The greedoid in the example above does not have the interval property. In fact, we will see later that interval greedoids are precisely those greedoids for which the rank feasible sets are also closure feasible, i.e. $\mathscr{C} = \mathscr{R}$ or, equivalently, for which all feasible sets are closure feasible.

4. Minors and Extensions

In the same way as the occurrence or nonoccurrence of certain submatroids reflects the structure of a given matroid, we can look at "sub-greedoids" or minors which are induced by greedoids. Therefore we extend the usual minor operations for matroids to greedoids (cf. Korte and Lovász [1983] and Björner, Korte and Lovász [1985].

The first and simplest operation is truncation. Let $(E, \mathscr{F})$ be a greedoid and $k \in \mathbb{N}$. It is easy to see that the **truncation (to rank k)**

$$\mathscr{F}^{(k)} = \{X \in \mathscr{F} : |X| \leq k\}$$

again gives rise to a greedoid $(E, \mathscr{F}^{(k)})$. Its rank function $r_k(X)$ is given by the relation

$$r_k(X) = \min(r(X), k) .$$

For the second operation consider any subset $T \subseteq E$. The **restriction of $(E, \mathscr{F})$** to T is the system $(T, \mathscr{F}_T)$, where

$$\mathscr{F}_T = \{X \in \mathscr{F} : X \subseteq T\}.$$

It is trivial to verify that $(T, \mathscr{F}_T)$ defines a greedoid on T. We say that the restriction of T is obtained by the **deletion** of its complement. We denote this by

$$\mathscr{F} \setminus (E \setminus T) = \mathscr{F}_T .$$

Obviously the rank function r_T of $\mathscr{F}_T$ is just the restriction of the rank function r to T. Similarly, the closure operator σ_T is given by

$$\sigma_T(X) = \sigma(X) \cap T \text{ for all } X \subseteq T .$$

This relation, however, does not carry over to the monotone closure operator. One readily verifies that $\mu(X) \cap T \subseteq \mu_T(X)$ for all $X \subseteq T$, but this inclusion may be strict. In fact, if equality holds for all $X \subseteq T$, then T is closure feasible.

Similarly, the systems of rank feasible sets and of closure feasible sets of a restriction greedoid do not directly relate to the corresponding systems of the original greedoid. Again, the inclusions

$$\{X \in \mathcal{R} : X \subseteq T\} \subseteq \mathcal{R}(T, \mathcal{F}_T)$$

and

$$\{X \in \mathcal{C} : X \subseteq T\} \subseteq \mathcal{C}(T, \mathcal{F}_T)$$

hold but strict inclusions may occur.

For the third operation, consider a feasible set $B \in \mathcal{F}$. The **contraction of B** is the system $(E \setminus B, \mathcal{F}/B)$, where

$$\mathcal{F}/B = \{X \subseteq E \setminus B : X \cup B \in \mathcal{F}\}.$$

Clearly, the contraction of a feasible set in a greedoid results in a greedoid. The following lemma relates the corresponding rank functions.

Lemma 4.1. *Let $(E, \mathcal{F})$ be a greedoid, r its rank function, $B \in \mathcal{F}$ and r_B the rank function of the contraction $(E \setminus B, \mathcal{F}/B)$. Then for all $X \subseteq E \setminus B$*

$$r_B(X) = r(X \cup B) - |B| .$$

Proof. If A is a basis of X in $(E \setminus B, \mathcal{F}/B)$, then $A \cup B \in \mathcal{F}$. Hence

$$r(X \cup B) \geq |A \cup B| = r_B(X) + |B| .$$

Conversely, let A be a basis of $X \cup B$ in $(E, \mathcal{F})$. Since $B \in \mathcal{F}$, we can assume $B \subseteq A$ and hence $A \setminus B \in \mathcal{F}/B$. Thus

$$r_B(X) \geq |A \setminus B| = r(X \cup B) - |B| . \qquad \square$$

As a consequence we obtain for the closure σ_B and the monotone closure μ_B of $\mathcal{F}/B$,

Corollary 4.2.
(i) X is closed in $(E \setminus B, \mathcal{F}/B)$ if and only if $X \cup B$ is closed in $(E, \mathcal{F})$.
(ii) $\sigma_B(X) = \sigma(X \cup B) \setminus B$ and $\mu_B(X) = \mu(X \cup B) \setminus B$. $\qquad \square$

Let $\mathcal{R}/B$ and $\mathcal{C}/B$ denote the system of rank feasible sets and the system of closure feasible sets of $\mathcal{F}/B$.

Corollary 4.3.
(i)   *For all subsets $X \subseteq E$ we have $B \cup X \in \mathcal{R}$ if and only if $X \in \mathcal{R}/B$.*
(ii)   *If $Y \in \mathcal{C}$, then $Y \setminus B \in \mathcal{C}/B$.*

Proof. (i) If $B \cup X \in \mathcal{R}$, then we have in particular for all $B \cup B' \in \mathcal{F}$ with $B' \subseteq E \setminus B$:

$$|(B \cup X) \cap (B \cup B')| \le r(B \cup X).$$

Hence $|X \cap B'| \le r_B(X)$ for all $B' \in \mathcal{F}/B$.

Conversely, if $B \cup X \notin \mathcal{R}$, then by Lemma 3.2 there exists a feasible set $B \cup B' \in \mathcal{F}$ such that $|(B \cup X) \cap (B \cup B')| > r(B \cup X)$. Hence $|X \cap B'| > r_B(X)$ for some $B' \in \mathcal{F}/B$.

(ii) If $Y \in \mathscr{C}$, then for any $A \subseteq E \setminus B$, $Y \subseteq \mu(A \cup B)$ whenever $Y \subseteq \sigma(A \cup B)$. The relation between the closure operators of $\mathcal{F}$ and $\mathcal{F}/B$ immediately gives $Y \in \mathscr{C}/B$.   □

We now extend the definition of contraction from feasible sets to rank feasible sets. For $U \in \mathcal{R}$ define a rank function on $E \setminus U$ as

$$r_U(X) = r(X \cup U) - r(U).$$

Clearly, conditions (1.1), (1.3) and (1.4) of a greedoid rank function carry over and condition (1.2) follows from Theorem 3.2. Hence r_U is the rank function of a greedoid $(E \setminus U, \mathcal{F}/U)$, which we call the **contraction** of U.

The third part of Theorem 3.2 shows that $r_U(X) = r_B(X)$ for any basis $B \subseteq U$. Combining this with the result of Lemma 4.1 we see that

(4.1) $\mathcal{F}/U = \mathcal{F}/B$ for any basis B of U.

This result suggests defining the contraction of an arbitrary set $U \subseteq E$ by contracting a basis B of U. We will show in Section 5 that for interval greedoids this operation turns out to be independent of the particular choice of the basis B. In general, however, this is not true, as the following example shows.

Example. Consider the greedoid $(E, \mathcal{F})$ on the ground set $E = \{a, b, c, d\}$ with $\mathcal{F} = \{\emptyset, a, b, \{a, c\}, \{b, c\}, \{a, b, c\}, \{a, c, d\}\}$. Let $U = \{a, b\}$, then U has two bases $\{a\}$ and $\{b\}$. But $(\mathcal{F}/a) \setminus b = \{\emptyset, c, \{c, d\}\}$ while $(\mathcal{F}/b) \setminus a = \{\emptyset, c\}$.

A fourth minor-producing operation has already been used in Chapter III, where we applied it to antimatroids. In general, it will not induce a greedoid. For $T \subseteq E$ define the **trace** of $(E, \mathcal{F})$ on T as

$$\mathcal{F} : T = \{A \cap T : A \in \mathcal{F}\} .$$

Example. Let $E = \{w, x, y, z\}$ and $\mathcal{F} = \{\emptyset, w, x, \{w, x\}, \{w, y\}, \{x, z\}\}$. Then $\mathcal{F} : (E \setminus w) = \{\emptyset, x, y, \{x, z\}\}$ is not a greedoid.

In Chapter IX we will relate the trace operation to other operations and study classes of greedoids for which the traces of the form $\mathcal{F} : (E \setminus x)$ are again greedoids.

Up to now we have only dealt with substructures of greedoids. We now turn to constructions which extend the ground set. These constructions are well-known in matroid theory and were extended to greedoids by Brylawski and Dieter [1986].

Given two greedoids $(E_1, \mathcal{F}_1)$ and $(E_2, \mathcal{F}_2)$, the **union** $(E_1 \cup E_2, \mathcal{F}_1(\cup)\mathcal{F}_2)$ is the set system

$$\mathcal{F}_1(\cup)\mathcal{F}_2 = \{X_1 \cup X_2 : X_1 \in \mathcal{F}_1, \; X_2 \in \mathcal{F}_2\}.$$

The union of two greedoids is an accessible set system but is in general not a greedoid. If $E_1 \cap E_2 = \emptyset$, then the union is the **direct sum** $(E_1 \cup E_2, \mathcal{F}_1 + \mathcal{F}_2)$.

Lemma 4.4. *Let $(E_1, \mathcal{F}_1)$, $(E_2, \mathcal{F}_2)$ be two greedoids with $E_1 \cap E_2 = \emptyset$. Then for all $A \subseteq E_1, B \subseteq E_2$,*

(i) $(E_1 \cup E_2, \mathcal{F}_1 + \mathcal{F}_2)$ *is a greedoid,*
(ii) $r_{12}(A \cup B) = r_1(A) + r_2(B)$,
(iii) $\sigma_{12}(A \cup B) = \sigma_1(A) \cup \sigma_2(B)$,
(iv) $\lambda_{12}(A \cup B) = \lambda_1(A) \cup \lambda_2(B)$,
(v) $\mu_{12}(A \cup B) = \mu_1(A) \cup \mu_2(B)$,
(vi) $\beta_{12}(A \cup B) = \beta_1(A) + \beta_2(B)$,
(vii) $\mathcal{R}_{12} = \mathcal{R}_1 \cup \mathcal{R}_2, \; \mathcal{C}_{12} = \mathcal{C}_1 \cup \mathcal{C}_2$.

Proof. Straightforward.   □

Let $\mathcal{B}$ be the set of bases of a greedoid $(E, \mathcal{F})$, and let $\phi : \mathcal{B} \to 2^E \setminus E$ be a mapping with the following properties:

(4.2) $B \subseteq \phi(B)$ for all $B \in \mathcal{B}$,
(4.3) $\phi(B_1) \neq \phi(B_2)$ implies $B_1 \nsubseteq \phi(B_2)$ for all $B_1, B_2 \in \mathcal{B}$.

Given such a mapping, define the **erection** $(E, \mathcal{F}')$ of $(E, \mathcal{F})$ by

$$\mathcal{F}' = \mathcal{F} \cup \{B \cup e : B \in \mathcal{B} , e \in E \setminus \phi(B)\}.$$

Lemma 4.5. *The erection $(E, \mathcal{F}')$ of a greedoid $(E, \mathcal{F})$ is a greedoid.*

Proof. Obviously $(E, \mathcal{F}')$ is accessible. Consider now two sets $X, Y \in \mathcal{F}'$ with $|Y| < |X|$. If $|X| \leq r(E)$, the augmentation property of $\mathcal{F}'$ follows from that of $\mathcal{F}$.

If $|X| = r(E) + 1$, then $X = B \cup e$ for some $B \in \mathcal{B}$ and $e \in E \setminus \phi(B)$. We distinguish three subcases.

(i) $|Y| < r(E)$. We can then augment Y in $\mathcal{F}$ from B.
(ii) $|Y| = r(E)$ and $\phi(Y) = \phi(B)$. Then $Y \cup e \in \mathcal{F}'$.
(iii) $|Y| = r(E)$ and $\phi(Y) \neq \phi(B)$. Then by (4.3) there exists an element $x \in B \setminus \phi(Y)$. Hence $Y \cup x \in \mathcal{F}'$.   □

For the particular mapping $\phi(B) = B$, we obtain the **free erection** of a greedoid.

Define the **lift** $(E, \mathcal{F}')$ of a greedoid $(E, \mathcal{F})$ as the set system

$$\mathcal{F}' = \{X \subseteq E : |X| - 1 \leq r(X) \leq |X|\} .$$

Denoting by $\mathcal{M}_1$ the 1-truncation of the free matroid on E, we see that

$$\mathcal{F}' = \mathcal{F}(\cup)\mathcal{M}_1.$$

Lemma 4.6. *The lift $(E, \mathcal{F}')$ of a greedoid $(E, \mathcal{F})$ is a greedoid.*

Proof. Straightforward.   □

5. Interval Greedoids

Recall from Chapter IV that a greedoid $(E, \mathcal{F})$ is an **interval greedoid** if for all feasible sets $A, B, C \in \mathcal{F}$ with $A \subseteq B \subseteq C$ and all $x \in E \setminus C$ the following holds:

(5.1) $A \cup \{x\} \in \mathcal{F}$ and $C \cup \{x\} \in \mathcal{F}$ implies $B \cup \{x\} \in \mathcal{F}$.

We will now investigate the structural properties of this class of greedoids. The first characterization of interval greedoids is a strengthening of the augmentation property (IV.1.3).

Lemma 5.1. *A hereditary language $(E, \mathcal{L})$ is an interval greedoid if and only if the following holds:*

(5.2)   *If $\alpha, \beta \in \mathcal{L}$ with $|\beta| > |\alpha|$, then there exists a substring β' of β with $|\beta'| \geq |\beta| - |\alpha|$ such that $\alpha\beta' \in \mathcal{L}$.*

(By a substring of β we mean any subset of $\tilde{\beta}$, ordered in the same way as β.)

Proof. Let $(E, \mathcal{L})$ be a hereditary language satisfying (5.2), and $(E, \mathcal{F})$ the corresponding accessible set system. Trivially, $(E, \mathcal{L})$ is a greedoid. Consider $A, B, C \in \mathcal{F}$ with $A \subseteq B \subseteq C$ and $A \cup x, C \cup x \in F$ for some $x \in E \setminus C$.
   Using (5.2) we can find words α, $\alpha\beta$ and $\alpha\beta\gamma$ in $\mathcal{L}$ such that $\tilde{\alpha} = A$, $\tilde{\beta} = B \setminus A$ and $\tilde{\gamma} = C \setminus B$. Now $\alpha x \in \mathcal{L}$ and $\alpha\beta\gamma x \in \mathcal{L}$ together with (5.2) imply $\alpha x\beta\gamma \in \mathcal{L}$, hence $\alpha x\beta \in \mathcal{L}$. Thus $B \cup x \in \mathcal{F}$.
   Conversely, let $(E, \mathcal{L})$ be an interval greedoid. We prove property (5.2) by induction on $|\alpha|$ and observe that (5.2) holds trivially for $\alpha = \emptyset$.
   Let $\alpha z, \beta \in \mathcal{L}$ be two feasible words with $|\beta| > |\alpha z|$. By induction, there exists a substring $\beta' = y_1 y_2 \ldots y_l$ of β such that $\alpha\beta' \in \mathcal{L}$. Now use the augmentation property to successively augment αz from $\alpha\beta'$ with the additional stipulation of always selecting a y_i with the smallest index possible. Suppose at step k we arrive at a word

$$\alpha z y_1 y_2 \ldots y_{j-1} y_{j+1} \ldots y_k y_j.$$

The interval property applied to $A = \tilde{\alpha} \cup \{y_1, \ldots, y_{j-1}\}, B = A \cup z$ and $C = A \cup \{z, y_{j+1}, \ldots, y_k\}$ and $x = y_j$ yields $\alpha z y_1 y_2 \ldots y_{j-1} y_j \in \mathcal{L}$, in contradiction to the selection rule. Thus αz can be augmented from $\alpha\beta'$ by some substring of β' which is again a substring of β.   □

If we reformulate the interval property in terms of the closure operator, we obtain the following condition.

Lemma 5.2. *A greedoid has the interval property if and only if for all $A \subseteq B \subseteq C$*

$$\sigma(B) \subseteq \sigma(A) \cup \sigma(C) .$$ □

It is clear that it suffices to require this for feasible sets A, B and C.

Lemma 5.3. *A hereditary language $(E, \mathscr{L})$ is an interval greedoid if and only if the following conditions hold:*

(5.3.1) $\alpha x y \beta \in \mathscr{L}$ *and* $\alpha y \in \mathscr{L}$ *implies* $\alpha y x \beta \in \mathscr{L}$,
(5.3.2) $\alpha x \in \mathscr{L}, \alpha y \in \mathscr{L}, \alpha x \beta \in \mathscr{L}$ *and* $\alpha x y \notin \mathscr{L}$ *implies* $\alpha y \beta \in \mathscr{L}$.

Proof. Let $(E, \mathscr{L})$ be an interval greedoid. Then (5.3.1) is an immediate consequence of the augmentation property applied to $\alpha x y \beta$ and αy.

To verify (5.3.2), let $\alpha x \in \mathscr{L}, \alpha y \in \mathscr{L}$ and $\alpha x \beta \in \mathscr{L}$ be feasible words and assume $\alpha x y \notin \mathscr{L}$. If we successively augment αy from $\alpha x \beta$, then either $\alpha y \beta \in \mathscr{L}$ or $\alpha y \beta_1 x \in \mathscr{L}$ for some substring β_1 of β. Suppose the second case holds. Then the interval property applied to $\alpha x \in \mathscr{L}$ and $\alpha y \beta_1 x \in \mathscr{L}$ implies that $\alpha y x \in \mathscr{L}$. Hence $\alpha x y \in \mathscr{L}$, in contradiction to our assumption. Thus $\alpha y \beta \in \mathscr{L}$.

Conversely, assume that (5.3) is satisfied and consider $\gamma \alpha \in \mathscr{L}$ and $\gamma \beta \in \mathscr{L}$, where $|\beta| > |\alpha|$ and γ is the longest common initial substring of both words. To verify (5.2), we have to show that $\gamma \alpha \beta' \in \mathscr{L}$ for some substring β' in β of appropriate length.

Since (5.2) is trivially fulfilled if $|\alpha| = 0$, we proceed by induction on the length of α. If $\alpha \neq \emptyset$, we may write $\gamma \beta = \gamma x \beta_1$ and $\gamma \alpha = \gamma y \alpha_1$ and distinguish two cases.

(i) $\gamma y x \in \mathscr{L}$. The induction hypothesis applied to $\gamma y x \in \mathscr{L}$ and $\gamma y \alpha_1$ yields a substring α_2 of α_1 so that $\gamma y x \alpha_2 \in \mathscr{L}$ and $|\gamma y x \alpha_2| = |\gamma \alpha|$.

Then by (5.3.1), $\gamma x y \alpha_2 \in \mathscr{L}$. Using induction, we obtain a substring β_2 of β_1 with $\gamma x y \alpha_2 \beta_2 \in \mathscr{L}$ and $|\gamma x y \alpha_2 \beta_2| = |\gamma \beta|$. Again by (5.3.1), we get $\gamma y x \alpha_2 \beta_2 \in \mathscr{L}$ and, applying induction to this word and $\gamma y \alpha_1$, we obtain a substring β' of $x \beta_2 \subseteq \beta$ such that $\gamma y \alpha_1 \beta' = \gamma \alpha \beta' \in \mathscr{L}$.

(ii) $\gamma y x \notin \mathscr{L}$. Then $\gamma y \beta_1 \in \mathscr{L}$ by (5.3.2). The induction hypothesis then yields a substring β' of β_1 such that $\gamma y \alpha_1 \beta' = \gamma \alpha \beta' \in \mathscr{L}$.

Thus in both cases we can find a substring β' of β with $\gamma \alpha \beta' \in \mathscr{L}$, which proves the assertion.   □

Lemma 5.4. *A greedoid $(E, \mathscr{F})$ has the interval property if and only if $X \subseteq \sigma(A)$ implies $X \subseteq \sigma(B)$ for any $X \in \mathscr{F}$ and any two sets $A, B \subseteq E$ with $A \subseteq B$.*

Proof. Let $(E, \mathscr{F})$ have the above property and consider feasible sets $A, B, C \in \mathscr{F}$ with $A \subseteq B \subseteq C$. Assume $A \cup x \in \mathscr{F}$ and $B \cup x \notin \mathscr{F}$ for some $x \notin B$. Then $A \cup x \subseteq \sigma(B)$ which implies $A \cup x \subseteq \sigma(C)$, i.e. $C \cup x \notin \mathscr{F}$. Thus $(E, \mathscr{F})$ is an interval greedoid.

Conversely, assume that $(E, \mathscr{F})$ is an interval greedoid and let C be a basis of $X \cap \sigma(B)$. If $C = X$ we are done, so suppose $C \neq X$. Then there exists an element $x \in X \setminus C$ such that $C \cup \{x\} \in \mathscr{F}$. Clearly $x \notin \sigma(B)$ and $x \notin \sigma(C)$. We have $C \subseteq \sigma(A) \cap \sigma(B) \subseteq \sigma(B)$ and thus by Lemma 5.2 $\sigma(\sigma(A) \cap \sigma(B)) \subseteq \sigma(C) \cup \sigma(B)$. Furthermore, $A \subseteq \sigma(A) \cap \sigma(B) \subseteq \sigma(A)$, hence by Lemma 2.3 $\sigma(\sigma(A) \cap \sigma(B)) = \sigma(A)$. Thus $\sigma(A) \subseteq \sigma(C) \cup \sigma(B)$, in contradiction to $x \in \sigma(A)$, $x \notin \sigma(B) \cup \sigma(C)$.   □

Observe that the previous statement does not imply the monotonicity of σ, i.e. the restriction to feasible sets is crucial.

We have seen in Lemma 3.8 that for arbitrary greedoids the system of closure feasible sets is contained in the system of rank feasible sets and that, in general, this inclusion is strict. The interval greedoids now turn out to be exactly those greedoids for which these two systems are the same.

Theorem 5.5. *Let $(E, \mathscr{F})$ be a greedoid, $\mathscr{C}$ its system of closure feasible sets and $\mathscr{R}$ its system of rank feasible sets. $(E, \mathscr{F})$ has the interval property if and only if $\mathscr{C} = \mathscr{R}$.*

Proof. Suppose $(E, \mathscr{F})$ has the interval property and let $U \in \mathscr{R}$ be a rank feasible set. We have to show that $U \in \mathscr{C}$. By Theorem 3.5, there exists a set $X \in \mathscr{F}$ such that $X \subseteq U \subseteq \mu(X)$. Now let $A \subseteq B$ and $U \subseteq \sigma(A)$. Then $X \subseteq \sigma(A)$, and by Lemma 5.4, $X \subseteq \sigma(B)$. The definition of the monotone closure operator implies $\mu(X) \subseteq \sigma(B)$. Hence $U \subseteq \sigma(B)$, and thus $U \in \mathscr{C}$.

To see the converse, suppose $(E, \mathscr{F})$ does not have the interval property. Then there exist sets $A \subseteq B \subseteq C$ such that $A, B, C \in \mathscr{F}$ and for some $x \in E \setminus C$ we have $A \cup \{x\}$, $C \cup \{x\} \in \mathscr{F}$ but $B \cup \{x\} \notin \mathscr{F}$. Then $A \cup \{x\} \subseteq \sigma(B)$; however, $A \cup \{x\} \not\subseteq \sigma(C)$. Hence $A \cup \{x\} \notin \mathscr{C}$, but $A \cup \{x\} \in \mathscr{F} \subseteq \mathscr{R}$.   □

Corollary 5.6. *For an interval greedoid, $(E, \mathscr{C}) = (E, \mathscr{R})$ is an antimatroid.*

Proof. By Lemmas 3.7 and 3.9, $\mathscr{C} = \mathscr{R}$ is accessible and closed under union.   □

While in general no inclusion relation between $\mathscr{F}$ and $\mathscr{C}$ holds, the previous theorems yield such a property for interval greedoids.

Corollary 5.7. *The greedoid $(E, \mathscr{F})$ has the interval property if and only if $\mathscr{F} \subseteq \mathscr{C}$.*

Proof. This is an immediate consequence of Theorem 5.5.   □

Recall that the collection $(E, \mathscr{A})$ of partial alphabets consists of all unions of feasible sets of a greedoid.

Corollary 5.8. *For a greedoid $(E, \mathscr{F})$ the following statements are equivalent:*

(i) $(E, \mathscr{F})$ *has the interval property.*
(ii) $\mathscr{A} \subseteq \mathscr{R}$.
(iii) $A, B, C \in \mathscr{F}$ *with* $A \cup B \subseteq C$ *implies* $A \cup B \in \mathscr{F}$ **(local union property).**

The local union property says, in other words, that the restriction of an interval greedoid to a feasible set is an antimatroid.

Proof. (i) $\Rightarrow$ (ii): If $(E, \mathscr{F})$ has the interval property, then by Theorem 5.5, $\mathscr{R}$ is closed under union, so, since $\mathscr{F} \subseteq \mathscr{R}$, we have $\mathscr{A} \subseteq \mathscr{R}$.

(ii) $\Rightarrow$ (iii): If $\mathscr{A} \subseteq \mathscr{R}$, then $A \cup B$ is rank feasible. Hence

$$r(A \cup B) = \beta(A \cup B) \geq |(A \cup B) \cap C| = |A \cup B|,$$

i.e. $A \cup B \in \mathscr{F}$.

(iii) $\Rightarrow$ (i): $A, B, C \in \mathscr{F}$ with $A \subseteq B \subseteq C$ and $A \cup x, C \cup x \in \mathscr{F}$ implies $(A \cup x) \cup B = B \cup x \in \mathscr{F}$. $\qquad\square$

In Section 2 we have seen that the kernel closure is not monotone in general. It turns out that interval greedoids are precisely those greedoids whose kernel closure is monotone.

Corollary 5.9. *The greedoid $(E, \mathscr{F})$ has the interval property if and only if its kernel closure λ is monotone.*

Proof. Assume that the kernel closure λ is monotone and consider three sets $A, B, C \in \mathscr{F}$ with $A \subseteq B \subseteq C$ and $A \cup x$, $C \cup x \in \mathscr{F}$. Suppose $B \cup x \notin \mathscr{F}$. Then B is a basis of $B \cup x$ and hence $\lambda(B) = \lambda(B \cup x)$. If $x \in \lambda(B)$, then by the monotonicity of λ, $x \in \lambda(C)$, contradicting $C \cup x \in \mathscr{F}$. On the other hand, if $x \notin \lambda(B) = \lambda(B \cup x)$, then by the monotonicity of λ, $x \notin \lambda(A \cup x)$, contradicting $A \cup x \in \mathscr{F}$. Thus $(E, \mathscr{F})$ has the interval property.

Conversely, let $(E, \mathscr{F})$ be an interval greedoid and $A \subseteq B$ be two arbitrary subsets. Consider any basis X of A and any basis Y of B with $X \subseteq Y$. Since the rank function r is submodular on $\mathscr{R}$ and $\mathscr{A} \subseteq \mathscr{R}$, we obtain for any $Z \in \mathscr{F}$ with $r(X \cup Z) = r(X)$:

$$\begin{aligned} r(X) + r(Y) &= r(X \cup Z) + r(Y) \\ &\geq r(Y \cup Z) + r(X \cup (Y \cap Z)) \\ &\geq r(Y \cup Z) + r(X) . \end{aligned}$$

Hence $r(Y) = r(Y \cup Z)$ and thus $\lambda(A) \subseteq \lambda(B)$, i.e. the kernel closure is monotone. $\qquad\square$

We say that two feasible sets $A, B \in \mathscr{F}$ are **equivalent** if $\mathscr{F}/A = \mathscr{F}/B$, i.e. if they have the same set of continuations. The equivalence classes $[A]$ are the **flats** of the greedoid. Note that equivalent sets are cospanning (since cospanning means that they have the same 1-element continuations).

Theorem 5.10. *A greedoid $(E, \mathscr{F})$ has the interval property if and only if cospanning sets are equivalent.*

Proof. Let $(E, \mathscr{F})$ be an interval greedoid and $A, B \in \mathscr{F}$ two bases of X. If $A \cup Y \in \mathscr{F}$, we can augment B from $A \cup Y$ to a set $B \cup Z \in \mathscr{F}$ with

$|B \cup Z| = |A \cup Y|$. Suppose $Z \neq Y$, i.e. Z contains some elements of A. Then $\beta(A \cup B) > |B| = r(B)$, in contradiction to $A \cup B \in \mathscr{A} \subseteq \mathscr{R}$. Hence $B \cup Y \in \mathscr{F}$, i.e. A and B are equivalent.

Conversely, assume that cospanning sets are equivalent. We show that the greedoid has the local union property of Corollary 5.8(iii). Let $A, B, C \in \mathscr{F}$ with $A, B \subseteq C$. Suppose $A \cup B \notin \mathscr{F}$. We augment A and B to bases $A \cup B'$ and $B \cup A'$ of $A \cup B$. Now $C \in \mathscr{F}$ implies $C \setminus (A \cup B') \in \mathscr{F}/(A \cup B')$ but $C \setminus (A \cup B') \notin \mathscr{F}/(B \cup A')$, contradicting the assumption that $A \cup B'$ and $B \cup A'$ are equivalent. Hence $A \cup B \in \mathscr{F}$ and $(E, \mathscr{F})$ has the interval property by Corollary 5.8.   □

For interval greedoids the concepts of flats and λ-closed sets are equivalent in the following sense.

Lemma 5.11. *Let $(E, \mathscr{F})$ be an interval greedoid. Then a set $A \in \mathscr{A}$ is λ-closed if and only if its bases form a flat.*

Proof. Let A be λ-closed and X be a basis of A. By Theorem 5.10, every other basis is equvalent to X. Moreover, if Y is a feasible set equivalent to X, then X and Y are cospanning and hence $Y \subseteq \sigma(A)$. But since A is λ-closed, $X \subseteq A$. Thus the bases of A form an equivalence class.

Conversely, let A be such that its bases form an equivalence class. Consider a set $X \in \mathscr{F}$ with $X \subseteq \sigma(A)$ and extend X to a basis X' of $\sigma(A)$. Consider any $Z \in \mathscr{F}/X'$. We can augment any basis Y of A from $X \cup Z$ to a basis $Y \cup Z'$ of $\sigma(A) \cup Z$. Clearly, $|Z'| = |Z|$ and Z' cannot contain any element of X' since $X' \cup Y \in \mathscr{R}$. So $Z' = Z$ and hence $\mathscr{F}/X' \subseteq \mathscr{F}/Y$. The reverse inclusion is shown similarly. Thus X' is equivalent to Y and hence $X \subseteq X' \subseteq A$. So A is λ-closed.   □

In Section 4 we remarked that for arbitrary greedoids only the contraction of rank feasible sets is meaningful. For interval greedoids, we may contract any subset U by contracting some basis of U. The resulting greedoid will be independent of the special choice of the basis by Theorem 5.10.

Chapter VI. Further Structural Properties

We continue to develop some basic tools for the study of the structure of greedoids. We relate greedoids to lattices and extend the notion of connectivity to greedoids.

The first section collects results about ordered sets associated with greedoids, some of which date back to Dilworth in 1940. We show that feasible sets, closed sets and flats of an interval greedoid when endowed with an appropriate order relation, induce semimodular lattices.

In Section 2 we extend the notion of connectivity and dominating sets in graphs to greedoids. We introduce an exchange property for bases which gives rise to an abstraction of the pivot operation in linear algebra. Depending on the connectivity of the greedoid, we derive upper bounds on the number of pivot steps necessary to transform one basis into another. While for arbitrary 2-connected greedoids this bound is an exponential function, it can be reduced to a quadratic expression in the case of branching greedoids.

1. Lattices Associated with Greedoids

Greedoids, and in particular antimatroids, have already been investigated implicitly within the framework of lattice theory by Dilworth [1940]. Here we review some of these results and outline further connections between greedoids and lattices which have been found more recently (cf. Edelmann [1980], Korte and Lovász [1983]).

For an accessible set system $(E, \mathcal{F})$, let $(\mathcal{F}, \subseteq)$ be the ordered set on $\mathcal{F}$ with respect to inclusion. It is immediate from Birkhoff's Theorem that $(E, \mathcal{F})$ is a poset antimatroid if and only if $(\mathcal{F}, \subseteq)$ is a distributive lattice.

Theorem 1.1. *For any set system $\mathcal{F} \subseteq 2^E$ with $\emptyset, E \in \mathcal{F}$ the following statements are equivalent:*

(i)   *$(E, \mathcal{F})$ is an antimatroid.*
(ii) *$(\mathcal{F}, \subseteq)$ is a locally free lattice of height $|E|$.*
(iii) *$(\mathcal{F}, \subseteq)$ is a semimodular lattice of height $|E|$.*

Proof. Since every locally free lattice is semimodular, the result follows from Lemma III.1.2.   □

Corollary 1.2. *A lattice is locally free if and only if it is isomorphic to the lattice of feasible sets of some antimatroid.*

Proof. The "if"-direction follows from the above theorem. To prove the "only if"-part we use the following constructions of Greene and Markowsky [1974]. Let L be a locally free lattice. Let E be the set of its meet-irreducible elements. Then for each x let $U_x = \{u \in E : x \not\leq u\}$ and let $\mathscr{F}$ be the family of all U_x, $x \in L$. We claim that $(E, \mathscr{F})$ is an antimatroid and that $x \to U_x$ is an isomorphism between L and $(\mathscr{F}, \subseteq)$. Note that U_x also determines x by $x = \bigwedge\{u : u \in E \setminus U_x\}$. It is trivial that $\mathscr{F}$ is union-closed since $U_{x \vee y} = U_x \cup U_y$. To show the accessibility of $(E, \mathscr{F})$ we need the following

Claim: Let $u, y, x, a \in L$ such that $u, y, x \geq a$, $x \not\leq u$, x covers a and u is meet-irreducible. Then either $y \leq u$ or $y \geq x$.

We use induction on the height of $[y \wedge u, u]$. If $y \geq u$, then consider $u \vee x$. It is clear that $u \vee x$ covers u by semimodularity. Moreover, $u \leq (u \vee x) \wedge y \leq u \vee x$. The first inequality must be strict since u is meet-irreducible. Hence the second inequality is equality and so $y \geq u \vee x \geq x$.

Now assume that $y \not\geq u$. Let $a_1 = u \wedge y$, let u_1 be any element in $[a_1, u]$ covering a_1, let y_1 be any element in $[a_1, y]$ covering a_1 and let $x_1 = a_1 \vee x$. Clearly x_1 also covers a_1. It is obvious that $x_1 \neq u_1$ and $y_1 \neq u_1$. If $x_1 = y_1$ then $y \geq x$. So we may assume that x_1, y_1 and u_1 are distinct. Hence $[a_1, x_1 \vee y_1 \vee u_1]$ is a Boolean algebra. This in particular implies that $y_2 = u_1 \vee y_1$ is not above x_1. But then we get a contradiction if we apply the induction hypothesis to u, y_2, x_1 and a_1. This proves the claim.

To show the accessibility of $\mathscr{F}$, take any element $x \in L, x \neq 0$. We want to show that we can delete an element of U_x and stay feasible. Let a be any element covered by x. Clearly U_a is properly contained in U_x, and it remains to show that $|U_x \setminus U_a| = 1$. Consider $u_1, u_2 \in U_x \setminus U_a$. Then, by the claim, $u_1 \geq u_2$ and $u_2 \geq u_1$. So $u_1 = u_2$.   □

A further consequence of the previous theorem is

Corollary 1.3. *A greedoid $(E, \mathscr{F})$ has the interval property if and only if all intervals $[\emptyset, X]$ in $(\mathscr{F}, \subseteq)$ are semimodular lattices.*

Proof. Follows from Theorem 1.1 and Corollary V.5.8.   □

For arbitrary greedoids, the ordered set $(\mathscr{F}, \subseteq)$ will not be a lattice in general. However, by formally adding E to $\mathscr{F}$ (if necessary) to get a family $\widehat{\mathscr{F}}$, we obtain a lattice in the case of interval greedoids.

Theorem 1.4. *For an interval greedoid, $(\widehat{\mathscr{F}}, \subseteq)$ is a lattice.*

Proof. Interval greedoids have the local union property. So the join of two sets $X, Y \in \widehat{\mathscr{F}}$ exists and is either $X \cup Y$ or E. We have seen in Chapter I that the existence of a minimal element in $\widehat{\mathscr{F}}$ and the meet operation

$$X \wedge Y = \bigcup \{Z \in \widehat{\mathscr{F}} : Z \subseteq X \cap Y\}$$

turn the join-semilattice $(\widehat{\mathscr{F}}, \subseteq)$ into a lattice. ☐

Since closure feasible sets are closed under union and since $\emptyset \in \mathscr{C}$, we get

Theorem 1.5. *The poset $(\mathscr{C}, \subseteq)$ of closure feasible sets of a greedoid is a lattice.* ☐

This lattice will in general not be semimodular. The collection of partial alphabets, however, is an antimatroid and hence a locally free lattice. For interval greedoids we know by Corollary V.5.6 that $(E, \mathscr{C}) = (E, \mathscr{R})$ is an antimatroid and hence $(\mathscr{C}, \subseteq) = (\mathscr{R}, \subseteq)$ is again a locally free lattice. For non-interval greedoids $(\mathscr{R}, \subseteq)$ is not necessarily a lattice. However, as a poset it has a nice property.

Lemma 1.6. *The poset $(\mathscr{R}, \subseteq)$ of a greedoid is graded with height function $h(X) = |X|$.*

Proof. Let $X, Y \in \mathscr{R}$ and assume that X covers Y. If $r(X) = r(Y)$, then for any $x \in X \setminus Y$ we have $r(Y \cup x) = r(X) = \beta(X) \geq \beta(Y \cup x) \geq r(Y \cup x)$ and hence $Y \cup x \in \mathscr{R}$ and $X = Y \cup x$. If $r(X) > r(Y)$ then there exists an element $x \in X \setminus Y$ such that $r(Y \cup x) \geq r(Y) + 1$.

Moreover, $r(Y \cup x) \leq \beta(Y \cup x) \leq \beta(Y) + 1 = r(Y) + 1 \leq r(Y \cup x)$. Hence, again $Y \cup x \in \mathscr{R}$ and $X = Y \cup x$. ☐

Let us denote by $\mathscr{Cl}$ the family of λ-closed sets of a greedoid $(E, \mathscr{F})$.

Theorem 1.7. *The poset $(\mathscr{Cl}, \subseteq)$ of λ-closed sets of an interval greedoid is a semi-modular lattice with height function equal to the rank function of the greedoid.*

Proof. First we show that for $A, B \in \mathscr{Cl}$, $\lambda(A \cup B)$ is the join of A and B in this poset. By the monotonicity of λ, $A = \lambda(A) \subseteq \lambda(A \cup B)$ and, similarily, $B \subseteq \lambda(A \cup B)$. Moreover, if $C \in \mathscr{Cl}$ and $A, B \subseteq C$, then $C = \lambda(C) \supseteq \lambda(A \cup B)$. Hence $(\mathscr{Cl}, \subseteq)$ is a lattice.

Next we show that if $A, B \in \mathscr{Cl}$ and B covers A, then $r(B) = r(A) + 1$. To this end, take $x \in B \setminus A$. Let $U \in \mathscr{F}$ be such that $x \in U \subseteq B$. Let V be a basis of A. Then $U \nsubseteq \sigma(V)$, since it is feasible and not contained in $\lambda(V) = A$. Hence there is a $y \in U$ such that $V \cup y \in \mathscr{F}$. Then $A = \lambda(V) \subseteq \lambda(V \cup y) \subseteq \lambda(B) = B$. Since $r(A) < r(V \cup y)$, $A \neq \lambda(V \cup y)$ and thus $B = V \cup y$.

So we know that the height function of $(\mathscr{Cl}, \subseteq)$ is the greedoid rank function. Since the greedoid rank function is submodular on partial alphabets, $(\mathscr{Cl}, \subseteq)$ is a semimodular lattice. ☐

We remark that an argument similar to the one used in the previous proof shows that the meet of closed sets is the λ-closure of their intersection.

We have seen that there is a natural one-to-one correspondence between λ-closed sets, closed sets and flats of an interval greedoid. Hence the lattice structure of the previous theorem can be carried over.

We close this section by also proving the converse of Theorem 1.7, i.e. that every semimodular lattice is isomorphic to the poset of λ-closed sets of an interval greedoid.

To this end, we say that a sequence $\alpha = x_1 \ldots x_n$ of distinct elements in a semimodular lattice L is a **generating sequence** if the joins $y_i = x_1 \vee \ldots \vee x_i$, $0 \leq i \leq n$, form a maximal chain from 0 to y_n in L.

Lemma 1.8. *Let $(E, \mathcal{L})$ be the language of generating sequences in a semimodular lattice. Then for every partial alphabet $A \in \mathcal{A}$ and every maximal word $x_1 \ldots x_n$ in $\mathcal{L}$ formed by elements of A,*

$$\vee x_i = \vee \{a : a \in A\}.$$

Proof. Let $x_1 \ldots x_n$ be a maximal word of elements of A and $y \in A$ be arbitrary. Suppose $y \not\leq x = x_1 \vee x_2 \vee \ldots \vee x_n$. Then let $y_1 \ldots y_k$ be a word in A with $y_k = y$ and y_i the first letter for which $y_i \not\leq x$. Then $y_1 \vee \ldots \vee y_i$ covers $y_1 \vee \ldots \vee y_{i-1} \leq x$. By semimodularity, $y_i \vee x$ covers x and hence $x_1 \ldots x_n y_i$ is a word in $\mathcal{L}$, contradicting the maximility of $x_1 \ldots x_n$. Thus $x \in A$ is an upper bound for A and hence $x = \vee \{a : a \in A\}$. $\qquad\square$

Theorem 1.9. *Any semimodular lattice L is isomorphic to the lattice of λ-closed sets of an interval greedoid.*

Proof. Let $(E, \mathcal{L})$ be the language of generating sequences. Obviously, $(E, \mathcal{L})$ is hereditary. An argument as used in the proof of Lemma 1.8 reveals the strong augmentation property, i.e. $(E, \mathcal{L})$ is an interval greedoid.

Again by Lemma 1.8, the rank of a set $A \in \mathcal{A}$ is equal to the rank of $\vee \{a : a \in A\}$ in the lattice. For two partial alphabets $A, B \in \mathcal{A}$, $r(A) = r(B)$ if and only if $\vee A = \vee B$. Thus

$$\lambda(A) = \{x \in L : x \leq \vee A\}$$

and the λ-closed sets form a lattice isomorphic to L. $\qquad\square$

2. Connectivity in Greedoids

In this section we introduce a notion of connectivity and domination in ordered sets and greedoids.

Consider a graded poset P of rank $r = r(P)$ and some element $x \in P$. A set A of upper covers is called x-**free** if A is the set of atoms of a Boolean algebra in P with minimal element x. P is called k-**connected** for some $1 \leq k \leq r$ if for every $x \in P$ there exists an x-free set of size at least $\min\{k, r(P) - r(x)\}$. Then every graded ordered set is 1-connected and every k-connected ordered set is $(k - 1)$-connected. In the case that the poset consists of the feasible sets of a greedoid $(E, \mathcal{F})$ ordered by inclusion, a set A is X-free ($X \in \mathcal{F}$) if and only if $X \cup A' \in \mathcal{F}$ for every $A' \subseteq A$.

A greedoid $(E, \mathcal{F})$ is k-**connected** if the poset $(\mathcal{F}, \subseteq)$ is k-connected. That is, for every $X \in \mathcal{F}$ there exists an X-free set $A \subseteq E \setminus X$ with $|A| \geq \min\{k, r(E) - |X|\}$. A set $X \in \mathcal{F}$ is **dominating** if there is an X-free set A with $|A| = r(E) - |X|$.

Example. In a matroid $(E, \mathcal{F})$ every independent set is dominating and hence $(E, \mathcal{F})$ is r-connected.

Example. The free sets of an antimatroid $(E, \mathcal{F})$ are precisely the X-free sets for some $X \in \mathcal{F}$. The set $X \in \mathcal{F}$ is dominating if $\Gamma(X) = E \setminus X$. $(E, \mathcal{F})$ is k-connected if $\Gamma(X) \geq k$ for all nondominating feasible sets. Hence $(E, \mathcal{F})$ is k-connected if and only if every circuit has at least $k + 1$ elements.

Example. In a directed branching greedoid $(E, \mathcal{F})$ a set A is X-free if and only if the edges in A connect $V(X)$ to different nodes in $V \setminus V(X)$. Hence $(E, \mathcal{F})$ is k-connected if and only if for all $X \in \mathcal{F}$ there are at least $\min\{k, |V \setminus V(X)|\}$ nodes in $V \setminus V(X)$ which can be reached by an edge leaving $V(X)$. That is, $(E, \mathcal{F})$ is k-connected if and only if no node in V can be separated by less than k nodes from the root. Moreover, X is dominating precisely when every node in $V \setminus V(X)$ can be reached from X by an edge, i.e. if X is dominating in the graph theoretical sense.

The following lemma yields a simple algorithm for checking whether a set is X-free.

Lemma 2.1. *Let $(E, \mathcal{F})$ be an interval greedoid and $X \in \mathcal{F}$ a feasible set. Then $A \subseteq E \setminus X$ is X-free if and only if $X \cup A \in \mathcal{F}$ and $A \subseteq \Gamma(X)$. Moreover, the X-free sets are the independent sets of a matroid.*

Proof. Let $A \subseteq E \setminus X$ be such that $X \cup A \in \mathcal{F}$ and $A \subseteq \Gamma(X)$. By Corollary V.5.8, $X \cup B \in \mathcal{F}$ for any subset $B \subseteq A$. Moreover, the greedoid $(\Gamma(X), \mathcal{M}_X)$ with $\mathcal{M}_X = \{B \subseteq \Gamma(X) : X \cup B \in \mathcal{F}\}$ is a matroid by Lemma IV.1.5. $\qquad\square$

Lemma 2.2. *Let $(E, \mathcal{F})$ be an interval greedoid and $A, B, C \in \mathcal{F}$ feasible sets with $A \subseteq B \subseteq C$. Then $C \subseteq A \cup \Gamma(A)$ implies $C \subseteq B \cup \Gamma(B)$.*

Proof. Consider an element $x \in C \setminus B$. By augmentation we obtain a set $D \in \mathcal{F}$ with $B \subseteq D \subseteq C \setminus x$ and $D \cup x \in \mathcal{F}$. The interval property then implies $B \cup x \in \mathcal{F}$, i.e. $x \in \Gamma(B)$. $\qquad\square$

Lemma 2.3. *Let $(E, \mathcal{F})$ be an interval greedoid and $A, B \in \mathcal{F}$ feasible sets with $A \subseteq B$. Then $\Gamma(A) \cap B$ is A-free.*

Proof. This is a simple consequence of Corollary V.5.8. $\qquad\square$

Lemma 2.4. *Let $(E, \mathcal{F})$ be an interval greedoid and let $A, B \in \mathcal{F}$ be feasible sets with $A \subseteq B$. Then B is dominating if A is dominating.*

Proof. Let $X \subseteq \Gamma(A)$ be such that $C = A \cup X$ is a basis. Augment B from C to obtain a basis $B \cup Y$. By Lemma 2.2, $Y \subseteq \Gamma(A)$ implies $Y \subseteq \Gamma(B)$, i.e. B is dominating. $\qquad \square$

Thus the dominating sets of an interval greedoid $(E, \mathscr{F})$ form a filter $\mathscr{D}$ in the lattice $(\widehat{\mathscr{F}}, \subseteq)$ of feasible sets. By adding $\emptyset$ and E to $\mathscr{D}$ we obtain the following:

Corollary 2.5. *The dominating sets of an interval greedoid together with $\emptyset$ and E ordered by inclusion form a lattice.* $\qquad \square$

The property of k-connectivity is preserved under truncations and contractions. For restrictions, however, the situation is different.

Lemma 2.6. *An interval greedoid $(E, \mathscr{F})$ is a matroid if and only if every restriction to a feasible set is 2-connected.*

Proof. Only the sufficiency requires proof. Proceeding by induction on $|E|$, we may assume that $(E, \mathscr{F})$ is normal. If the greedoid has more than one basis, we may again apply induction to the restriction to the basis and obtain $x \in \mathscr{F}$ for all $x \in E$. Thus $(E, \mathscr{F})$ is a matroid by Lemma IV.1.5.

So we may assume that E is the only basis. Then the 2-connectivity implies that there are two elements $x, y \in E$ such that $E \setminus x, E \setminus y \in \mathscr{F}$. Then induction and Lemma IV.1.5 give the desired result. $\qquad \square$

Lemma 2.7. *An interval greedoid $(E, \mathscr{F})$ is k-connected if and only if its lattice $\mathscr{C}\ell$ of flats is k-connected.*

Proof. Straightforward. $\qquad \square$

Let $(E, \mathscr{F})$ be a k-connected interval greedoid and $X \in \mathscr{F}$ with $|X| \leq r(E) - k$. Then the X-free sets form a matroid and hence every $x \in \Gamma(X)$ is contained in an X-free set of size k. However, for $k = 2$ this holds without the interval property.

Lemma 2.8. *Let $(E, \mathscr{F})$ be a 2-connected greedoid and consider a set $X \in \mathscr{F}$ with $|X| \leq r(E) - 2$. Then the X-free sets of size ≤ 2 form a matroid.*

Proof. Obviously X-free sets form a hereditary set system. To show the exchange property, let $\{x\}, \{y, z\} \subseteq \Gamma(X)$ be X-free. Then $X \cup x$ can be augmented from $X \cup \{y, z\}$ to $X \cup \{x, y\}$, say, and $\{x, y\}$ is X-free. $\qquad \square$

Lemma 2.9. *Let $(E, \mathscr{F})$ be a 2-connected greedoid and $a, b \in \Gamma(\emptyset)$. Then either there exists a basis containing both a and b or there exists a $Y \in \mathscr{F}$ such that $Y \cup a$ and $Y \cup b$ are bases.*

Proof. Consider a maximal set $Y \in \mathscr{F}$ such that $a, b \in \Gamma(Y)$. If $\{a, b\}$ is Y-free then $Y \cup \{a, b\}$ can be extended to a basis containing a and b. If $|Y| = r - 1$ then $Y \cup a$ and $Y \cup b$ are bases. Suppose $|Y| \leq r - 2$. Then by 2-connectivity and Lemma 2.8 there exists a $c \in \Gamma(Y)$ such that $\{a, c\}$ and $\{b, c\}$ are Y-free. But then $a, b \in \Gamma(Y \cup c)$, a contradiction. $\square$

The pivot operation is one of the fundamental algorithmic tools in linear algebra. Korte and Lovász [1985e] introduced an analoguous operation for greedoids, which we will study in the remainder of this section.

We say that two bases B_1 and B_2 of a greedoid $(E, \mathcal{F})$ are **adjacent** if

$$|B_1 \setminus B_2| = 1 \quad \text{and} \quad B_1 \cap B_2 \in \mathcal{F}.$$

B_1 can then be obtained from B_2 by a pivot operation where the element $y = B_2 \setminus B_1$ is pivoted out and the element $x = B_1 \setminus B_2$ is pivoted in. Similarly, two basic words α and β are **adjacent** if α arises from β by interchanging two consecutive letters or by exchanging the last letters.

The adjacency relations give rise to a **basis graph** $G(\mathcal{F})$ and a **basic word graph** $G(\mathcal{L})$ where the former can be obtained from the latter by identifying basic words consisting of the same letters. We will see that these graphs are connected for very large classes of greedoids.

From now on it will be convenient to consider all greedoids of rank 1 to be 2-connected.

For two bases B, B' let $d(B, B')$ be the length of a shortest path from B to B' in $G(\mathcal{F})$, and let

$$\text{diam } G(\mathcal{F}) = \max\{d(B, B') : B, B' \in \mathcal{B}\}$$

be the **diameter** of the basis graph. For $\{x\} \in \mathcal{F}$ and $B \in \mathcal{B}$ let $d(x, B)$ be the minimum distance of B to a basis containing x. Set

$$d(x, \mathcal{B}) = \max\{d(x, B) : B \in \mathcal{B}\}$$

and denote by $\delta(r)$ the maximum of $d(x, B)$ taken over all 2-connected greedoids $(E, \mathcal{F})$ with rank r and over all singletons $\{x\} \in \mathcal{F}$.

Lemma 2.10. $\delta(r) \leq 2^{r-1}$.

Proof. Use induction on r. The statement is true for $r = 1$. Let $r \geq 2$ and let $B \in \mathcal{B}$ and $\{x\} \in \mathcal{F}$. Let $y \in B$ be such that $\{y\} \in \mathcal{F}$. If $\{x, y\} \in \mathcal{F}$, then there is a path of length 2^{r-2} connecting $B \setminus y$ to some basis containing x in the graph $G(\mathcal{F}/y)$, and hence $d(x, B) \leq 2^{r-2}$. If $\{x, y\} \notin \mathcal{F}$, then by Lemma 2.8 there exists an element u such that $\{x, u\}, \{y, u\}$ are $\emptyset$-free. Then by the induction hypothesis, there is a path of length at most 2^{r-2} in $G(\mathcal{F}/y)$ connecting $B \setminus y$ to some basis B_1 of $\mathcal{F}/y$ containing u, and a path of length at most 2^{r-2} in $G(\mathcal{F}/u)$ connecting $B_1 \cup y \setminus u$ to some basis B_2 of $\mathcal{F}/u$ containing x. This shows that $d(x, B) \leq 2 \cdot 2^{r-2} = 2^{r-1}$. $\square$

Theorem 2.11. *Let $(E, \mathcal{F})$ be a 2-connected greedoid with basis graph $G(\mathcal{F})$ and basic word graph $G(\mathcal{L})$. Then these graphs are connected and*

(i)   diam $G(\mathscr{F}) \le 2^{r(E)} - 1$,

(ii)   diam $G(\mathscr{L}) \le 3 \cdot 2^{r(E)} - 2r(E) - 3$.

Proof. Again by induction on the rank. For $r = 1$, the assertion is trivial. Let B, B' be two bases and $x \in B$ with $\{x\} \in \mathscr{F}$. Then by Lemma 2.10,

$$d(B, B') \le d(x, B') + \text{diam } G(\mathscr{F}/x) \le 2^{r-1} + 2^{r-1} - 1 = 2^r - 1 .$$

A similar argument works for the basic word graph $G(\mathscr{L})$.   □

An alternative induction proof can be based on Lemma 2.9. Ziegler [1988] has constructed greedoids showing that these bounds are tight.

Note that we have used a slightly weaker property than 2-connectivity. We only use the existence of an X-free set of size two if $|\Gamma(X)| \ge 2$. One class of greedoids that satisfy this weaker assumption are truncations of antimatroids. And so we obtain

Lemma 2.12. *The basic word graph and the basis graph of truncations of antimatroids are connected.*   □

The following relates the diameters of the basis graphs of a greedoid and its truncations. It turns out that some of the following results can be considerably strengthened for greedoids which behave locally like poset antimatroids. This class of so-called local poset greedoids will be studied in more detail in the next chapter. What we need here is the local intersection property: if $A, B, C \in \mathscr{F}$ and $A, B \subseteq C$, then $A \cap B \in \mathscr{F}$.

Theorem 2.13. *Let $(E, \mathscr{F})$ be a 2-connected greedoid of rank r. Then*

(i)   diam $G(\mathscr{F}) \le 2 \cdot \text{diam } G(\mathscr{F}^{(r-1)}) + 1$,

(ii)   diam $G(\mathscr{F}^{(r-1)}) \le (r-1)(\text{diam } G(\mathscr{F}) + 1)$ *if $(E, \mathscr{F})$ is an interval greedoid,*

(iii) diam $G(\mathscr{F}^{(r-1)}) \le \text{diam } G(\mathscr{F}) + 1$ *if $(E, \mathscr{F})$ is a local poset greedoid.*

Proof. (i) Let B and B' be two bases of $(E, \mathscr{F})$. Let C and C' be any two feasible subsets of B and B' respectively of size $r - 1$. Let $C_0 = C, C_1, \ldots, C_k = C'$ be a shortest path in $G(\mathscr{F}^{(r-1)})$. Applying Lemma 2.10 to $\mathscr{F}/(C_0 \cap C_1)$, we obtain a path in $G(\mathscr{F})$ of length at most 2 connecting B to a basis B_1 containing C_1. Similarly we find a path of length at most 2 connecting B_1 to B_2. In $2k$ steps we get a basis B_k containing C_k which is adjacent to B' in $G(\mathscr{F})$.

(ii) Let C and C' be two bases of the truncation $\mathscr{F}^{(r-1)}$. We extend C and C' to bases B and B' of $\mathscr{F}$. Let $B = B_0, B_1, \ldots, B_k = B'$ be a shortest path in $G(\mathscr{F})$. The sets $B_i \cap B_{i-1}$ and $B_{i+1} \cap B_i$ are bases of the truncated antimatroid obtained by restricting $\mathscr{F}^{(r-1)}$ to B_i. By Lemma 2.12 this truncation is connected. It has at most r bases and its basis graph has diameter at most $r - 1$. Similarly the distance of C from $B_0 \cap B_1$ as well as the distance of C' from $B_k \cap B_{k-1}$ are both at most $r - 1$. This yields a path of length at most $(r - 1)(k + 1)$.

(iii) follows by the same argument and the observation that $B_{i-1} \cap B_i \cap B_{i+1}$ is feasible.   □

Note that this theorem yields a different proof of Theorem 2.11(i).

Theorem 2.11 has some implications for 2-connected greedoids, which we summarize below.

Corollary 2.14. *Let $(E, \mathcal{F})$ be a 2-connected greedoid, $e \in E$, and let ℓ be the minimal size of a feasible word containing e. Then for every $\ell \leq k \leq r(E)$, there exists a basic word whose k-th letter is e.*

Proof. Let α be a basic word with e at its ℓth position and γ a maximal feasible word not containing e. Then $|\gamma| \geq r(E) - 1$, since otherwise γ could be extended by at least one letter different from e by 2-connectivity. If $|\gamma| = r(E)$, set $\beta = \gamma$, otherwise set $\beta = \gamma e \in \mathcal{F}$. In the path connecting α to β the position of e changes by at most one and e can only drop out from the last position. Hence for every $\ell \leq k \leq r(E)$, there is some basic word with e at its kth position.   □

By an observation of Frank (cf. Lovász [1977]) for any 2-connected graph with specified vertices $v_1, v_2 \in V$ and integers n_1, n_2 with $n_1 + n_2 = |V|$ there exists a partition $V_1 \cup V_2$ of V into two connected subgraphs so that $v_i \in V_i$. This result extends to local poset greedoids. Before we can prove this, we need a lemma.

Lemma 2.15. *Let $(E, \mathcal{F})$ be a local poset greedoid, $X \subseteq E$, and B_1, B_2 two adjacent bases. Then $|r(B_1 \cap X) - r(B_2 \cap X)| \leq 1$.*

Proof. Let A be a basis of $B_1 \cap X$. Then the local intersection property applied to $A, B_1 \cap B_2 \subseteq B_1$ gives $A \cap B_2 = A \cap (B_1 \cap B_2) \in \mathcal{F}$. Hence

$$r(B_2 \cap X) \geq |A \cap B_2| \geq |A| - 1 = r(B_1 \cap X) - 1 .$$

Interchanging B_1 and B_2 proves the lemma.   □

Corollary 2.16. *Let $(E, \mathcal{F})$ be a 2-connected local poset greedoid and $X \subseteq E$. Then the set of numbers $r(B \cap X)$, $B \in \mathcal{B}$, consists of consecutive integers.*

Proof. Let $B_1, B_2 \in \mathcal{B}$ be two bases with $r(B_i \cap X) = n_i$, $i = 1, 2$. By Theorem 2.11, B_1 and B_2 are connected by a path of adjacent bases along which, by the lemma above, the value $r(B \cap X)$ changes by at most one. Hence all integers between n_1 and n_2 occur.   □

To see that the previous corollary generalizes the aforementioned result on 2-connected graphs $G = (V, E)$, observe that undirected branching greedoids are local poset greedoids. Add a root vertex r to V with edges $(r, v_1), (r, v_2)$. For B_0 a spanning tree of G, $B_1 = B_0 \cup (r, v_1)$ and $B_2 = B_0 \cup (r, v_2)$ are bases of the undirected branching greedoid in the enlarged graph. For $X = E \cup (r, v_1)$, we get $r(B_1 \cap X) = |V|$ and $r(B_2 \cap X) = 0$. Hence for all integers $0 < n_1 < |V|$, there exists a rooted spanning tree B with $r(B \cap X) = r(B \setminus (r, v_2)) = n_1$. Then the branches of this tree rooted at v_1 and v_2 induce the desired partition.

If we assume higher connectivity, we get sharper bounds on the diameter but we have to assume the interval property. A different kind of extension of the previous results to higher connectivity will be discussed in the framework of homotopy in Chapter XII.

Lemma 2.17. *Let* $(E, \mathcal{F})$ *be an interval greedoid,* X *and* Y *two bases and assume that* Y *is an* $\emptyset$*-free set. Then* $d(X, Y) \leq r(E)$.

Proof. By a straightforward induction on r. $\square$

Lemma 2.18. *Let* $(E, \mathcal{F})$ *be an* r*-connected interval greedoid of rank* $r = r(E)$. *Then* diam $G(\mathcal{F}) \leq 2r - 1$.

Proof. Let $B_1, B_2 \in \mathcal{B}$ be two bases. Let $x \in B_1$ be a feasible singleton. By Lemma 2.1 and by the assumption, there exists a $\emptyset$-free basis B_3 containing x. Hence $d(B_1, B_2) \leq d(B_1, B_3) + d(B_3, B_2) \leq r - 1 + r$. $\square$

Theorem 2.19. *Let* $(E, \mathcal{F})$ *be a* k*-connected interval greedoid. Then* diam $G(\mathcal{F}) \leq k \cdot 2^{r-k+1} - 1$.

Proof. By induction on $r - k$, where the case $r = k$ is proved in Lemma 2.18. The induction step follows from Lemma 2.9. $\square$

Ziegler [1988] reduced the bounds on the diameter to a quadratic expression in the rank in the case of directed and undirected branching greedoids. We need a preparatory lemma for this result.

Lemma 2.20. *Let* $\mathcal{F}$ *be a* 2*-connected branching greedoid and* $\{x_1\}, \{y_1\} \in \mathcal{F}$ *with* $\{x_1, y_1\} \in \mathcal{F}$. *Then there exist sets* $X_i = \{x_1, ..., x_i\} \in \mathcal{F}$ *and* $Y_i = \{y_1, ..., y_i\} \in \mathcal{F}$ *such that* $X_i \cup Y_{r-i}$ *is a basis for all* $1 \leq i \leq r - 1$.

Proof. We first construct sets $X_i = \{x_1, ..., x_i\} \in \mathcal{F}$ and $Z_1, ..., Z_{r-1} \in \mathcal{F}$ inductively such that $X_i \cup Z_{r-i}$ is a basis and $y_1 \in Z_i$ for all i.

For $i = 1$, we know from Lemma 2.9 that there exists a set $A \in \mathcal{F}$ such that either $A \cup x_1$ and $A \cup y_1$ are bases or $A \cup \{y_1, x_1\}$ is a basis. The first case cannot hold. Hence we may set $X_1 = \{x_1\}$ and $Z_{r-1} = A \cup y_1$.

Suppose now that for some $i \leq r - 2$, X_i and Z_{r-i} have been constructed and consider the contraction $\mathcal{F}/X_i$. We distinguish two cases.

(i) Every set $A \in \mathcal{F}/X_i$ containing y_1 is also feasible in $\mathcal{F}$. By 2-connectivity there exists an $a \in \mathcal{F}/X_i, a \neq y_1$. By induction on r, there exist sets $A_j = \{a, a_2, ..., a_{j-1}\}$ and $B_j = \{y_1, b_2, ..., b_{j-1}\}$ such that $A_j \cup B_{r-i-j}$ is a basis in $\mathcal{F}/X_i$ for all $1 \leq j \leq r - i - 1$. Then setting $X_{i+j} = X_i \cup A_j$ and $Z_j = B_j$ finishes the construction.

(ii) There exists a smallest set $A \cup a \in \mathcal{F}/X_i$ containing y_1 such that $A \cup a \notin \mathcal{F}$. Then there exist singletons (we call them **red**) $a \in \mathcal{F}/X_i$ with $a \notin \mathcal{F}$ such that the endnodes of a and y_1 are distinct. Choose a basis B of $\mathcal{F}/X_i$ containing a red element. Since by Theorem 2.11 the basis graph of $\mathscr{F}/X_i$ is connected and Z_{r-i} is a basis of $\mathscr{F}/X_i$, there exists a path in $G(\mathscr{F}/X_i)$ from Z_{r-i} to B. Let A be the first basis on this path containing a red element a. Since the endnodes of y_1 and a are distinct, we may assume that $y_1 \in A$. Then, by construction, $Z_{r-i-1} = A \backslash a \in \mathscr{F}/X_i$ and $Z_{r-i-1} \in \mathscr{F}$. Setting $X_{i+1} = X_i \cup a$ finishes the first part of the proof.

Now let $Y_1 = \{y_1\}$, and successively augment Y_{i-1} in $\mathscr{F}/X_i$ from Z_i. Then $Y_i = \{y_1, ..., y_i\} \in \mathscr{F}$ and $X_i \cup Y_{r-i}$ is a basis for $1 \leq i \leq r-1$. $\square$

For the proof of the upper bound on the diameter of branching greedoids it seems convenient to work with **subbases**, i.e. with sets $X \in \mathscr{F}$ of rank $r(E) - 1$. We say that two subbases are **adjacent** if their union is a basis. Denoting by $\bar{d}(A, B)$ the distance between two subbases and by $\overline{G}(\mathscr{F})$ the **subbasis graph**, we have

$$(2.1) \qquad \qquad \left| \text{diam } G(\mathscr{F}) - \text{diam } \overline{G}(\mathscr{F}) \right| \leq 1.$$

Theorem 2.21. *Let $(E, \mathscr{F})$ be a 2-connected branching greedoid of rank r. Then*

(i)   diam $G(\mathscr{F}) \leq r^2 - r + 1$,
(ii)   diam $G(\mathscr{L}) \leq \binom{r+2}{3} - 2r + 1$.

Proof. We only show (i) by proving diam $\overline{G}(\mathscr{F}) \leq r^2 - r$. The proof of (ii) follows along the same lines.

Let A_{r-1} and B_{r-1} be two subbases, and label their elements so that $A_i = \{a_1, ..., a_i\}$ and $B_i = \{b_1, ..., b_i\}$ are feasible sets for $1 \leq i \leq r-1$. We distinguish five cases:

(i) a_1 and b_1 are identical. Consider $\mathscr{F}/a_1$. By induction on r we have $\bar{d}(A_{r-1}, B_{r-1}) \leq (r-1)^2 - r + 1 \leq r^2 - r$.

(ii) $A_i \cup B_{r-i}$ is a basis for all $1 \leq i \leq r-1$. Then $A_{r-1}, A_{r-2} \cup B_1, A_{r-3} \cup B_2, ..., B_{r-1}$ is a path in $\overline{G}(\mathscr{F})$, i.e. $\bar{d}(A_{r-1}, B_{r-1}) \leq r-1$.

(iii) $X_i \cup B_{r-i}$ is a basis for some sets $X_i = \{a_1, x_2, ..., x_i\} \in \mathscr{F}, 1 \leq i \leq r-1$. Let $\bar{d}_1(r)$ be the maximal distance in $\overline{G}(\mathscr{F})$ between two subbases for which this case applies. Consider the subbases $Z_i = X_i \cup B_{r-1-i}, 0 \leq i \leq r-1$. We claim that $a_2 \notin \sigma(Z_j)$ for at least one $j \geq 0$. Suppose not. The only node not covered by the branching Z_0 is the endnode of arc a_1. Similarly, the only node covered by the branching Z_i is the endnode of arc b_{r-i}. Hence the endnode of a_2 is none of the endpoints of $a_1, b_1, ..., b_{r-1}$. This, however, contradicts $A_2 \in \mathscr{F}$ and $A_1 \cup B_{r-1}$ a basis.

Observe that $Z_0 \cup a_2 \notin \mathscr{F}$. Hence $Z_j \cup a_2 \in \mathscr{F}$ for some $1 \leq j \leq r-1$. In other words, $(Z_j \backslash a_1) \cup (A_2 \backslash a_2)$ is a basis in $\mathscr{F}/a_1$ and we may argue by induction that $\bar{d}_1(r) \leq \bar{d}_1(r-1) + r - 1$ and thus $\bar{d}_1(r) \leq \frac{1}{2}r(r-1)$.

(iv) a_1 and b_1 are parallel. We use Lemma 2.20 to construct feasible sets $X_i = \{a_1, x_2, ..., x_i\}, X_i' = \{b_1, x_2, ..., x_i\}$ and $Y_i = \{y_1, y_2, ..., y_i\}$ such that $X_i \cup Y_{r-i}$ and $X_i' \cup Y_{r-i}$ are bases. Then $\bar{d}(A, B) \leq \bar{d}_1(A, Y) + \bar{d}_1(Y, B) \leq r^2 - r$.

(v) If none of the above cases applies, we apply Lemma 2.20 to construct sets $X_i = \{a_1, x_2, \ldots, x_i\} \in \mathscr{F}$ and $Y_i = \{b_1, y_2, \ldots, y_i\} \in \mathscr{F}$ such that $X_i \cup Y_{r-i}$ is a basis for all $1 \leq i \leq r-1$. As in case (ii), we find indices $1 \leq i, j \leq r-1$ such that $(X_i \backslash a_1) \cup Y_{r-1-i} \cup a_2$ is a basis in $\mathscr{F}/a_1$ and $(X_j \backslash a_1) \cup Y_{r-1-j} \cup b_2$ is a basis in $\mathscr{F}/b_1$. Hence $\bar{d}(A_{r-1}, B_{r-1}) \leq \bar{d}_1(r-1) + r - 1 + \bar{d}_1(r-1) \leq r^2 - 2r + 1$, which finishes the proof.   □

Chapter VII. Local Poset Greedoids

In this chapter we study the class of local poset greedoids. Their name is motivated by the fact that their restrictions to feasible sets are poset antimatroids. These greedoids have the interval property. Local poset greedoids cover a broad range of examples: directed and undirected branchings, poset antimatroids and polymatroid greedoids.

The first section presents some results on polymatroid greedoids (which contain the undirected branching greedoids as a special case). We show for example, that polymatroid greedoids have a representation as an intersection of a matroid with some antimatroid – a construction principle which we will investigate more closely in Chapter IX.

In Section 2 we introduce local poset greedoids, derive some alternative characterizations, and relate them to polymatroid greedoids.

Section 3 describes three simple greedoids which obviously do not have the local poset property. Conversely, we verify that it is enough to exclude these three types of greedoids as induced minors in order to obtain local poset greedoids.

The last two sections are motivated by another important class of local poset greedoids, namely branching greedoids. In Section 4 we introduce the notion of a path to a given element in a feasible set, and show that local poset greedoids can be characterized by the uniqueness of this path.

In the last section we prove an excluded minor characterization of undirected branching greedoids due to Schmidt [1988]. (We shall return to the characterization problem of directed branching greedoids in Chapter IX.)

1. Polymatroid Greedoids

Recall that given a polymatroid rank function f, we define a hereditary language $(E, \mathscr{L})$ by

$$\mathscr{L} = \{x_1 x_2 \ldots x_k : f(\{x_1, \ldots, x_i\}) = i \text{ for all } 1 \le i \le k\} .$$

The language $(E, \mathscr{L})$ is called a **polymatroid greedoid**.

Theorem 1.1. it Every polymatroid greedoid is an interval greedoid.

Proof. For the greedoid axioms only the augmentation property needs a proof. Let $x_1 x_2 \ldots x_k, y_1 y_2 \ldots y_j \in \mathscr{L}$ be two feasible words with $k > j$. By the monotonicity of f, there exists a smallest index i such that

$$f(\{y_1,\ldots,y_j,x_1,\ldots,x_i\}) > j .$$

Set $Y = \{y_1,\ldots,y_j\}$ and $X = \{x_1,\ldots x_i\}$. The submodularity of f implies

$$j < f(Y \cup X) \le f(X) + f(Y \cup (X \setminus x_i)) - f(X \setminus x_i) \le j+1 ,$$

thus $f(X \cup Y) = j+1$. Using monotonicity and submodularity we get

$$j+1 = f(Y \cup X) \ge f(Y \cup x_i) \ge f(Y) + f(Y \cup X) - f(Y \cup (X \setminus x_i)) \ge j+1.$$

Hence $f(Y \cup x_i) = j+1$ and therefore $Y \cup x_i \in \mathscr{L}$. So $(E,\mathscr{L})$ is a greedoid.

It remains to show that $(E,\mathscr{L})$ has the interval property. Let $A,B,C \in \mathscr{F} = \mathscr{F}(\mathscr{L})$ be feasible sets with $A \subseteq B \subseteq C$ and assume that $A \cup x$ and $C \cup x \in \mathscr{F}$ for some $x \in E \setminus C$. The submodularity of f yields

$$1 = f(C \cup x) - f(C) \le f(B \cup x) - f(B) \le f(A \cup x) - f(A) = 1 ,$$

hence $B \cup x \in \mathscr{F}$.   □

Various classes of greedoids can be represented as polymatroid greedoids.

Example 1. Consider the undirected branching greedoid as defined in IV.2.9. Let $G = (V,E)$ with root $r \in V$. For any subset $X \subseteq E$ let $f(X)$ count the number of vertices in $V \setminus r$ covered by X. It is straightforward to check that f is submodular and that the polymatroid greedoid induced by f consists of the subtrees of G covering the root r. Hence the undirected branching greedoid is a polymatroid greedoid.

Example 2. Let $(E,\le)$ be an ordered set. For every subset $X \subseteq E$ let $f(X) = |I(X)|$ denote the cardinality of the ideal generated by X. It is easily verified that f is a polymatroid rank function. Moreover, $X \subseteq E$ is feasible in the associated greedoid $(E,\mathscr{F})$ if and only if X is an ideal. Thus $(E,\mathscr{F})$ is the poset antimatroid.

If in Example 1, C is a cycle not covering the root r, then $f(C) = |C|$. This shows that the condition $f(X) = |X|$ is necessary but not sufficient for feasible sets. However, an inductive argument shows that in any polymatroid greedoid, if $f(X) = |X|$ for some $X \subseteq E$ and $X \setminus x \in \mathscr{F}$ for some $x \in X$, then $X \in \mathscr{F}$.

A similar statement is true for subsets of feasible sets:

Lemma 1.2. *Let $X \in \mathscr{F}$ be a feasible set and $A \subseteq X$. Then $f(A) \ge |A|$, where equality holds if and only if $A \in \mathscr{F}$.*

Proof. We use induction on $|A|$. Let B be a maximal feasible subset of X such that $A \nsubseteq B$. Then there exists an $x \in X \setminus B$ with $B \cup x \in \mathscr{F}$. The choice of B then implies $A \subseteq B \cup x$ and hence $A \cup B = B \cup x \in \mathscr{F}$.

Using the induction hypothesis for $A \cap B$ and the submodularity of f we get:

$$f(A) \geq f(A \cap B) + f(A \cup B) - f(B) \geq |A \cap B| + |A \cup B| - |B| = |A|.$$

Clearly if $A \in \mathscr{F}$, then equality holds. Conversely, if equality holds, then

$$f(A \cap B) = f(A \setminus x) = |A| - 1.$$

Hence $A \setminus x \in \mathscr{F}$ and thus $A \in \mathscr{F}$.   □

Lemma 1.3. *Let* $(E, \mathscr{F})$ *be a normal polymatroid greedoid. Then* $f(E) = r(E)$.

Proof. Let B be a basis of $(E, \mathscr{F})$. Then clearly $f(E) \geq f(B) = |B| = r(E)$.

Choose E' to be a maximal set with $B \subseteq E' \subseteq E$ and $f(E') = f(B)$. Suppose $E' \neq E$. Since $(E, \mathscr{F})$ is normal, there exists a feasible set $A \not\subseteq E'$. We can choose A such that $A \setminus E' = \{a\}$ for some $a \in A$. Applying Lemma 1.2 and using the fact that for any superset of E' the polymatroid rank exceeds the greedoid rank, we get:

$$\begin{aligned} |B| + 1 \leq f(E' \cup a) = f(E' \cup A) &\leq f(A) + f(E') - f(E' \cap A) \\ &\leq |A| + |B| - |E' \cap A| \\ &= |B| + 1. \end{aligned}$$

Similarly,

$$\begin{aligned} f(B \cup a) \leq f(E' \cup a) &\leq f(B \cup a) + f(E') - f(B) \\ &= f(B \cup a). \end{aligned}$$

Hence $f(B \cup a) = |B| + 1$, i.e. $B \cup a \in \mathscr{F}$, which contradicts the assumption that B is a basis.   □

Lemma 1.4. *Let* $(E, \mathscr{F})$ *be a polymatroid,* $A \in \mathscr{F}$ *with a feasible ordering* $a_1 a_2 \ldots a_k$ *and* $x \in E \setminus A$ *such that* $f(x) > 0$ *and* $f(A \cup x) = f(A)$. *Then* $\{a_1, \ldots, a_i\} \cup x \in \mathscr{F}$ *for some index* $0 \leq i \leq k - 1$.

Proof. Let $a_1 a_2 \ldots a_k$ be a feasible ordering of A. Then the submodularity of f implies that

$$g(i) = f(\{a_1, \ldots, a_i, x\}) - f(\{a_1, \ldots, a_i\})$$

is monotone decreasing in i. Furthermore, it decreases by at most one, since

$$g(i) - g(i+1) = f(\{a_1, \ldots, a_i, x\}) - f(\{a_1, \ldots, a_{i+1}, x\}) + 1 \leq 1.$$

As by assumption $g(k) = 0$ and $g(0) > 0$, there must exist an index i with $g(i) = 1$ and hence $\{a_1, \ldots, a_i, x\} \in \mathscr{F}$.   □

Theorem 1.5. *Let* $(E, \mathscr{F})$ *be a normal polymatroid greedoid. Then* $\mathscr{A} = \mathscr{R}$.

Proof. Since $(E, \mathscr{F})$ is an interval greedoid, $\mathscr{A} \subseteq \mathscr{R}$ by Corollary V.5.8.

Conversely, let $X \in \mathscr{R}$ be a rank feasible set, A a basis of X, B a basis of E with $A \subseteq B$ and $x \in X \setminus A$. $X \in \mathscr{R}$ then implies $x \in X \setminus B$. Since $(E, \mathscr{F})$ is normal, there exists a $Y \in \mathscr{F}$ such that $Y \cup x \in \mathscr{F}$, i.e.

$$f(x) \geq f(Y \cup x) - f(Y) = 1 .$$

Moreover, by Lemma 1.3,

$$|B| = f(B) \leq f(B \cup x) \leq f(E) = r(E) = |B| .$$

Let $a_1 a_2 \ldots a_k b_{k+1} \ldots b_m$ be a feasible ordering of B. By Lemma 1.4 there exists an index i such that either $a_1 \ldots a_i x \in \mathcal{F}$ or $a_1 a_2 \ldots a_k b_{k+1} \ldots b_i x \in \mathcal{F}$. The second possibility is ruled out since A is a basis of a rank feasible set, which implies that x is contained in a feasible subset of X. Hence $X \in \mathcal{A}$.   □

Lemma 1.6. *Let* $(E, \mathcal{F})$ *be a polymatroid greedoid. Then for every* $X \subseteq E$

$$\beta(X) \leq f(X),$$

and equality holds if and only if $X \in \mathcal{R}$.

Proof. Let B be a basis of E with $|X \cap B| = \beta(X)$. Lemma 1.2 then yields

$$f(X) \geq f(X \cap B) \geq |X \cap B| = \beta(X) .$$

If equality holds, then $f(X \cap B) = |X \cap B|$, and hence, again by Lemma 1.2, $X \in \mathcal{R}$. Conversely if $X \in \mathcal{R} = \mathcal{A}$, then we can apply Lemma 1.3 to the restriction of $(E, \mathcal{F})$ to X.   □

Trivially, this lemma remains valid if β is replaced by r.

We know that the system $\mathcal{R}$ of rank feasible sets of an interval greedoid is closed under unions and the rank function is submodular on pairs of sets from $\mathcal{R}$. For polymatroid greedoids we can say somewhat more:

Lemma 1.7. *Let* $(E, \mathcal{F})$ *be a polymatroid greedoid and* $X, Y \in \mathcal{R}$, *and suppose that* $r(X \cap Y) + r(X \cup Y) = r(X) + r(Y)$. *Then* $X \cap Y \in \mathcal{R}$.

Proof. We have

$$r(X) + r(Y) = r(X \cup Y) + r(X \cap Y)$$
$$\leq f(X \cup Y) + f(X \cap Y) \leq f(X) + f(Y) = r(X) + r(Y).$$

Hence $r(X \cap Y) = f(X \cap Y)$, i.e. $X \cap Y \in \mathcal{R}$.   □

A polymatroid is not uniquely determined by its associated greedoid. The next theorem in fact shows that there is a whole set of submodular functions defining the same polymatroid greedoid.

Theorem 1.8. *Let* f *be a polymatroid rank function,* f_k *its* k-*truncation for some* $k \geq 2$ *and* $(E, \mathcal{F})$, $(E, \mathcal{F}_k)$ *the associated polymatroid greedoids. Then* $\mathcal{F} = \mathcal{F}_k$.

Proof. Consider some set $X \in \mathcal{F}$. By applying Lemma 1.2 we get

$$|X| = f(X) \geq f_k(X) = \min_{Y \subseteq X}\{f(Y) + k|X \setminus Y|\}$$
$$\geq \min_{Y \subseteq X}\{|Y| + k|X \setminus Y|\} \geq |X|.$$

Since $X \setminus x \in \mathcal{F}$ for some $x \in X$, we can use induction on the cardinality of X to prove that $X \in \mathcal{F}_k$.

Conversely, assume $X \in \mathcal{F}_k$. Then again

$$|X| = f_k(X) = \min_{Y \subseteq X}\{f(Y) + k|X \setminus Y|\} \geq \min_{Y \subseteq X}\{f_k(Y) + k|X \setminus Y|\}$$
$$\geq \min_{Y \subseteq X}\{|Y| + k|X \setminus Y|\}$$
$$\geq |X|.$$

In view of $k \geq 2$, the minimum can only be attained for $Y = X$. Then $X \in \mathcal{F}$ follows as before. $\square$

We have seen that if $(E, \leq)$ is a poset, then $f(X) = |I(X)|$ represents the poset antimatroid as a polymatroid greedoid. Clearly this polymatroid function is not a 2-polymatroid function in general. In fact it is the largest among all submodular functions defining the poset antimatroid: for any other such function g we have $g(X) \leq g(I(X)) = |I(X)|$.

Note that Theorem 1.8 does not cover the case $k = 1$. We will now give an explicit construction relating a polymatroid greedoid to its 1-truncation (the induced matroid).

Theorem 1.9. *Let f be a polymatroid rank function, $\mathcal{R}$ the rank feasible sets of the associated polymatroid greedoid $(E, \mathcal{F})$, and $(E, \mathcal{M})$ the induced matroid. Then $\mathcal{F} = \mathcal{R} \cap \mathcal{M}$.*

Proof. By Lemma 1.2, $\mathcal{F} \subseteq \mathcal{M}$ and hence $\mathcal{F} \subseteq \mathcal{R} \cap \mathcal{M}$. To prove the reverse inclusion, we first show that if $X, Y \in \mathcal{F}$ and $X \cup Y \in \mathcal{M}$, then $X \cup Y \in \mathcal{F}$.

Let $y \in Y$ be such that $Y \setminus y \in \mathcal{F}$. By induction we may assume that $X \cup (Y \setminus y) \in \mathcal{F}$. The assumptions then imply

$$|X| + |Y| = |X \cup Y| + |X \cap Y| \leq f(X \cup Y) + f(X \cap Y) \leq f(X) + f(Y)$$
$$= |X| + |Y|.$$

Hence $|X \cup Y| = f(X \cup Y)$. Since $X \cup (Y \setminus y) \in \mathcal{F}$, we obtain $X \cup Y \in \mathcal{F}$.

Now let $X \in \mathcal{R} \cap \mathcal{M}$. In view of Theorem 1.5, $X = X_1 \cup \ldots \cup X_k$ for some $X_i \in \mathcal{F}$. By induction, $X_1 \cup \ldots \cup X_i \in \mathcal{F}$ for all $i = 1, \ldots, k$, and in particular $X \in \mathcal{F}$. $\square$

We have seen that undirected branchings are particular polymatroid greedoids. Their directed analogues do not fit directly into this framework. However, in Korte and Lovász [1985c] a polymatroid generalization of directed branchings is given, which we will now describe.

Let f be a polymatroid rank function defined on some ground set E' and let $(E', \mathcal{L}')$ be the associated polymatroid greedoid. Let $E \subseteq E'$ and assume that for each $x \in E$ we have an associated element $x' \in E'$.

Define a hereditary language $(E, \mathcal{L})$ by

$$\mathcal{L} = \{x_1 \ldots x_k \in \mathcal{L}' : f(\{x_1, \ldots, x_{j-1}, x_j'\}) = j - 1 \text{ for all } 1 \le j \le k\} .$$

We call this language $(E, \mathcal{L})$ a **polymatroid branching greedoid**.

Lemma 1.10. *Every polymatroid branching greedoid is a greedoid.*

Proof. Let $x_1 \ldots x_k \in \mathcal{L}$ and $y_1 \ldots y_j \in \mathcal{L}$ be feasible strings with $k > j$. As in the proof of Theorem 1.1, there exists an index i such that $f(\{y_1, \ldots, y_j, x_i\}) = j + 1$ but $f(\{y_1, \ldots, y_j, x_1, \ldots, x_{i-1}\}) = j$.
Using the submodularity and monotonicity of f, we obtain

$$
\begin{aligned}
f(\{y_1, \ldots, y_j, x_i'\}) &\le f(\{y_1, \ldots, y_j, x_1, \ldots, x_{i-1}, x_i'\}) \\
&\le f(\{y_1, \ldots, y_j, x_1, \ldots, x_{i-1}\}) + f(\{x_1, \ldots, x_{i-1}, x_i'\}) \\
&\quad - f(\{x_1, \ldots, x_{i-1}\}) \\
&= j + (i-1) - (i-1) = j .
\end{aligned}
$$

Hence $y_1 \ldots y_j x_i \in \mathcal{L}$. This proves that $(E, \mathcal{L})$ is a greedoid. $\square$

It is not hard to see that every polymatroid greedoid is also a polymatroid branching greedoid: consider a polymatroid rank function f on some ground set E. We enlarge E by a new element e to $E' = E \cup e$ and extend f to E' by setting

$$f(X) = f(X \setminus e)$$

for any set $X \subseteq E'$. If for every $x \in E$ we set $x' = e$, then the polymatroid branching greedoid $(E, \mathcal{F})$ is precisely the polymatroid greedoid $(E', \mathcal{F}')$.

We have defined polymatroid branching greedoids as sublanguages of polymatroid greedoids. The next result shows that they are in a certain sense dense in the polymatroid greedoid.

Lemma 1.11. *Let $(E', \mathcal{F}')$ be a polymatroid greedoid, $(E, \mathcal{F})$ an associated polymatroid branching greedoid with $f(\{e, e'\}) = f(e)$, and $X \in \mathcal{F}$. Then*

$$\mathcal{F}'_X = \mathcal{F}_X .$$

Proof. Clearly $\mathcal{F}_X \subseteq \mathcal{F}'_X$. To see the converse, consider $Y \subseteq X$, $Y \in \mathcal{F}'$. Let $x_1 \ldots x_k$ be a feasible ordering of X in $(E, \mathcal{F})$ and $y_1 \ldots y_j$ a feasible ordering of Y in $(E', \mathcal{F}')$. Let $y_j = x_{\ell+1}$. We may assume by induction that $y_1 \ldots y_{j-1} \in \mathcal{F}$. Using monotonicity and submodularity of f, we get

$$
\begin{aligned}
j - 1 \le f((Y \setminus y_j) \cup y_j') & \\
&\le f(Y \cup y_j') + f(\{x_1, \ldots, x_\ell\} \cup y_j') - f(\{x_1, \ldots, x_\ell\} \cup \{y_j, y_j'\}) \\
&\le f(Y) + f(\{y_j, y_j'\}) - f(y_j) + \ell - (\ell + 1) = j - 1 .
\end{aligned}
$$

Hence also $y_1 \ldots y_j \in \mathcal{F}$ and thus $\mathcal{F}'_X = \mathcal{F}_X$. $\square$

Corollary 1.12 *Every polymatroid branching greedoid has the interval property.* □

In Chapter VIII we will discuss two more examples of polymatroid branching greedoids.

2. Local Properties of Local Poset Greedoids

We call a greedoid $(E, \mathscr{F})$ a **local poset greedoid** if

(2.1) $A, B, C \in \mathscr{F}$ and $A, B \subseteq C$ imply $A \cup B \in \mathscr{F}$ and $A \cap B \in \mathscr{F}$.

In other words, the restriction of a local poset greedoid to a feasible set is a poset antimatroid. Hence $(E, \mathscr{F})$ is a local poset greedoid if and only if all intervals $[\emptyset, X]$ in the poset $(\mathscr{F}, \subseteq)$ are distributive lattices.

The purpose of this section is to give a structural characterization of local poset greedoids and to relate them to polymatroid greedoids.

Consider the following properties of a greedoid $(E, \mathscr{F})$:

(2.2)   **Local intersection property**
$A, A \setminus x, A \setminus y \in \mathscr{F}$ implies $A \setminus \{x, y\} \in \mathscr{F}$,

(2.3)   **Local union property**
$A, A \cup x, A \cup y, A \cup \{x, y, z\} \in \mathscr{F}$ implies $A \cup \{x, y\} \in \mathscr{F}$.

The subdiagrams of the Boolean algebra in Figure 7 illustrate these properties. If the black points are feasible, then the light points must also be feasible.

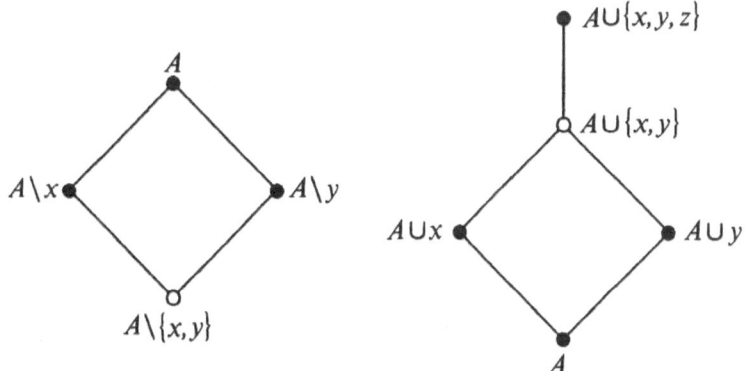

Fig. 7. Forbidden subdiagrams for local poset greedoids

Clearly, if $(E, \mathscr{F})$ is a local poset greedoid, then the above properties hold. But the converse is also true, as we will see.

We first show that (2.2) and (2.3) imply the interval property.

Lemma 2.1. *Every greedoid* $(E, \mathscr{F})$ *satisfying* (2.2) *and* (2.3) *is an interval greedoid.*

Proof. Let $A, B, C \in \mathscr{F}$ be feasible sets with $A \subseteq B \subseteq C$ and assume $A \cup x$, $C \cup x \in \mathscr{F}$ for some $x \in E \setminus C$. We may furthermore assume that $B = A \cup b$ and that C is the smallest feasible superset of B with $C \cup x \in \mathscr{F}$.

Let $C = B \cup \{c_1, \ldots, c_k\}$ such that $C_i = B \cup \{c_1, \ldots, c_i\} \in \mathscr{F}$ for $1 \leq i \leq k$. Consider the largest i for which $A \cup \{x, c_1, \ldots, c_i\} \in \mathscr{F}$. We show that $i = k$. Assume $i < k$. Then we can augment this set from C_{i+1} by either b or c_{i+1}. The first possibility is ruled out since C was chosen to be minimal, and the second since i was maximal. Hence $i = k$ and so $C \cup x \setminus b \in \mathscr{F}$.

Now $C \setminus b \in \mathscr{F}$ follows from (2.2) and $C \cup x$, $C, C \cup x \setminus b \in \mathscr{F}$. Similarly, since $C \setminus c_k \in \mathscr{F}$, we get $C \setminus \{b, c_k\} \in \mathscr{F}$. Hence we have the configuration in Figure 8.

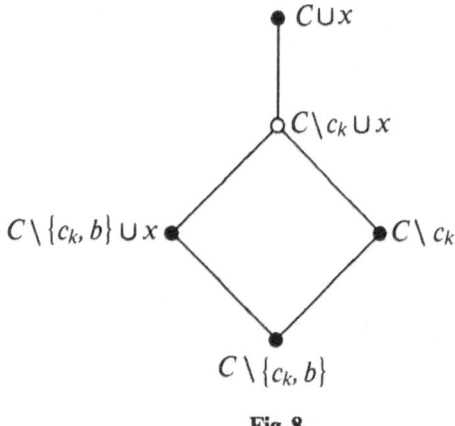

$C \cup x$

$C \setminus c_k \cup x$

$C \setminus \{c_k, b\} \cup x$ $C \setminus c_k$

$C \setminus \{c_k, b\}$

Fig. 8

Thus $C \setminus c_k \cup x \in \mathscr{F}$ by (2.3), contradicting the choice of C.   □

We are now ready to give a characterization of local poset greedoids.

Theorem 2.2. *For a greedoid* $(E, \mathscr{F})$ *the following statements are equivalent:*

(i) $(E, \mathscr{F})$ *is a local poset greedoid.*
(ii) $(E, \mathscr{F})$ *is an interval greedoid with property* (2.2).
(iii) $(E, \mathscr{F})$ *satisfies* (2.2) *and* (2.3).

Proof. It remains to show that (ii) implies (i). For this purpose let A, B and C be feasible sets with $A, B \subseteq C$. By Corollary V.5.8, $A \cup B \in \mathscr{F}$.

To prove $A \cap B \in \mathscr{F}$, we use induction on $|A \cup B|$. If either $A \subseteq B$ or $B \subseteq A$, then there is nothing to show. Otherwise, augment A from $A \cup B$ to a feasible set X with $X = A \cup (B \setminus b)$ for some $b \in B \setminus A$. Similarly, we obtain a feasible set Y with $Y = B \cup (A \setminus a)$. Property (2.2) implies that $X \cap Y \in \mathscr{F}$.

By the induction hypothesis $A \cap (X \cap Y) = A \setminus a \in \mathscr{F}$ and hence, again using induction, $(A \setminus a) \cap B = A \cap B \in \mathscr{F}$.   □

Corollary 2.3. *Every polymatroid branching greedoid is a local poset greedoid.*

Proof. In view of Lemma 1.11 we need to prove this fact only for polymatroid greedoids. By Theorem 1.1 and the above result we only have to show that (2.2) holds. But this follows directly from Lemma 1.2:

$$|A| - 2 \le f(A \setminus \{x, y\}) \le f(A \setminus x) + f(A \setminus y) - f(A) = |A| - 2 ,$$

hence $A \setminus \{x, y\} \in \mathscr{F}$.   □

For a full greedoid the concepts of poset antimatroid, polymatroid greedoid, and local poset greedoid coincide:

Corollary 2.4. *For a full greedoid the following properties are equivalent:*

(i) $(E, \mathscr{F})$ *is a poset antimatroid.*
(ii) $\mathscr{F}$ *is closed under union and intersection.*
(iii) $(E, \mathscr{F})$ *has properties (2.2) and (2.3).*
(iv) $(E, \mathscr{F})$ *is a polymatroid greedoid.*

Proof. We have already seen that (i) ⇒ (iv) ⇒ (iii) ⇒ (ii), and we noted in Chapter III, Example 2.3, that (ii) ⇒ (i).   □

We have just seen that polymatroid greedoids are examples of local poset greedoids. We can formulate a further property similar to the local union and local intersection properties that is satisfied by polymatroid greedoids:

(2.4)   If $A, A \cup x, A \cup y, A \cup \{x, y\}, A \cup \{y, z\} \in \mathscr{F}$, then either $A \cup z \in \mathscr{F}$ or $A \cup \{x, z\} \in \mathscr{F}$ or $A \cup \{x, y, z\} \in \mathscr{F}$.

It is not difficult to see that this property is equivalent to excluding the following greedoid as a minor:

(2.5) $E = \{x, y, z\}, \ \mathscr{F} = \{\emptyset, x, y, \{x, y\}, \{y, z\}\}.$

Lemma 2.5. *Every polymatroid greedoid has property (2.4).*

Proof. By Lemma 1.2, $f(A \cup z) \ge |A| + 1$, and, since $A \in \mathscr{F}$, equality holds if and only if $A \cup z \in \mathscr{F}$. Suppose $A \cup z \notin \mathscr{F}$, then in particular $f(A \cup \{x, z\}) \ge |A| + 2$. Again, if equality holds, then $A \cup \{x, z\} \in \mathscr{F}$. Suppose not, then

$$\begin{aligned} |A| + 3 &\le f(A \cup \{x, y, z\}) \\ &\le f(A \cup \{y, z\}) + f(A \cup \{x\}) - f(A) \\ &= |A| + 3 , \end{aligned}$$

i.e. $A \cup \{x, y, z\} \in \mathscr{F}$.   □

Local poset greedoids with property (2.4) have many properties in common with polymatroid greedoids. For example, every rank feasible set is a partial alphabet both for polymatroid greedoids and local poset greedoids with property (2.4). In fact, it was conjectured for a while that both concepts are equivalent. This conjecture, however, was disproved by an elaborate counterexample described in Korte and Lovász [1986a].

The problem of characterizing polymatroid greedoids by exchange properties analogous to (2.2)–(2.4) is an interesting open problem.

3. Excluded Minors for Local Posets

Observing that local poset greedoids are closed under taking minors, we derive in this section a characterization by forbidden minors (cf. Korte and Lovász [1985c]).

The first two minors are defined on a 3-element ground set $E = \{x, y, z\}$ as follows:

(3.1) $$\mathcal{F} = 2^E \setminus \{z\},$$

(3.2) $$\mathcal{F} = 2^E \setminus \{x, y\}.$$

For the third type of minor, let $(E', \mathcal{F}')$ be a poset antimatroid, $x, y \notin E'$ and $E = E' \cup \{x, y\}$. Then

(3.3) $$\mathcal{F} = \{\emptyset, E\} \cup \{\{A \cup x\}, \{A \cup y\} : A \in \mathcal{F}'\}.$$

Theorem 3.1. *A greedoid* $(E, \mathcal{F})$ *is a local poset greedoid if and only if it does not contain a greedoid of type (3.1), (3.2) or (3.3) as a minor.*

Proof. If $(E, \mathcal{F})$ is a local poset greedoid, then in the case of (3.1) and (3.3) $E, E \setminus x, E \setminus y \in \mathcal{F}$ but $E \setminus \{x, y\} \notin \mathcal{F}$, contradicting property (2.2). In the case of (3.2), we have $x, y, \{x, y, z\} \in \mathcal{F}$ but $\{x, y\} \notin \mathcal{F}$, which contradicts (2.3).

Conversely, let $(E, \mathcal{F})$ be a greedoid without minors of type (3.1) - (3.3). Suppose $(E, \mathcal{F})$ is not a local poset greedoid. We may assume that all proper minors are local poset greedoids.

If property (2.3) is violated, there exists a feasible set $A \in \mathcal{F}$ with

$$A \cup x, \quad A \cup y, \quad A \cup \{x, y, z\} \in \mathcal{F} \text{ but } A \cup \{x, y\} \notin \mathcal{F}.$$

Contraction of A and restriction to $\{x, y, z\}$ yields a minor $\mathcal{F}'$ on $E' = \{x, y, z\}$. Since $x, y, \{x, y, z\} \in \mathcal{F}$ and $\{x, y\} \notin \mathcal{F}$, both $\{x, z\}$ and $\{y, z\}$ are feasible. So either

$$\mathcal{F}' = 2^{E'} \setminus \{x, y\}$$

or

$$\mathcal{F}' = 2^{E'} \setminus \{z, \{x, y\}\}.$$

The first case is ruled out by (3.2) and the second is a minor of type (3.3) constructed from the one-element poset antimatroid $\{\emptyset, z\}$.

Suppose property (2.2) is violated , i.e. $A, A \backslash x, A \backslash y \in \mathscr{F}$ for some $x, y \in A$ and $A \backslash \{x, y\} \notin \mathscr{F}$. We may restrict the greedoid to A and contract every feasible element $z \in E \backslash \{x, y\}$. Thus, $E = A$ and $r(E \backslash \{x, y\}) = 0$. But since $E \backslash y \in \mathscr{F}$, it must contain a feasible element. So $x \in \mathscr{F}$, and similarly $y \in \mathscr{F}$. We distinguish two cases:

(i) there exists an element $z \in E \backslash \{x, y\}$ such that $E \backslash z \in \mathscr{F}$. Then

$$E \backslash \{x, z\} \in \mathscr{F},$$

since otherwise E, x and z also violate (2.2) and the contraction of y would yield a smaller counterexample. Similarly $E \backslash \{y, z\} \in \mathscr{F}$. Moreover,

$$E \backslash \{x, y, z\} \in \mathscr{F},$$

for if not, then the deletion of z would result in a smaller counterexample. In view of $r(E \backslash \{x, y\}) = 0$ we now get

$$E = \{x, y, z\}$$

and

$$\mathscr{F} = 2^E \backslash \{z\} ,$$

i.e. a minor of type (3.1).

(ii) $E \backslash z \notin \mathscr{F}$ for all $z \in E \backslash \{x, y\}$. Then for each $X \in \mathscr{F}, X \neq \emptyset, E,$

(3.4) $$|X \cap \{x, y\}| = 1 ,$$

since if $|X \cap \{x, y\}| = 0$, then $r(E \backslash \{x, y\}) > 0$, and if $|X \cap \{x, y\}| = 2$, then we could augment X to some set $E \backslash z$ with $z \neq x, y$. Set

$$E' = E \backslash \{x, y\}$$
$$\mathscr{F}' = (\mathscr{F}/x) \backslash y$$
$$\mathscr{F}'' = (\mathscr{F}/y) \backslash x .$$

We claim that $\mathscr{F}' = \mathscr{F}''$. For let $X \in \mathscr{F}'$, i.e. $X \cup x \in \mathscr{F}$, $y \notin X$. Augment y from $X \cup x$ to a set $Y \cup y$ with $|X \cup x| = |Y \cup y|$. Then $Y \cup y \neq E$ by (3.4), so $x \notin Y \cup y$ and thus $X = Y$. Hence $X \in \mathscr{F}''$.

By the minimality of $(E, \mathscr{F})$, $(E', \mathscr{F}')$ is a local poset antimatroid. Moreover, it is full since $E \backslash y \in \mathscr{F}$. Hence it is a poset antimatroid by Corollary 2.4. Let

$$\overline{\mathscr{F}} = \{\emptyset, E\} \cup \{\{A \cup x\}, \{A \cup y\} : A \in \mathscr{F}'\} .$$

We claim $\mathscr{F} = \overline{\mathscr{F}}$. Consider $X \in \mathscr{F}$. If $X = \emptyset$ or $X = E$, then clearly $X \in \overline{\mathscr{F}}$. Otherwise $|X \cap \{x, y\}| = 1$ by (3.4). Suppose $x \in X$, then $X \backslash x \in \mathscr{F}'$, and hence $X = (X \backslash x) \cup x \in \overline{\mathscr{F}}$.

Conversely, let $X \in \overline{\mathscr{F}}$. Again, $X = \emptyset$ or $X = E$ are trivial. So let $X = A \cup x$, say, for some $A \in \mathscr{F}'$. Then since $\mathscr{F}' = (\mathscr{F}/x) \backslash y$, we get $X \in \mathscr{F}$, and $(E, \overline{\mathscr{F}})$ is a minor of type (3.3). □

Observe that the greedoids (3.2) and (3.3) do not have the interval property. We therefore get the following:

Corollary 3.2. *An interval greedoid is a local poset greedoid if and only if it does not contain the greedoid (3.1) as a minor.*   □

4. Paths in Local Poset Greedoids

In this section we summarize some properties of paths in local poset greedoids. Most of these properties appeared in Korte and Lovász [1985c], Schmidt [1985a] and Goecke and Schrader [1990]. The motivation comes from paths in branching greedoids. Indeed, these results will be used in the next section to characterize the latter.

Recall that a feasible set A is a **path** if for some element $x \in A$ no proper feasible subset of A contains x. In this case x is called the **endpoint** of A, and the set A is an x-**path.** The collection of paths is denoted by $\mathscr{P}$.

Lemma 4.1. *An interval greedoid $(E, \mathscr{F})$ is a local poset greedoid if and only if every $A \in \mathscr{F}$ contains exactly one x-path for every $x \in A$.*

Proof. If $(E, \mathscr{F})$ is a local poset greedoid, then by Theorem 2.2,

$$\bigcap \{B \subseteq A : x \in B, \ B \in \mathscr{F}\}$$

is the unique x-path.

Conversely, let $A, A \setminus x, A \setminus y \in \mathscr{F}$. Then for each $z \in A \setminus \{x, y\}$, A contains a unique z-path which is also the unique z-path in $A \setminus x$ and in $A \setminus y$. Hence

$$A \setminus \{x, y\} = \bigcup \{X : X \text{ is a } z\text{-path}, \ z \in A \setminus \{x, y\}\},$$

i.e. $A \setminus \{x, y\} \in \mathscr{F}$ by Corollary V.5.8. Thus $(E, \mathscr{F})$ is a local poset greedoid.   □

The paths $\mathscr{P}/x$ in a contraction $\mathscr{F}/x$ relate to the paths $\mathscr{P}$ in the following way.

Lemma 4.2. *Let $(E, \mathscr{L})$ be a local poset greedoid, α a feasible word, and $\{x\} \in \mathscr{F}$ with $x \notin \tilde{\alpha}$.*

(i)   *If $\alpha \in \mathscr{P}$ and $\alpha \in \mathscr{L}/x$, then $\alpha \in \mathscr{P}/x$.*
(ii)   *If $\alpha \in \mathscr{P}/x$, then either $\alpha \in \mathscr{P}$ or $x\alpha \in \mathscr{P}$.*

Proof. Let u be the last letter of α.

(i) Choose a u-path $\alpha' u$ in α with respect to $\mathscr{L}/x$. Denote by A and A' the sets underlying α and α', respectively. Then $A, A' \cup \{u, x\} \subseteq A \cup x$ and the local intersection property implies $\alpha' u \in \mathscr{L}$. Hence by Lemma 4.1, $A' \cup u = A$ and $\alpha \in \mathscr{P}/x$.

(ii) Choose a u-path $\alpha'u$ in $x\alpha$ with respect to $\mathscr{L}$. If $x \in \alpha'$, then $A' \cup u \setminus x$ is a path in $\mathscr{L}/x$, thus $A' \cup u \setminus x = A$ and $x\alpha \in \mathscr{P}$. If $x \notin \alpha'u$, then $x\alpha'u \in \mathscr{L}$ since $x, A' \cup u \subseteq A \cup x \in \mathscr{F}$. This implies that $\alpha'u \in \mathscr{L}/x$, hence $A' \cup u = A$ and so $\alpha \in \mathscr{P}/x$. □

It is natural to ask when paths have a unique feasible ordering. It turns out that this is equivalent to a local exchange property analogous to property (2.4) of polymatroid greedoids.

(4.1) Local Forest Property

If $A, A \cup x, A \cup y, A \cup \{x, y\}, A \cup \{x, y, z\} \in \mathscr{F}$, then either $A \cup \{x, z\} \in \mathscr{F}$ or $A \cup \{y, z\} \in \mathscr{F}$.

A local poset greedoid with this property is called a **local forest greedoid**. It is easy to see that local forest greedoids may be characterized by the following forbidden minor:

(4.2) $E = \{x, y, z\}, \mathscr{F} = \{\emptyset, x, y, \{x, y\}, \{x, y, z\}\}$.

The name of property (4.1) is explained by the following lemma.

Lemma 4.3. *A full greedoid is a local forest greedoid if and only if it is the poset antimatroid of a forest poset.*

Proof. The "if"-part is obvious. For the converse, consider a full local forest greedoid $(E, \mathscr{F})$. Then by Corollary 2.4, it is a poset antimatroid associated with a poset $(E, \leq)$. Assume that this is not a forest poset. Then there exists an element z that covers two elements x and y. Let $A = E \setminus (F(x) \cup F(y))$. Then A, x, y and z violate (4.1). □

Theorem 4.4. *A local poset greedoid has the local forest property if and only if every path has a unique feasible ordering.*

Proof. If the greedoid has the local forest property, then its restriction to any path is a forest poset antimatroid. Since it is a path, this forest poset is a single chain. Hence it has a unique feasible ordering. Suppose conversely that every path has a unique feasible ordering. Consider the restriction of the greedoid to any feasible set. By Corollary 2.4, this restriction is a poset antimatroid. It suffices to prove that this poset is a forest poset. Suppose it contains an element z covering two elements x and y. Then the ideal generated by z has at least two feasible orderings. □

For later reference we formulate a characterization of local forest greedoids in terms of the closure operator.

Lemma 4.5. *A local poset greedoid has the local forest property if and only if $\sigma(X) \cap \sigma(Y) \subseteq \sigma(X \cup Y)$ for all $X, Y \in \mathscr{R}$.*

Proof. Assume that we have a local forest greedoid and that it contains two sets $X, Y \in \mathcal{R}$ such that $\sigma(X) \cap \sigma(Y) \not\subseteq \sigma(X \cup Y)$. Choose such a pair with $|X \cup Y|$ minimal, and from among these one with $r(X \cap Y)$ maximal, and again from among these one with $|X \cap Y|$ maximal.

Let A be a basis of $X \cap Y$. If $\sigma(A) \supseteq X$, then $\sigma(Y) \supseteq X$ since $X \in \mathcal{C} = \mathcal{R}$. Then $\sigma(X \cup Y) = \sigma(Y)$, which contradicts the assumption that $\sigma(X) \cap \sigma(Y) \not\subseteq \sigma(X \cup Y)$. Hence we can find an element $x \in X$ such that $A \cup x \in \mathcal{F}$. Let $z \in (\sigma(X) \cap \sigma(Y)) \setminus \sigma(X \cup Y)$. Since $A \cup x \subseteq X \subseteq X \cup Y$ and $z \notin \sigma(X \cup Y)$ but $z \in \sigma(X)$, the interval property implies that $z \in \sigma(A \cup x)$.

Note that $Y \cup x = Y \cup (A \cup x) \in \mathcal{R}$. Hence if $z \in \sigma(Y \cup x)$, then the pair X and $Y \cup x$ contradicts the extremal choice of X and Y. So $z \notin \sigma(Y \cup x)$. But then the pair $A \cup x$ and Y contradicts the extremal choice of X and Y unless $X = A \cup x$.

Similarly it follows that $Y = A \cup y \in \mathcal{F}$. Since $z \notin \sigma(A \cup x \cup y)$ but $z \in \sigma(A \cup x), A \cup x \cup y$ is not cospanning with $A \cup x$ and hence $A \cup x \cup y \in \mathcal{F}$. Now A, x, y and z violate (4.1).

The converse is obvious.   □

5. Excluded Minors for Undirected Branching Greedoids

The aim of this section is to characterize undirected branching greedoids. This characterization is given in Schmidt [1988]. The shorter proof presented here follows Goecke and Schrader [1990].

Consider the following four greedoids defined on the ground set $E = \{x, y, z\}$.

(5.1) $\mathcal{F} = 2^E \setminus \{z\}$,
(5.2) $\mathcal{F} = \{\emptyset, x, y, \{x, y\}, \{x, y, z\}\}$,
(5.3) $\mathcal{F} = \{\emptyset, x, y, \{x, y\}, \{x, z\}\}$,
(5.4) $\mathcal{F} = 2^E \setminus E$.

Recall from Section 2 that (5.3) is excluded for all polymatroid greedoids, while the exclusion of (5.1) characterizes local posets among interval greedoids and the exclusion of (5.2) characterizes local forest greedoids among local poset greedoids.

Theorem 5.1. *A greedoid is an undirected branching greedoid if and only if it is an interval greedoid without minors* (5.1)–(5.4).

The necessity is of course obvious. As a preparation for the proof of the sufficiency, consider an interval greedoid $(E, \mathcal{F})$ and a basic word $a_1 \ldots a_r$. We associate a graph G with $(E, \mathcal{F})$ as follows. Let $V(G) = \{0, 1, \ldots, r\}$. For each $e \in E$, let $i(e)$ be the first index i such that $a_1 \ldots a_i e \in \mathcal{L}$. Let $j(e)$ be the largest index such that $a_1 \ldots a_{j-1} e \in \mathcal{L}$. If no t exists such that $a_1 \ldots a_t e \in \mathcal{L}$, then put $i(e) = j(e) = 0$. Connect $i(e)$ to $j(e)$ for each $e \in E$ by an edge.

We will show that if $(E, \mathscr{F})$ does not have any of the given minors, then the undirected branching greedoid of this graph with respect to root 0 is just $(E, \mathscr{F})$. For this purpose we need a series of lemmas.

Lemma 5.2. *Let $(E, \mathscr{F})$ be an interval greedoid without minors (5.1), (5.2) and (5.3). Suppose that $\alpha = x_1 \ldots x_\ell \in \mathscr{P}$, $x_0 \in \mathscr{F}$ and that $\{x_0, x_1, \ldots, x_{k-1}\} \in \mathscr{F}$, $\{x_0, x_1, \ldots, x_k\} \notin \mathscr{F}$ for some $k \in \{1, \ldots, \ell\}$. Then $x_0 x_k x_{k-1} \ldots x_2 \in \mathscr{P}$, $x_0 x_{k+1} \ldots x_\ell \in \mathscr{P}$ and*

$$\{x_0, x_1, \ldots, x_\ell\} \setminus \{x_i, \ldots, x_j\} \in \mathscr{F} \text{ for } 1 \le i \le j \le k.$$

Proof. For $k = 1$ it suffices to prove that $x_0 x_2 \ldots x_\ell \in \mathscr{P}$. To see this, successively augment $x_0 \in \mathscr{F}$ from $x_1 \ldots x_\ell \in \mathscr{F}$. Since $\{x_0, x_1\} \notin \mathscr{F}$ and by the interval property, $x_0 x_2 \ldots x_\ell \in \mathscr{P}$.

For $k \ge 2$ consider $\mathscr{F}/x_1$. Then $x_2 \ldots x_\ell, x_0$ satisfy the requirements of our lemma, hence by induction we get

$$(5.5) \qquad\qquad x_0 x_k x_{k-1} \ldots x_3 \in \mathscr{P}/x_1 ,$$

$$(5.6) \qquad\qquad x_0 x_{k+1} \ldots x_\ell \in \mathscr{P}/x_1 ,$$

$$(5.7) \qquad \{x_0, x_2, \ldots, x_\ell\} \setminus \{x_i, \ldots, x_j\} \in \mathscr{F}/x_1 \text{ for } 2 \le i \le j \le k.$$

Since $\{x_0, x_1, \ldots, x_{k-1}\} \in \mathscr{F}$ and $x_1 \ldots x_{k-1} \in \mathscr{P}$, Lemma 4.2 implies

$$(5.8) \qquad\qquad x_1 \ldots x_{k-1} \in \mathscr{P}/x_0 .$$

We now claim that $x_k \in \mathscr{P}/x_0$. For $k \ge 3$ this immediately follows from (5.5). For $k = 2$ this also holds, since otherwise $\{x_0, x_1, x_2\}$ induces a minor of type (5.2) in $\mathscr{F}$.

From (5.7) we deduce that

$$\{x_k, x_1, \ldots, x_{k-2}\} \in \mathscr{F}/x_0 \text{ and } \{x_k, x_1, \ldots, x_{k-1}\} \notin \mathscr{F}/x_0 .$$

Hence the induction hypothesis implies $x_k x_{k-1} \ldots x_2 \in \mathscr{P}/x_0$ and

$$(5.9) \quad \{x_k, x_1, \ldots, x_{k-1}, x_{k+1}, \ldots, x_\ell\} \setminus \{x_1, \ldots, x_j\} \in \mathscr{F}/x_0 \text{ for } 1 \le j \le k-1.$$

Thus, using Lemma 4.2, $x_0 x_k x_{k-1} \ldots x_2 \in \mathscr{P}$. Moreover, (5.6) and Lemma 4.2 yield $x_0 x_{k+1} \ldots x_\ell \in \mathscr{P}$, and finally (5.7) and (5.9) imply $\{x_0, x_1, \ldots, x_\ell\} \setminus \{x_i, \ldots, x_j\} \in \mathscr{F}$ for $1 \le i \le j \le k$. □

We say that two elements x, y are **parallel** (denoted by $x \parallel y$) in $\mathscr{F}$ if $\alpha x \beta \in \mathscr{F}$ if and only if $\alpha y \beta \in \mathscr{F}$ for all words α and β.

Lemma 5.3. *Let $(E, \mathscr{L})$ be an interval greedoid without minors (5.1)–(5.4). Then $x \parallel y$ in $\mathscr{L}$ if and only if either x and y are loops or there exist paths $\alpha x, \alpha y$ such that $\alpha x y \notin \mathscr{L}$.*

Proof. If $x \parallel y$ in $\mathscr{L}$ and x, y are not loops, then for any x-path αx we have $\alpha y \in \mathscr{P}$ and $\alpha x y \notin \mathscr{L}$.

Conversely if x and y are loops, then clearly they are parallel. Suppose now that $\alpha x, \alpha y \in \mathscr{P}$ and $\alpha x y \notin \mathscr{L}$. If $\alpha = \emptyset$, then the interval property immediately implies $x \parallel y$ in $\mathscr{L}$.

Now let $\alpha = a\alpha'$ and suppose that x, y are not parallel in $\mathscr{L}$. Choose minimal words β, γ such that $\beta x y \in \mathscr{L}$, $\beta y \gamma \notin \mathscr{L}$. Then $\beta x \in \mathscr{P}$, for if $\beta' x \in \mathscr{P}$ with $\beta = \beta'\beta''$, then $\beta'x\beta''\gamma \in \mathscr{L}$. Hence by the minimality of β, $\beta'y\beta''\gamma \in \mathscr{L}$ and, by the interval property, $\beta y \gamma \in \mathscr{L}$. Clearly $\beta \neq \emptyset$.

Furthermore, we claim that we may assume $\gamma = \emptyset$ or $\gamma = y$. If $\gamma = \gamma'b$, then by the minimality of γ, $\beta y \gamma' \in \mathscr{L}$ and, since $\beta y \gamma \notin \mathscr{L}$, $\beta y \gamma'x \in \mathscr{L}$. By the interval property this implies that $\beta x y \in \mathscr{L}$. Since $\beta y y \notin \mathscr{L}$, we thus may assume $\gamma = y$.

We now distinguish two cases.

(i) $\beta x \in \mathscr{P}/a$ or $\beta y \in \mathscr{P}/a$. Then both $\beta x \in \mathscr{P}/a$ and $\beta y \in \mathscr{P}/a$ since $x \parallel y$ in $\mathscr{L}/a$ by induction. By Lemma 4.2, $\beta x \in \mathscr{P}$ and $\beta y \in \mathscr{P}$, in particular $\gamma = y$. But then $\{a, x, y\}$ induces a minor (5.4) in $\mathscr{L}/\beta$.

(ii) $\beta x, \beta y \notin \mathscr{P}/a$. From Lemma 4.2 we get $a\beta x, a\beta y \notin \mathscr{L}$. Let B be the set underlying β.

For $\gamma = y$, augment $a \in \mathscr{L}$ from $\beta x y \in \mathscr{L}$. Then necessarily

$$B \cup \{a, x, y\} \setminus b \in \mathscr{L}$$

for some $b \in B$, contradicting $x \parallel y$ in $\mathscr{L}/a$. Thus $\gamma = \emptyset$. Let β' be the longest initial substring of β such that $\beta'a \in \mathscr{L}$. If $\beta' = \beta$, apply Lemma 5.2 to $\beta x \in \mathscr{P}$, $a \in \mathscr{L}$, $\beta x a \notin \mathscr{L}$. Then $a x \delta \in \mathscr{P}$, where δ is β in reverse order without the first letter b of β. Since $x \parallel y$ in $\mathscr{L}/a$, also $a y \delta \in \mathscr{P}$. A second application of Lemma 5.2 to $a y \delta \in \mathscr{P}$, $b \in \mathscr{L}$, $a y \delta b \notin \mathscr{L}$ yields $\beta y \in \mathscr{P}$, contradicting $\gamma = \emptyset$.

Suppose now that $\beta = \beta'b\beta''$. In $\mathscr{L}/\beta'$, $a \parallel b$. Hence $\beta'b\beta''x \in \mathscr{L}$ implies $\beta'a\beta''x \in \mathscr{L}$. Since $x \parallel y$ in $\mathscr{L}/a$, we get $\beta'a\beta''y \in \mathscr{L}$. Again, by $a \parallel b$ in $\mathscr{L}/\beta'$, we finally conclude $\beta'b\beta''y = \beta y \in \mathscr{L}$, a contradiction.   □

Lemma 5.4. *Let* $(E, \mathscr{F})$, $(E, \mathscr{F}')$ *be interval greedoids without minors* (5.1)–(5.4). *Suppose that* $\mathscr{P} = \mathscr{P}'$ *and that* $x \parallel y$ *in* $\mathscr{F}$ *if and only if* $x \parallel y$ *in* $\mathscr{F}'$. *Then* $\mathscr{F} = \mathscr{F}'$.

Proof. Suppose $\mathscr{F} \neq \mathscr{F}'$. Let A be a minimal set in $\mathscr{F} \setminus \mathscr{F}'$. By Lemma 4.3, the restriction of $\mathscr{F}$ to A is a forest poset antimatroid.

(i) If A has a unique maximal element, then it is a path and so $A \in \mathscr{F}'$ by hypothesis.

(ii) If A has exactly two maximal elements x and y and every other element is below x and y, then $x \parallel y$ in $(E, \mathscr{F}')$ by Lemma 5.3. But clearly $x \not\parallel y$ in $(E, \mathscr{F})$, a contradiction.

(iii) If A has exactly two maximal elements y and z and there exists an element x below y but not below z, then we may assume tht z covers x. Then contracting $A \setminus \{x, y, z\}$ in $(E, \mathscr{F}')$, we get a minor (5.3), again a contradiction.

(iv) If A has at least three maximal elements, then let x, y, z be thre of these. Contracting $A \setminus \{x, y, z\}$ in $(E, \mathscr{F}')$ we get a minor (5.4).   □

Proof of Theorem 5.1. As we have already remarked, the necessity is trivial. To prove the sufficiency, we show that an interval greedoid without minors (5.1)–(5.4) is the branching greedoid of the graph G as defined above. In view of Lemma 5.4, it suffices to show that the paths and the parallel elements in $\mathcal{F}$ and G coincide.

We may assume that the assertion is true for $\mathcal{F}/a_1$ and G/a_1. For paths of length one and paths containing a_1, the claim is easily verified.

So let $k \geq 2$ and assume that the paths shorter than k coincide in $\mathcal{F}$ and G. Let $\alpha = x_1 \ldots x_k$ be a path in G. If α is a path in G/a_1, i.e. $\alpha \in \mathcal{P}/a_1$, then by Lemma 4.2, $\alpha \in \mathcal{P}$ or $a_1\alpha \in \mathcal{P}$. However, the latter case cannot occur since $x_1 \in \mathcal{F}$.

Therefore we can now assume that α is not a path in G/a_1. We thus have the subgraph of G/a_1 shown in of Figure 9:

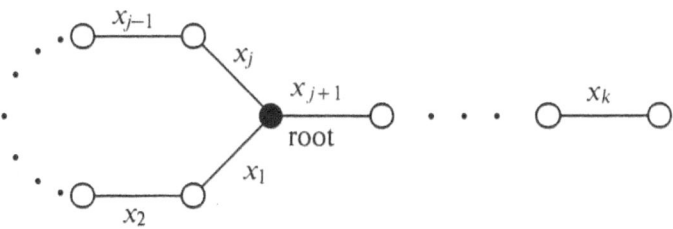

Fig. 9

where $1 \leq j \leq k$. Since the branchings in G/a_1 are the feasible sets in $\mathcal{F}/a_1$, we know that $\{a_1, x_1, \ldots, x_k\} \setminus \{x_j\} \in \mathcal{F}$. Augmenting $\{x_1, \ldots, x_{k-1}\}$ from the latter gives that either $\{a_1, x_1, \ldots, x_{k-1}\} \in \mathcal{F}$ or $\{x_1, \ldots, x_k\} \in \mathcal{F}$.

If $\{x_1, \ldots, x_k\} \in \mathcal{F}$, it contains an x_k-path. However, this path has length k since strictly shorter paths coincide in $\mathcal{F}$ and G, thus $\alpha \in \mathcal{P}$.

If $\{a_1, x_1, \ldots, x_{k-1}\} \in \mathcal{F}$,then clearly $j = k$. Applying Lemma 5.2 to $x_1 \in \mathcal{F}$, $a_1 x_k x_{k-1} \ldots x_2 \in \mathcal{P}$ (since $x_k x_{k-1} \ldots x_2 \in \mathcal{P}/a_1$) and $\{a_1, x_1, \ldots x_k\} \notin \mathcal{F}$, which yields $\alpha = x_1 x_2 \ldots x_k \in \mathcal{P}$.

Conversely, let $\alpha = x_1 \ldots x_k \in \mathcal{P}$ with $k \geq 2$. If $\alpha \in \mathcal{P}/a_1$, then α or $a_1\alpha$ is a path in G. However, $a_1\alpha$ cannot be a path in G since x_1 is incident with the root. If $\alpha \notin \mathcal{P}/a_1$, then by Lemma 4.2, $\alpha \notin \mathcal{F}/a_1$. Choose the index j such that $a_1 x_1 \ldots x_{j-1} \in \mathcal{F}$, $a_1 x_1 \ldots x_j \notin \mathcal{F}$. Lemma 5.2 implies that $x_j x_{j-1} \ldots x_2 \in \mathcal{P}$ and $\{a_1, x_1, \ldots, x_k\} \setminus x_j \in \mathcal{F}$, i.e. $\{x_1, \ldots, x_k\} \setminus x_j \in \mathcal{F}/a_1$. Hence $\{x_1, \ldots, x_k\} \setminus x_j$ is a branching in G/a_1 and, by construction, $\{a_1, x_1, \ldots, x_k\} \setminus x_j$ is a branching in G. To see that α is a path in G, we proceed as before.

It remains to show that $a \parallel b$ in $\mathcal{F}$ if and only if $a \parallel b$ in G. For $a_1 = a$ or $a_1 = b$ this is obvious from the construction. If $a \parallel b$ in $\mathcal{F}$, then, trivially, $a \parallel b$ in $\mathcal{F}/a_1$. By induction $a \parallel b$ in G/a_1. Now if a and b are not parallel in G, then, by construction, exactly one of a, b is incident to the root. This, however, contradicts $a \parallel b$ in $\mathcal{F}$. Conversely, let $a \parallel b$ in G. Then $a \parallel b$ in G/a_1 and hence $a \parallel b$ in $\mathcal{F}/a_1$. By Lemma 5.3, there exist $\alpha a, \alpha b \in \mathcal{P}/a_1$, $\alpha ab \notin \mathcal{F}/a_1$. Thus $\alpha a \in \mathcal{P}$ or

$a_1 \alpha a \in \mathscr{P}$ by Lemma 5.2. If $\alpha a \in \mathscr{P}$ and $a_1 \alpha b \in \mathscr{P}$ or $a_1 \alpha a \in \mathscr{P}$ and $\alpha b \in \mathscr{P}$, then a and b are not parallel in G since the paths of G and $\mathscr{F}$ coincide. If $\alpha a, \alpha b \in \mathscr{P}$, then $\alpha ab \notin \mathscr{F}$ since otherwise $\{a, b, a_1\}$ induces a minor (5.4) in $\mathscr{F}/\alpha$. Hence by Lemma 5.3, $a \parallel b$ in $\mathscr{F}$. If $a_1 \alpha a, a_1 \alpha b \in \mathscr{P}$, then $a \parallel b$ in $\mathscr{F}$ since $a_1 \alpha ab \notin \mathscr{F}$.

Thus the requirements of Lemma 5.4 are fulfilled and $\mathscr{F}$ is the branching greedoid of G.   □

Chapter VIII. Greedoids on Partially Ordered Sets

Basically there have been two attempts to generalize classical matroids to ordered sets. The first one is the notion of supermatroids introduced by Dunstan, Ingleton and Welsh [1972]. Here the independent sets correspond to nodes in a partially ordered set so that if x is independent, then the whole principal ideal of x consists of independent nodes and for any y, any two maximal independent nodes below y have the same height.

In a different approach, Faigle [1980] introduced his ordered geometries as hereditary structures on ordered sets endowed with an exchange property. He gave several characterizations of these structures, for example in terms of closed sets, hyperplanes, the rank function or the closure operator.

Section 1 collects basic facts about supermatroids. We exhibit greedoids as a subclass of supermatroids with submodular rank functions. In Section 2 we introduce the notion of an ordered geometry as a hereditary language defined on a poset, whose words are compatible with the order and satisfy a certain exchange property. We show that ordered geometries are interval greedoids.

Section 3 is devoted to a characterization of those interval greedoids which are ordered geometries. For ordered geometries we can define a new rank function called the ideal rank. This is always a polymatroid rank function, but does not determine the ordered geometry uniquely. We show in Section 4 that among all ordered geometries with the same rank function, there is a unique minimal and a unique maximal one. Minimal and maximal ordered geometries are polymatroid branching greedoids, while ordered geometries in general are not even local posets.

1. Supermatroids

Dunstan, Ingleton and Welsh [1972] introduced a common generalization of matroids and polymatroids to partially ordered sets.

Let $P = (E, \leq)$ be a poset with a unique minimal element 0 and height function h. For any subset $B \subseteq E$ we denote by $I(B)$ the ideal generated by B, i.e.

$$I(B) = \{x \in E : x \leq b \text{ for some } b \in B\}.$$

Dually, $F(B)$ is the filter generated by B.

A collection $\mathscr{F}$ of elements of P is a **supermatroid** if

(1.1) $0 \in \mathscr{F}$,
(1.2) $x \in \mathscr{F}$ and $y \leq x$ implies $y \in \mathscr{F}$,
(1.3)   For $x \in E$, all maximal elements $y \in I(x) \cap \mathscr{F}$ have the same height.

If P is a Boolean algebra 2^E, then the supermatroids on P are exactly the matroids on E. The definitions of basis, circuit, spanning and closed elements naturally extend to supermatroids. In particular, the **rank** $r(x)$ of an element $x \in E$ is the height $h(y)$ of a maximal element $y \in \mathscr{F}$ in the principal ideal $I(x)$.

A supermatroid is **strong** if its rank function is unit-increasing and locally submodular, i.e. satisfies

(1.4)   if x covers y, then $r(y) \leq r(x) \leq r(y) + 1$,
(1.5)   if x and x' cover y, and z covers x and x', then $r(y) = r(x) = r(x')$ implies $r(y) = r(z)$.

For semimodular lattices local submodularity is equivalent to submodularity in the following sense:

Lemma 1.1. *Let L be a semimodular lattice and f a monotone increasing function on L with the unit increase property. Then f is submodular if and only if f is locally submodular.*

Proof. If f is submodular, then it is easy to see that f is also locally submodular. Conversely, let f be locally submodular. We show that $f(x) + f(y) \geq f(x \vee y) + f(x \wedge y)$ for all x, y. We use induction on $h(x) + h(y) - 2h(x \wedge y)$. If $h(x) \leq h(x \wedge y) + 1$ and $h(y) \leq h(x \wedge y) + 1$, then the assertion follows from the local submodularity of f and the semimodularity of L. Assume e.g. that $h(x) > h(x \wedge y) + 1$. Then there exists an element x' with $x \wedge y < x' < x$. Set $a = x' \vee y$. Then $x' \wedge y = x \wedge y$ and hence by the induction hypothesis, $f(x') + f(y) \geq f(x' \vee y) + f(x' \wedge y) = f(a) + f(x \wedge y)$. Moreover, $x \vee a = x \vee y$ and $x \wedge a \geq x'$, and hence by the induction hypothesis again, $f(x) + f(a) \geq f(x \vee a) + f(x \wedge a) \geq f(x \vee y) + f(x')$. Adding up, we obtain $f(x) + f(y) \geq f(x \wedge y) + f(x \vee y)$.   □

This result immediately implies

Theorem 1.2. *The rank function of a strong supermatroid on a semimodular lattice is submodular.*   □

For more details on supermatroids in general, see Dunstan, Ingleton and Welsh [1972]. Here we are more concerned with the following question. Both supermatroids and greedoids are generalizations of matroids; how are these two classes related? This is (partially) answered by the next results.

Theorem 1.3. *Every interval greedoid $(E, \mathscr{F})$ is a strong supermatroid on a locally free semimodular lattice of subsets of E with the same rank function.*

Proof. $(E, \mathscr{R})$ is an antimatroid, hence the poset $(\mathscr{R}, \subseteq)$ is a locally free semimodular lattice with height function $h(X) = |X|$. Trivially, $\mathscr{F}$ is an ideal in $(\mathscr{R}, \subseteq)$.

For every principal ideal $I(X)$ in $(\mathcal{R}, \subseteq)$, the maximal elements in $\mathcal{F}$ below X are the bases of X. Since they have the same cardinality, they have the same height in the lattice. Hence $\mathcal{F}$ is a supermatroid on $(\mathcal{R}, \subseteq)$ with corresponding rank function $\rho = r$. Since r is unit-increasing on $\mathcal{R}$ and locally submodular, $\mathcal{F}$ is a strong supermatroid. $\qquad\Box$

Conversely, with any strong supermatroid with rank function r on a semi-modular lattice we may associate an interval greedoid by copying the definition of polymatroid greedoids. Let

$$\mathcal{L} = \{x_1 \ldots x_k : r(x_1 \vee \ldots \vee x_i) = i, 1 \leq i \leq k\}.$$

Clearly $\mathcal{L}$ is hereditary. Consider two words $x_1 \ldots x_k$, $y_1 \ldots y_m \in \mathcal{L}$ and set $X = x_1 \vee \ldots \vee x_k$. Since r is submodular,

$$r(X \vee y_1 \vee \ldots \vee y_i) - r(X \vee y_1 \vee \ldots \vee y_{i-1}) \leq 1,$$

and since $r(X \vee y_1 \vee \ldots \vee y_k) \geq r(y_1 \vee \ldots \vee y_k)$, there must exist an index i such that $r(X \vee y_1 \vee \ldots \vee y_{i-1}) = r(X)$ and $r(X \vee y_1 \vee \ldots \vee y_i) = r(X) + 1$. Again, by submodularity

$$r(X) + 1 = r(X \vee y_1 \vee \ldots \vee y_i) \geq r(X \vee y_i)$$
$$\geq r(X \vee y_1 \vee \ldots \vee y_i) + r(X) - r(X \vee y_1 \vee \ldots \vee y_{i-1}) = r(X) + 1.$$

Thus $x_1 \ldots x_k y_i \in \mathcal{L}$.

Now let $A \subseteq B \subseteq C$ be feasible sets such that $A \cup x, C \cup x \in \mathcal{F}$. Then

$$r(\vee B) + 1 = r(\vee A \vee x) + r(\vee B) - r(\vee A) \geq r(\vee B \vee x)$$
$$\geq r(\vee B) + r(\vee C \vee x) - r(\vee C) = r(\vee B) + 1,$$

i.e. $B \cup x \in \mathcal{F}$.

Similarly, using Lemma VI.1.6 we obtain

Theorem 1.4. *Every greedoid is a strong supermatroid on a graded poset.* $\qquad\Box$

Of particular interest is the case of supermatroids on distributive lattices. By Birkhoff's theorem such a **distributive supermatroid** may be viewed as a collection $\mathcal{F}$ of ideals of some partially ordered set such that the following properties hold:

(1.6) $\quad \emptyset \in \mathcal{F}$,

(1.7) $\quad$ if $X \in \mathcal{F}$ and $Y \subseteq X$ is an ideal, then $Y \in \mathcal{F}$,

(1.8) $\quad$ for $X, Y \in \mathcal{F}$ with $|Y| < |X|$ there exists an $x \in X \setminus Y$ such that $Y \cup x \in \mathcal{F}$.

Hence distributive supermatroids are in particular interval greedoids. They will turn out to be precisely those interval greedoids which arise as intersections of matroids and poset antimatroids (cf. Chapter IX). They also give rise to a min-max-theorem (cf. Chapter X).

2. Ordered Geometries

Generalizing distributive supermatroids, Faigle [1980] introduced "matroids on posets". Let $P = (E, \leq)$ be a partially ordered set and let $(E, \mathcal{L})$ be a simple greedoid over E.

The triple $(E, \leq, \mathcal{L})$ is called an **ordered geometry** if it satisfies the following conditions:

(2.1)   If $x_1 x_2 \ldots x_k \in \mathcal{L}$ and $x_i \leq x_j$, then $i \leq j$,
(2.2)   if $X \subseteq Y$ are ideals and $x \in X$ such that $r(X \setminus x) = r(X)$, then $r(Y \setminus x) = r(Y)$.

Distributive supermatroids as introduced in the previous section are ordered geometries. The converse is not true. Consider the poset P in Figure 10 and the greedoid with basic words ab, ba, ac, bc. This is an ordered geometry but not a distributive supermatroid. This greedoid is an ordered geometry with respect to the order P' as well. So the order P is not uniquely determined.

Axiom (2.1) states that the ordering of a word in $\mathcal{L}$ must respect the underlying order. The following result gives equivalent axioms in terms of the closure operators.

Lemma 2.1. *Let $(E, \leq)$ be a partial order and $(E, \mathcal{F})$ a greedoid. Then (2.1) is equivalent to either of the following conditions:*

(2.1′) $I(A) \subseteq \sigma(A)$ *for every* $A \in \mathcal{F}$,
(2.1″) $I(A) \subseteq \mu(A)$ *for every* $A \in \mathcal{F}$.

Proof. (2.1) $\Rightarrow$ (2.1′): Let $A \in \mathcal{F}$ and $\alpha \in \mathcal{L}$ be a feasible ordering of A. Assume $x \notin \sigma(A)$. Then $\alpha x \in \mathcal{L}$, and hence $x \notin I(A)$ by (2.1).

(2.1′) $\Rightarrow$ (2.1″): Let $A \in \mathcal{F}$ be a feasible word and C a closed superset of A. Extend A to a basis B of C. Then $I(B) \subseteq \sigma(B) = C$, hence $I(A) \subseteq I(B) \subseteq C$ and thus $I(A) \subseteq \bigcap \{C : A \subseteq C, \sigma(C) = C\} = \mu(A)$.

(2.1″) $\Rightarrow$ (2.1): Let $x_1 x_2 \ldots x_k \in \mathcal{L}$ and $1 \leq i < j \leq k$. Then $A = \{x_1, \ldots, x_{j-1}\} \in \mathcal{F}, x_j \notin \mu(A)$, hence $x_j \notin I(A)$ and, in particular, $x_j \not< x_i$. $\square$

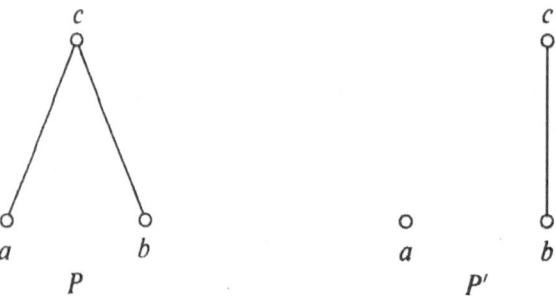

Fig. 10

The next lemma gives conditions equivalent to (2.2).

Lemma 2.2. *Let $(E, \le)$ be a partial order and $(E, \mathcal{F})$ a greedoid satisfying (2.1). Then (2.2) is equivalent to either of the following conditions:*

(2.2') *every ideal is closure feasible,*
(2.2'') *every ideal is rank feasible.*

Proof. (2.2) $\Rightarrow$ (2.2'): Let $(E, \mathcal{F})$ be an ordered geometry on the poset $(E, \le)$ and suppose that X is an ideal but $X \notin \mathcal{C}$, i.e. there exist feasible sets $A, B \in F$ with $A \subseteq B$ such that $X \subseteq \sigma(A)$ but $X \not\subseteq \sigma(B)$.

We can assume without loss of generality that $|X \setminus \sigma(B)| = 1$, since otherwise we could replace X by some suitable subideal. So let $\{p\} = X \setminus \sigma(B)$. We now claim that the ideals $I(A) \cup X$ and $I(B) \cup X$ violate (2.2). To see this, observe that by definition and by (2.1') $X \cup I(B) \setminus p \subseteq \sigma(B)$. Hence

$$r(X \cup I(B) \setminus p) \le r(B) = |B|.$$

On the other hand, $B \cup p$ is a feasible subset of $X \cup I(B)$, thus

$$r(X \cup I(B)) \ge |B| + 1 > r(X \cup I(B) \setminus p).$$

Similarly, $X \cup I(A) \subseteq \sigma(A)$. So $r(X \cup I(A)) \le r(\sigma(A)) = |A| \le r(X \cup I(A) \setminus p)$, in contradiction to (2.2).

(2.2') $\Rightarrow$ (2.2''): Clear, since by Lemma V.3.8, $\mathcal{C} \subseteq \mathcal{R}$.

(2.2'') $\Rightarrow$ (2.2): Let A be a basis of $X \setminus x$. Augment A to some basis B of Y. Then $|B \cap X| \le \beta(X) = r(X) = r(X \setminus x) = |A|$, and hence $x \notin B$. So $r(Y \setminus x) = r(Y)$. $\qquad\square$

We will show next that ordered geometries are in particular interval greedoids:

Lemma 2.3. *Every ordered geometry has the interval property.*

Proof. By Corollary V.5.7 we have to show that $\mathcal{F} \subseteq \mathcal{C}$. Let $X \in \mathcal{F}$, $A \in \mathcal{F}$ be feasible sets with $X \subseteq \sigma(A)$ and consider some superset B of A. Since $X \subseteq \sigma(A)$, we have $\mu(X) \subseteq \sigma(A)$ by the definition of the monotone closure. Lemma 2.1 now yields $I(X) \subseteq \sigma(A)$. Since $I(X)$ is closure feasible we obtain $X \subseteq I(X) \subseteq \sigma(B)$. $\qquad\square$

A detailed exposition of order geometries is given in Faigle [1980], where results on closed sets, hyperplanes and the rank function of these geometries can also be found. In the following, we concentrate on greedoid aspects and characterize ordered geometries as a subclass of interval greedoids.

3. Characterization of Ordered Geometries

Lemma 2.3 raises the question whether for every interval greedoid $(E, \mathcal{F})$ there is some appropriate partial order P on the ground set such that $(E, \mathcal{F})$ is an ordered geometry with respect to this partial order. The following example gives a negative answer to this question. Consider the double shelling antimatroid on the chain $a < b < c$. Suppose that a second order can be defined on the ground set $E = \{a, b, c\}$ which turns the double shelling antimatroid into an ordered geometry. Since abc and cba are basic words, the ideal $I(b)$ generated by b with repect to this second order can only consist of b. Thus $r(I(b)) = 0$, but $|I(b) \cap \{a, b, c\}| = 1$, contradicting the rank feasibility of ideals.

Then the question is, how to characterize ordered geometries among interval greedoids. Before we answer this question we need some terminology.

For every $x \in E$, we let $N(x)$ be the set of all elements in E which never appear after x, i.e.

$$(3.1) \qquad\qquad N(x) = \bigcap \{\sigma(A) : x \in A \in \mathcal{F}\}.$$

For $X \subseteq E$ we set $N(X) = \bigcup \{N(x) : x \in X\}$.

Lemma 3.1. *If $(E, \mathcal{F})$ is an ordered geometry with respect to some order $(E, \leq)$, then for every $X \subseteq E$*

$$(3.2) \qquad\qquad \beta(X) \leq r(N(X)).$$

Proof. Since by axiom (2.1) $I(x) \subseteq N(x)$, we get $I(X) \subseteq N(X)$. This, together with (2.2″), implies $\beta(X) \leq \beta(I(X)) = r(I(X)) \leq r(N(X))$.   □

We will show in the sequel that condition (3.2) of the above lemma is also sufficient to characterize ordered geometries among interval greedoids.

For this purpose we have to construct an appropriate order on the ground set. We first impose a preorder on E by writing

$$x \leq y \quad \text{if} \quad N(x) \subseteq N(y)$$

and an equivalence relation by

$$x_i \equiv x_j \quad \text{if} \quad N(x_i) = N(x_j).$$

Clearly, if $x_1 x_2 \ldots x_k \in \mathcal{L}$, then $x_i \leq x_j$ implies $i \leq j$, and $x_i \equiv x_j$ implies $i = j$.

We denote by $Q(X)$ the ideal generated by X with respect to the preorder $(E, \leq)$. Obviously $Q(X) \subseteq N(X)$. While equality need not always hold, the following converse relationship is true.

Lemma 3.2. *If $A \subseteq N(x)$ for some $A \in \mathcal{F}$ and $x \in E$, then $A \subseteq Q(x)$.*

Proof. We show that $N(a) \subseteq N(x)$ for every $a \in A$. Consider an arbitrary feasible set $U \in \mathcal{F}$ with $x \in U$. Then $A \subseteq N(x)$ implies $A \subseteq \sigma(U)$.

Augment A to a basis A' of $\sigma(U)$. Then $\sigma(A') = \sigma(U)$ and hence $N(a) \subseteq \sigma(U)$. Thus $N(a) \subseteq \bigcap \{\sigma(U) : x \in U \in \mathcal{F}\} = N(x)$.   □

Recall from Chapter V that the feasible kernel $\nu(Q(X))$ is the union of all feasible subsets in $Q(X)$. We now define an order on E by setting $x \leq y$ if one of the following possibilities occurs:

(3.3.1) $x = y$,

(3.3.2) $x \leq y$ and $x \not\equiv y$,

(3.3.3) $x \equiv y$ and $x \in v(Q(x))$, but $y \notin v(Q(y))$.

Clearly, $P = (E, \leq)$ is a partially ordered set and $x \leq y$ implies $x \leq y$. In particular, $(E, \leq, \mathscr{F})$ is a greedoid satisfying (2.1).

We are now ready to prove

Theorem 3.3. *A greedoid $(E, \mathscr{F})$ is an ordered geometry with respect to some order on E if and only if it has the interval property and*

$$\beta(X) \leq r(N(X)) \text{ for every } X \subseteq E .$$

Proof. Lemma 3.1 shows that this condition is necessary. For the converse, we claim that in particular the order given by (3.3) is appropriate. It remains to show that every ideal of $P = (E, \leq)$ is rank feasible. Denote by $I(X)$ the ideal generated by X with respect to the order (3.3).

Since $(E, \mathscr{F})$ is an interval greedoid, we know from Corollary V.5.6 that the rank feasible sets of $\mathscr{F}$ are closed under union. It therefore suffices to show that every ideal of the form $I(x)$, $x \in E$, is rank feasible.

By the definition of the partial order we have

$$I(x) = \begin{cases} Q(x) \setminus \{y : y \equiv x, y \neq x\} & \text{if } x \in v(Q(x)) \\ Q(x) \setminus \{y : y \equiv x, y \neq x, y \notin v(Q(y))\} & \text{if } x \notin v(Q(x)). \end{cases}$$

In the first case, consider any feasible subset of $Q(x)$ containing x and augment this to a basis B of $Q(x)$. Since B contains no two equivalent elements, it is also a basis of $I(x)$. In the second case, every basis of $Q(x)$ is contained in $I(x)$. Thus $r(I(x)) = r(Q(x))$. Next, in view of Lemma 3.2 and since $Q(x) \subseteq N(x)$, we get $r(Q(x)) = r(N(x))$. Hence by hypothesis,

$$r(I(x)) \leq \beta(I(x)) \leq r(N(I(x))) = r(N(x)) = r(I(x)),$$

i.e. the ideals are rank feasible. Thus $(E, \leq, \mathscr{F})$ is an ordered geometry.   □

Example. Let $E = \{a, b, c, x\}$ and $\mathscr{F}$ consist of all one- or two-element subsets of $\{a, b, c\}$ together with ax. Then the construction above will give the partial order P in Figure 11.

As we have seen, there may be more than one order underlying an ordered geometry. A natural question is, then, whether the order given by Theorem 3.3 is minimal, where minimal means that there is no other poset $P' = (E, \leq')$ such that $(E, \mathscr{F})$ is an ordered geometry with respect to P' and $x \leq' y$ implies $x \leq y$.

It can easily be seen, however, that the poset P in Figure 11 is not a minimal order representing the ordered geometry. For example, P' in Figure 11 is also an appropriate poset on E. More generally, let $(E, \leq)$ be any poset such that $(E, \mathscr{F}, \leq)$ is an ordered geometry, and $a < b$ a covering pair in $(E, \leq)$ such that $r(I(b) \backslash a) = r(I(b))$. Then the partial order obtained from $(E, \leq)$ by deleting the comparability between a and b is also an appropriate order on the ground set.

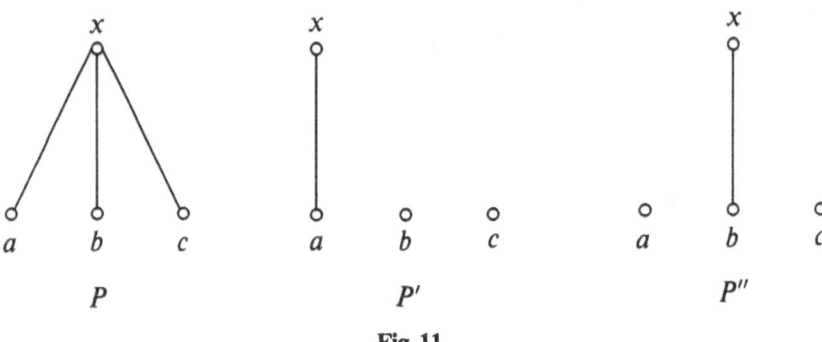

Fig. 11

In our example, P' can be obtained from P by omitting two covering relations according to the rule above. Note however, that even minimal orders need not be unique, since P'' in Figure 11 is also an appropriate order on E.

4. Minimal and Maximal Ordered Geometries

Following Faigle [1980], we consider a third rank function ρ, which we call **ideal rank** and define as $\rho(X) = r(I(X))$.

Theorem 4.1. Let $(E, \mathscr{F}, \leq)$ be an ordered geometry. Then ρ is submodular, monotone and $\rho(\emptyset) = 0$, i.e. a polymatroid rank function.

Proof. We show only submodularity. Since the rank function r is submodular on $\mathscr{R}$ in general and since by (2.2″) every ideal is rank feasible, ρ is submodular on the lattice of ideals.
    Consider $X, Y \subseteq E$. Then

$$\begin{aligned}
\rho(X \cup Y) + \rho(X \cap Y) &= r(I(X \cup Y)) + r(I(X \cap Y)) \\
&\leq r(I(X) \cup I(Y)) + r(I(X) \cap I(Y)) \\
&\leq r(I(X)) + r(I(Y)) \\
&= \rho(X) + \rho(Y).
\end{aligned}$$

$\square$

By definition, the ideal rank of X coincides with the rank if X is an ideal. The same is true if X is feasible.

Lemma 4.2. $\rho(X) = r(X) = |X|$ for every $X \in \mathscr{F}$.

Proof. By (2.1'), $X \subseteq I(X) \subseteq \sigma(X)$ and hence, since ρ is monotone, $\rho(X) = r(I(X)) = r(X)$.

$\square$

This identity, however, is no longer true for $X \in \mathcal{R}$. This can be seen from the greedoid with the basic words ab, ba and bc on the poset P in Figure 9 and the set $\{a, c\} \in \mathcal{R}$, where $r(\{a, c\}) = \beta(\{a, c\}) = 1$ but $r(I(\{a, c\})) = r(\{a, b, c\}) = 2$.

Moreover, the function ρ does not determine the ordered geometry uniquely. Consider the ordered set $x < y$ and the set systems $\mathscr{F}_1 = \{\emptyset, x\}$, $\mathscr{F}_2 = \{\emptyset, x, y\}$. Then $\mathscr{F}_1$ and $\mathscr{F}_2$ define ordered geometries with the same ideal rank.

We will show in the following, however, that for any ordered geometry there exist a unique largest and a unique smallest ordered geometry with the same ideal rank.

For this purpose let $(E, \mathscr{L}, \leq)$ be an ordered geometry. A word $\alpha = x_1 x_2 \ldots x_k$ in $\mathscr{L}$ is called **lexicographically minimal** if

(4.1) $x_1 x_2 \ldots x_{i-1} y \notin \mathscr{L}$ for all $y < x_i$ and $1 \leq i \leq k$.

Let $\mathscr{L}_{\min}$ be the set of lexicographically minimal words in $\mathscr{L}$.

Lemma 4.3. $(E, \mathscr{L}_{\min})$ is a greedoid.

Proof. We only have to show the augmentation property (VI.1.3). Let $\alpha = x_1 \ldots x_k$, $\beta = y_1 \ldots y_m \in \mathscr{L}_{\min}$ be lexicographically minimal words with $k < m$. Since α and β are in $\mathscr{L}$, there exists a smallest index $j \geq 1$ such that $\alpha y_j \in \mathscr{L}$. We show that $\alpha y_j \in \mathscr{L}_{\min}$.

Suppose not. Then there exists a $z < y_j$ with $\alpha z \in \mathscr{L}$. Set $Y = \{y_1, \ldots, y_{j-1}\}$ and $A = I(Y \cup z)$.

Suppose $y_1 \ldots y_{j-1} a \in \mathscr{L}$ for some $a \in A$. Then $a \in I(Y)$ or $a \in I(z)$. The first case is impossible in view of (2.1) and the second contradicts the lexicographic minimality of $y_1 y_2 \ldots y_j$. Hence $A \subseteq \sigma(Y)$. By the choice of j we have $Y \subseteq \sigma(\{x_1, \ldots, x_k\})$. Since $x_1 \ldots x_k z \in \mathscr{L}$, $A \nsubseteq \sigma(\{x_1, \ldots, x_k\})$, contradicting (3.2'). $\qquad \square$

Let $r_{\min}$ and $\beta_{\min}$ be the rank and basis rank functions of $(E, \mathscr{L}_{\min})$ respectively.

Lemma 4.4. *For every ideal I of P we have $r_{\min}(I) = r(I)$.*

Proof. Clearly $r(I) \geq r_{\min}(I)$. To see the converse, let $\alpha = x_1 x_2 \ldots x_k$ be a lexicographically minimal basis of I. It suffices to show that $\alpha \in \mathscr{L}_{\min}$.

Suppose not, then for some $i \in \{1, \ldots, k\}$ there exists a $y < x_i$ such that $\beta = x_1 x_2 \ldots x_{i-1} y \in \mathscr{L}$. Thus $y \in I$, since I is an ideal. Augment β from α to a basis α', which is then lexicographically smaller than α, contradicting the choice of α. $\qquad \square$

Theorem 4.5. $(E, \mathscr{L}_{\min}, \leq)$ *is an ordered geometry with the same ideal rank as* $(E, \mathscr{L}, \leq)$.

Proof. In view of Lemma 4.4, it remains to show that $(E, \mathscr{L}_{\min}, \leq)$ is an ordered geometry. But this is clear since axiom (2.1) is obviously satisfied and (2.2'')

follows immediately from Lemma 4.4, as for any ideal I, $r_{min}(I) = r(I) = \beta(I) \geq \beta_{min}(I) \geq r_{min}(I)$. $\square$

Example. Consider the poset P in Figure 11 and the greedoid $(E, \mathscr{F})$ in the associated example. Then $\mathscr{F}_{min}$ is the 2-truncation of the free matroid on $\{a, b, c\}$. Observe also that the minimal ordered geometry depends on the underlying order, since for the order P' we have $\mathscr{F}_{min} = \mathscr{F}$.

We have defined an ordered geometry $(E, \mathscr{F}_{min}, \leq)$ which is contained in $(E, \mathscr{F}, \leq)$ such that the ideal rank is preserved. In the following we describe a reverse process which imbeds an ordered geometry in a larger one leaving the ideal rank unchanged.

Let $\mathscr{L}_{max}$ be the polymatroid greedoid induced by ρ, i.e.

$$\mathscr{L}_{max} = \{x_1 x_2 \ldots x_k : \rho(\{x_1, \ldots, x_i\}) = i \text{ for all } 1 \leq i \leq k\}.$$

Theorem 4.6. $(E, \mathscr{L}_{max}, \leq)$ *is an ordered geometry with the same ideal rank function as* $(E, \mathscr{L}, \leq)$.

Proof. By Theorem 4.1, ρ is a polymatroid rank function. Theorem VII.1.1 then implies that $(E, \mathscr{L}_{max})$ is an interval greedoid. Let r_{max} and β_{max} denote the rank and the basis rank functions of $(E, \mathscr{L}_{max})$. We show that $(E, \mathscr{L}_{max}, \leq)$ satisfies (2.1'). Let $A \in \mathscr{F}$ and $x \in I(A) \backslash A$. Then $\rho(A \cup x) \leq \rho(I(A)) = \rho(A) = |A| < |A \cup x|$, so $A \cup x \notin \mathscr{F}$. This shows that $x \in \sigma(A)$. To prove (2.2") consider an ideal I. By Lemma VII.1.6, $\beta_{max}(I) \leq \rho(I) = r(I) \leq r_{max}(I)$.

This argument also implies that $r_{max}(I) = r(I)$, i.e. $(E, \mathscr{L}_{max}, \leq)$ has ideal rank function ρ. $\square$

In the remaining part of this chapter we will show that $\mathscr{L}_{min}$ and $\mathscr{L}_{max}$ are respectively the smallest and the largest ordered geometries with the same ideal rank as $\mathscr{L}$.

Lemma 4.7. $\mathscr{L} \subseteq \mathscr{L}_{max}$ *and for every* $x_1 x_2 \ldots x_k \in \mathscr{L}_{max}$ *and* $1 \leq i \leq k$ *there exist elements* $y_i \in I(x_i)$ *such that* $y_1 y_2 \ldots y_k \in \mathscr{L}$.

Proof. Let $x_1 x_2 \ldots x_k \in \mathscr{L}$. By Lemma 4.2 we have $\rho(\{x_1, \ldots, x_i\}) = i$ for every $1 \leq i \leq k$, and hence $x_1 x_2 \ldots x_k \in \mathscr{L}_{max}$.

On the other hand, let $x_1 x_2 \ldots x_k \in \mathscr{L}_{max}$. Assume by induction that $y_j \in I(x_j)$, $1 \leq j \leq i-1$, have been chosen so that $y_1 y_2 \ldots y_{i-1} \in \mathscr{L}$. Then

$$r(I(\{x_1, \ldots, x_{i-1}\})) = \rho(\{x_1, \ldots, x_{i-1}\}) = i - 1 = r(I(\{x_1, \ldots, x_i\})) - 1 ,$$

i.e. $y_1 y_2 \ldots y_{i-1}$ is a basis of $I(\{x_1, \ldots, x_{i-1}\})$ and can be augmented by some $y_i \in I(x_i)$ to a basis of $I(\{x_1, \ldots, x_i\})$. $\square$

Lemma 4.8. $\mathscr{L}_{min} = (\mathscr{L}_{max})_{min}$ *for every ordered geometry* $(E, \mathscr{L}, \leq)$.

Proof. Let $x_1 x_2 \ldots x_k \in \mathscr{L}_{min} \subseteq \mathscr{L}_{max}$ be a lexicographically minimal word and suppose $x_1 x_2 \ldots x_k \notin (\mathscr{L}_{max})_{min}$. Then there exists an index $1 \leq i \leq k$ and an element $y < x_i$ such that $x_1 x_2 \ldots x_{i-1} y \in \mathscr{L}_{\max}$. By Lemma 4.7 there are $z_1 \leq x_1, \ldots, z_{i-1} \leq x_{i-1}$ and $z_i \leq y$ such that $z_1 z_2 \ldots z_i \in \mathscr{L}$, contradicting the minimality of $x_1 x_2 \ldots x_i$.

Conversely, let $x_1 x_2 \ldots x_k \in (\mathscr{L}_{\max})_{\min} \subseteq \mathscr{L}_{\max}$. Again by Lemma 4.7, there exist $y_i \leq x_i$, $1 \leq i \leq k$, such that $y_1 y_2 \ldots y_k \in \mathscr{L} \subseteq \mathscr{L}_{\max}$, hence by the minimality of $x_1 x_2 \ldots x_k$, $y_i = x_i$ for $i = 1, \ldots, k$. Thus $x_1 x_2 \ldots x_k \in \mathscr{L}$, and since it is lexicographically minimal, it is also in $\mathscr{L}_{\min}$. $\square$

We are now ready to prove

Theorem 4.9. *Let* $(E, \mathscr{L}, \leq)$ *be an ordered geometry. Then for every other ordered geometry* $(E, \mathscr{S}, \leq)$ *on the same partial order with the same ideal rank function,*

$$\mathscr{L}_{\max} = \mathscr{S}_{\max} \quad \text{and} \quad \mathscr{L}_{\min} = \mathscr{S}_{\min} ,$$

and therefore $\mathscr{L}_{\min} \subseteq \mathscr{S} \subseteq \mathscr{L}_{\max}$.

Proof. Since the enlargement $\mathscr{L}_{\max}$ depends only on ρ, we get $\mathscr{S}_{\max} = \mathscr{L}_{\max}$. Lemma 4.8 then implies

$$\mathscr{L}_{\min} = (\mathscr{L}_{\max})_{\min} = (\mathscr{S}_{\max})_{\min} = \mathscr{S}_{\min} .$$ $\square$

An ordered geometry $(E, \mathscr{L}, \leq)$ is called **maximal (minimal)** if $\mathscr{L}_{\max} = \mathscr{L}$ $(\mathscr{L}_{\min} = \mathscr{L})$.

Let us remark that for the collection of rank feasible sets $\mathscr{R}_{\min}$ of a minimal ordered geometry and the collection of rank feasible sets $\mathscr{R}_{\max}$ of the corresponding maximal ordered geometry no inclusion relation holds.

Both maximal and minimal ordered geometries are local poset greedoids. More specifically, we show

Lemma 4.10. *Maximal and minimal ordered geometries are polymatroid branching greedoids.*

Proof. By definition, maximal ordered geometries are polymatroid greedoids and hence polymatroid branching greedoids.

Now let $(E, \mathscr{F}_{\min})$ be a minimal ordered geometry. By Theorem 4.1, the ideal rank function ρ is a polymatroid rank function on E. For each $x \in E$, we take a new element x' and consider the extension of ρ parallel to $I(x) \backslash x$ and define

$$E' = E \cup \{x' : x \in E\}.$$

Let us extend ρ to E' by

$$f(X) = \rho((X \cap E) \cup (\bigcup \{I(x) \backslash x : x' \in X\})).$$

Repeated application of Lemma II.4.3 gives that this extension f of ρ to E' is again a polymatroid rank function and it defines a polymatroid branching greedoid $(E, \mathscr{F}')$. Consider a feasible word $x_1 \ldots x_k \in \mathscr{F}_{\min}$. Then for any $1 \leq i \leq k$,

$$r(I(x_1,\ldots,x_i) \setminus x_i) = i - 1 \,,$$

since otherwise $x_1 \ldots x_{i-1}$ can be augmented by some $y \in I(x_1,\ldots,x_i) \setminus x_i$ to a feasible word $x_1 \ldots x_{i-1} y \in \mathscr{F}$. But obviously $y < x_i$, contradicting the lexicographical minimality of $x_1 \ldots x_i$. Hence

$$\begin{aligned}
f(\{x_1,\ldots,x_{i-1},x_i'\}) &= \rho(\{x_1,\ldots,x_{i-1}\} \cup (I(x_i) \setminus x_i)) \\
&= r(I(x_1,\ldots,x_i) \setminus x_i) \\
&= i - 1 \,,
\end{aligned}$$

i.e. $x_1 \ldots x_i \in \mathscr{F}'$ for all $1 \leq i \leq k$.

Conversely, let $x_1 \ldots x_k \in \mathscr{F}'$. Assume by induction that $x_1 \ldots x_{k-1} \in \mathscr{F}_{\min}$. Then $\rho(\{x_1,\ldots,x_k\}) = k$ and for every $y < x_k$ we have

$$\rho(\{x_1,\ldots,x_{k-1},y\}) \leq \rho(\{x_1,\ldots,x_{k-1}\} \cup (I(x_k) \setminus x_k)) \leq k - 1 \,.$$

Hence $x_1 \ldots x_k \in \mathscr{F}_{\min}$ and thus $\mathscr{F}_{\min} = \mathscr{F}'$.   □

Surprisingly enough, although maximal and minimal ordered geometries are polymatroid branching greedoids, such a statement is no longer true for arbitrary ordered geometries. Consider the poset $(E, \leq)$ in Figure 12.

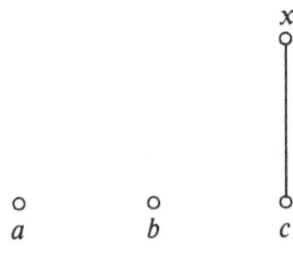

Fig. 12

Let $\mathscr{F} = 2^{\{a,b,c\}} \cup \{\{a,x\},\{b,x\},\{a,b,x\}\}$, where $2^{\{a,b,c\}}$ is the free matroid on $\{a,b,c\}$. Then $(E,\mathscr{F})$ is an ordered geometry but not a local poset greedoid, since

$$\{a,x\},\{b,x\},\{a,b,x\} \in \mathscr{F}$$

but

$$x = \{a,x\} \cap \{b,x\} \notin \mathscr{F} \,.$$

Hence, in particular, $(E,\mathscr{F})$ is not a polymatroid branching greedoid.

We will come back to ordered geometries in the next chapter, where we relate them to intersections of matroids and antimatroids.

Chapter IX. Intersection, Slimming and Trimming

Intersections of matroids and antimatroids have come up earlier in various ways, for example as polymatroid greedoids or ordered geometries. It is natural to try to find such a representation for other classes of greedoids. The intersection of a matroid and an antimatroid is not always a greedoid; but there is a slightly different way of combining a matroid with an antimatroid, which always results in a greedoid. We call this construction the "meet", which will be the main topic of this chapter.

Section 1 contains some technical preparations, in particular some conditions under which the intersection of a matroid and an antimatroid results in a greedoid. We will also strengthen these conditions in several ways in order to preserve the interval and local poset properties.

In Section 2 we introduce the "meet" of a matroid with an antimatroid. Various classes of greedoids have natural representations using this concept. We relate the meet operation to the results in Section 1 by showing that a greedoid representable as the intersection of a matroid with an antimatroid is also representable as a meet, and vice versa. We exhibit representations of polymatroid branching greedoids and ordered geometries using the meet operation.

We specialize our results further in Section 3 where intersections of antimatroids with partition matroids are analysed. We show that a closure condition characterizing partition matroids also extends to their meet with antimatroids and in fact characterizes these meets. We will prove in the next chapter that these so-called balanced interval greedoids are languages induced by certain elimination processes. Important examples of balanced interval greedoids are directed branching greedoids, and we derive two characterizations of these in terms of their circuits and their closure operator.

Brylawski and Dieter [1986] introduced exchange systems as greedoids with the property that the bases of any restriction form the bases of a matroid. Examples of such systems are antimatroids, undirected branching greedoids, and ear decomposition greedoids. Exchange systems are seen to be precisely those greedoids which are closed under some minor operations and some extensions. A subclass of exchange systems was investigated by Goecke [1986] under the name of Gauss greedoids. They are related to Gaussian elimination for matrices and have the property that the truncations induce a chain of matroids which are in strong map relation. These results are presented in the last section.

1. Intersections of Greedoids and Antimatroids

Let $(E, \mathcal{L}_1)$ and $(E, \mathcal{L}_2)$ be two greedoids. We define their **intersection** as the collection of common words, i.e. the language $(E, \mathcal{L}_1 \cap \mathcal{L}_2)$. Note that the underlying sets of this intersection need not be the same as the collection of common feasible sets.

The intersection of two greedoids is not necessarily a greedoid even if one is a matroid and the other an antimatroid. Consider the poset antimatroid of the poset in Figure 13a and the graphical matroid in Figure 13b.

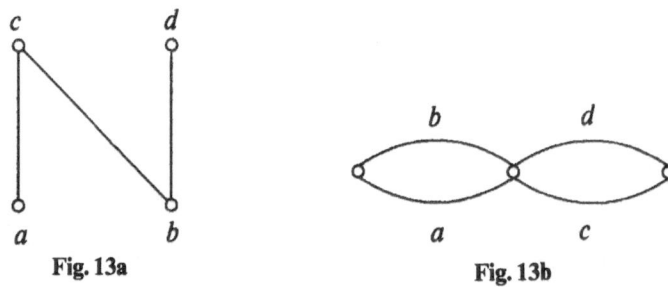

Fig. 13a   Fig. 13b

Then the intersection consists only of $\{\emptyset, a, b, bd\}$, which is not a greedoid.

For a given greedoid $(E, \mathcal{F})$ with closure operator σ, we call a set system $(E, \mathcal{S})$ **quasimodular** (with respect to $(E, \mathcal{F})$) if it is a filter in 2^E and

(1.1) $$\sigma(Y) \in \mathcal{S} \text{ implies } Y \in \mathcal{S}.$$

We say that an antimatroid $(E, \mathcal{A})$ is a **trimming antimatroid** (with respect to $(E, \mathcal{F})$) if for each $e \in E$ the system

(1.2) $$\mathcal{S}_e = \{X \subseteq E : \text{the basis of } X \cup e \text{ contains } e\}$$

is quasimodular. We now obtain

Theorem 1.1. *The intersection of a greedoid $(E, \mathcal{F})$ and a trimming antimatroid $(E, \mathcal{A})$ is a greedoid. Moreover, $\mathcal{L}(\mathcal{A}) \cap \mathcal{L}(\mathcal{F}) = \mathcal{L}(\mathcal{A} \cap \mathcal{F})$.*

Proof. Consider $x_1 \ldots x_k \in \mathcal{L}(\mathcal{F}) \cap \mathcal{L}(\mathcal{A})$ and $y_1 \ldots y_j \in \mathcal{L}(\mathcal{F}) \cap \mathcal{L}(\mathcal{A})$ with $j < k$. Since both words are feasible in $\mathcal{L}(\mathcal{F})$, there exists an index i such that $\{x_1, \ldots, x_{i-1}\} \subseteq \sigma(\{y_1, \ldots, y_j\})$ and $y_1 \ldots y_j x_i \in \mathcal{L}(\mathcal{F})$.

Since $x_1 \ldots x_i \in \mathcal{L}(\mathcal{A})$, we have that $\{x_1, \ldots, x_{i-1}\} \in \mathcal{S}_{x_i}$. By (1.1) we then get $\{y_1, \ldots, y_j\} \in \mathcal{S}_{x_i}$ and hence $y_1 \ldots y_j x_i \in \mathcal{L}(\mathcal{A}) \cap \mathcal{L}(\mathcal{F})$.

Let $X \in \mathcal{F} \cap \mathcal{A}$. Let A be a maximal subset of X that has an ordering $\alpha \in \mathcal{L}(\mathcal{F}) \cap \mathcal{L}(\mathcal{A})$. Let $\alpha y_1 \ldots y_k \in \mathcal{L}(\mathcal{A})$ be an ordering of X. Consider the first index j such that $\alpha y_j \in \mathcal{L}(\mathcal{F})$. Then $\{y_1, \ldots, y_{j-1}\} \subseteq \sigma(A)$. Hence $\sigma(A) \in \mathcal{S}_{y_j}$. By quasimodularity, $A \in \mathcal{S}_{y_j}$. Hence A contains a subset B such that $B \cup y_j \in \mathcal{A}$. But then $A \cup y_j = A \cup (B \cup y_j) \in \mathcal{A}$ and hence $\alpha y_j \in \mathcal{L}(\mathcal{A})$. This contradicts the maximality of A. $\square$

The interval property of a greedoid is preserved under intersections as the following result shows.

Corollary 1.2. *The intersection of an interval greedoid* $(E, \mathscr{F})$ *with a trimming antimatroid* $(E, \mathscr{A})$ *is an interval greedoid.*

Proof. Clearly, if $A, B, C \in \mathscr{F} \cap \mathscr{A}$ are feasible sets in the intersection with $A \subseteq B \subseteq C$ and

$$A \cup x, \ C \cup x \in \mathscr{F} \cap \mathscr{A},$$

then $B \cup x \in \mathscr{F} \cap \mathscr{A}$, since both $\mathscr{F}$ and $\mathscr{A}$ have the interval property.   □

It is easily seen that in contrast to the interval property, the local poset property need not be preserved, since the trimming antimatroid does not have this property in general. Local poset greedoids, however, are closed under intersections with trimming antimatroids if their alternative precedences are also locally closed under intersections.

For this purpose we call a set system $\mathscr{S} \subseteq 2^E$ **modular** if in addition to (1.1) it also satisfies:

(1.3)   Let $A \subseteq E$ and $x, y \in E \setminus A$ with $x \neq y$. Then $A, A \cup x, A \cup y, A \cup \{x, y\} \in \mathscr{F}$ and $A \cup x, A \cup y \in \mathscr{S}$ implies $A \in \mathscr{S}$.

We call a trimming antimatroid **modular** if the set systems $\mathscr{S}_e$ as defined in (1.2) are modular.

Theorem 1.3. *The intersection of a local poset greedoid* $(E, \mathscr{F})$ *with a modular trimming antimatroid* $(E, \mathscr{A})$ *is a local poset greedoid.*

Proof. Since a local poset greedoid is in particular an interval greedoid, the intersection greedoid $(E, \mathscr{F} \cap \mathscr{A})$ also has the interval property by Corollary 1.2.

In view of Corollary VII.3.2 it suffices to verify that the minor (VII.3.1) is excluded, i.e. that

$$A, A \cup \{x, z\}, \ A \cup \{y, z\}, \ A \cup \{x, y, z\} \in \mathscr{F} \cap \mathscr{A} \text{ implies } A \cup z \in \mathscr{F} \cap \mathscr{A}.$$

Since $A \cup z \in \mathscr{F}$, it remains to show that these conditions imply $A \cup z \in \mathscr{A}$. Suppose $A \cup z \notin \mathscr{A}$. Then $A \cup x \in \mathscr{A}$ and hence $A \cup x \in \mathscr{S}_z$. Similarly $A \cup y \in \mathscr{S}_z$.

Moreover, $A \cup z \in \mathscr{F}$ and $A \cup y \in \mathscr{F}$ since $\mathscr{F} \cap \mathscr{A}$ is a greedoid. The local poset property of $\mathscr{F}$ then yields $A \cup \{x, y\} \in \mathscr{F}$. Now the conditions of (1.3) are satisfied and thus $A \in \mathscr{S}_z$, i.e. $A \cup z \in \mathscr{A}$.   □

It may be interesting to point out that if $(E, \mathscr{F})$ is a matroid, then a modular set system is equivalent to a **modular cut** as defined by Crapo [1965], i.e. a system $\mathscr{T}$ of closed sets such that

(1.4) $X \in \mathscr{T}$ and $X \subseteq Y$ implies $Y \in \mathscr{T}$, and

(1.5)   if $X, Y \in \mathscr{T}$ with $r(X) + r(Y) = r(X \cap Y) + r(X \cup Y)$, then $X \cap Y \in \mathscr{T}$.

Lemma 1.4. *Let* $(E, \mathcal{M})$ *be a matroid.*

(i)   *If* $(E, \mathcal{T})$ *is a modular cut, then* $\mathcal{S} = \{X \subseteq E : \sigma(X) \in \mathcal{T}\}$ *is modular.*

(ii)   *If* $(E, \mathcal{S})$ *is a modular set system, then* $\mathcal{T} = \{\sigma(X) : X \in \mathcal{S}\}$ *is a modular cut.*

Proof. (i) is straightforward to check. To see (ii), observe that the condition (1.4) is obviously satisfied. Suppose (1.5) fails to hold. Choose two closed sets $X, Y \in \mathcal{T}$ violating (1.5), where $X \cap Y$ is maximal.

Let $x \in X \setminus Y$, $y \in Y \setminus X$ be two arbitrary elements and set $Z = \sigma(Y \cup x)$. Then

$$r(X) + r(Z) \geq r(X \cup Z) + r(X \cap Z) \geq r(X \cup Y) + r(X \cap Y) + 1$$
$$= r(X) + r(Y) + 1 = r(X) + r(Z) .$$

Since by (1.4), $Z \in \mathcal{T}$ and $|X \cap Z| > |X \cap Y|$, we have by the choice of X and Y that $X \cap Z \in \mathcal{T} \subseteq \mathcal{S}$.

Let A be a basis of $X \cap Y$. Then by the above, $A \cup x$ is a basis of $X \cap Z$. Moreover, by (1.1) $A \cup x \in \mathcal{S}$. Similarly one shows that $A \cup y \in \mathcal{S}$. Finally, since $y \notin \sigma(A \cup x)$, we have $A \cup \{x, y\} \in \mathcal{M}$. Now (1.3) implies that $A \in \mathcal{S}$ and hence $\sigma(A) = X \cap Y \in \mathcal{T}$, contradicting our assumptions. $\qquad \square$

A special instance of an intersection of a greedoid with an antimatroid is the following construction. Let $(E, \mathcal{F})$ be a greedoid and for each $e \in E$ let $\mathcal{T}_e \subseteq \mathcal{C}$ be a family of closure feasible sets. Define $(E, \mathcal{L}(\mathcal{T}))$ by

(1.6) $\mathcal{L}(\mathcal{T}) = \{x_1 \ldots x_k \in \mathcal{L} : \text{ for all } 1 \leq j \leq k \text{ there exists a } T \in \mathcal{T}_{x_j} \text{ with } T \subseteq \sigma(\{x_1, \ldots, x_{j-1}\})\}$.

We call the hereditary language $(E, \mathcal{L}(\mathcal{T}))$ the **strong trimming** of $(E, \mathcal{L})$ (with respect to $\mathcal{T}$). In particular, $(E, \mathcal{L}(\mathcal{T}))$ is a greedoid, as is implied by the following lemma.

Lemma 1.5. *Every strong trimming of a greedoid can be obtained as the intersection with a trimming antimatroid.*

Proof. For each $e \in E$ let

$$\mathcal{H}_e = \{X \subseteq E : \text{ there exists a } T \in \mathcal{T}_e \text{ with } T \subseteq \sigma(X)\} .$$

To verify condition (1.1), assume $X \in \mathcal{H}_e$ and $X \subseteq \sigma(Y)$. Then there exists a set $T \in \mathcal{T}_e$ with $T \subseteq \sigma(X)$. Since T is closure feasible, we have $T \subseteq \sigma(Y)$, i.e. $Y \in \mathcal{H}_e$ and (1.1) holds.

Let $(E, \mathcal{A})$ be the trimming antimatroid induced by the alternative precedences $\mathcal{H}_e$. Then one readily verifies that $\mathcal{F}(\mathcal{L}(\mathcal{T})) = \mathcal{F} \cap \mathcal{A}$. $\qquad \square$

For interval greedoids, however, both concepts turn out to be equivalent.

Lemma 1.6. *The intersection of an interval greedoid $(E, \mathcal{F})$ with a trimming anti-matroid $(E, \mathcal{A})$ is a strong trimming.*

Proof. For each $e \in E$ define

$$\mathcal{T}_e = \mathcal{H}_e \cap \mathcal{F} .$$

By Corollary V.5.7, $\mathcal{T}_e \subseteq \mathcal{C}$. Again it is easy to verify that $\mathcal{F}(\mathcal{L}(\mathcal{T})) = \mathcal{F} \cap \mathcal{A}$.

$\square$

For strong trimmings there is a simple condition which ensures that the local poset property is preserved.

Lemma 1.7. *If $(E, \mathcal{L})$ is a local poset greedoid and $|\mathcal{T}_e| = 1$ for all $e \in E$, then $(E, \mathcal{L}(\mathcal{T}))$ is also a local poset greedoid.*

Proof. Assume $\mathcal{T}_e = \{T_e\}$ and consider the system

$$\mathcal{H}_e = \{X \subseteq E : T_e \subseteq \sigma(X)\}$$

as in Lemma 1.5. We verify (1.3).

Assume $A, A \cup x, A \cup y, A \cup \{x, y\} \in \mathcal{F}$ and $A \cup x, A \cup y \in \mathcal{H}_e$. Suppose $A \in \mathcal{H}_e$, i.e. $T_e \subseteq \sigma(A)$. Then there exists an element $z \in T_e$ such that $A \cup z \in \mathcal{F}$. If we augment $A \cup z$ from $A \cup \{x, y\}$, we obtain a set $A \cup z \cup x \in \mathcal{F}$, say. Hence $z \notin \sigma(A \cup x)$, contradicting the assumption.

$\square$

Many classes which can be interpreted as intersections of greedoids and trimming antimatroids already turn out to be intersections of matroids with antimatroids. We therefore concentrate in the next section on this special case.

2. The Meet of a Matroid and an Antimatroid

Let $(E, \mathcal{M})$ be a matroid and $(E, \mathcal{A})$ an antimatroid on the same ground set. For each word $\alpha = x_1 x_2 \ldots x_m$ of the antimatroid define a subword $\pi(\alpha)$ by the following projection process: define indices $1 \le i_1 < \ldots < i_k \le m$ so that i_j is the first index such that

(2.1) $$r_{\mathcal{M}}(x_1, \ldots, x_{i_j}) = j \text{ for all } 1 \le j \le k .$$

Then set $\pi(\alpha) = x_{i_1} x_{i_2} \ldots x_{i_k}$. The set of all subwords derived in this way is called the **meet** of $\mathcal{M}$ and $\mathcal{A}$ and is denoted by $\mathcal{L}(\mathcal{M} \wedge \mathcal{A})$.

Lemma 2.1. *The meet of a matroid and an antimatroid is a greedoid.*

Proof. It is clear that the meet is a hereditary language. Consider $\gamma, \delta \in \mathcal{L}(\mathcal{M} \wedge \mathcal{A})$ with $|\gamma| < |\delta|$. Let $\gamma = \pi(\alpha)$ and $\delta = \pi(\beta)$, where $\alpha, \beta \in \mathcal{L}(\mathcal{A})$. Let x be the first element of β not in $\sigma_{\mathcal{M}}(\alpha)$ and write $\beta = \beta' x \beta''$. Note that $x \in \delta$ since

$r_{\mathcal{M}}(\beta'x) > r_{\mathcal{M}}(\beta')$. Then $\tilde{\alpha} \cup (\tilde{\beta}' \cup x) \in \mathcal{A}$ and also $\tilde{\alpha} \cup \tilde{\beta}' \in \mathcal{A}$. Hence $\tilde{\alpha} \cup \tilde{\beta}' \cup x$ has a feasible ordering $\alpha\beta_1 x$, where $\tilde{\beta}_1 \subseteq \tilde{\beta}'$. Then $\pi(\alpha\beta_1 x) = \gamma x$. So $\gamma x \in \mathcal{L}(\mathcal{M} \wedge \mathcal{A})$. $\square$

We will denote by $\mathcal{M} \wedge \mathcal{A}$ the system of feasible sets in $(E, \mathcal{L}(\mathcal{M} \wedge \mathcal{A}))$. The projection operation can be described in the unordered version, where as before $\mathscr{S}_x = \{X \subseteq E : \text{the } A\text{-basis of } X \cup x \text{ contains } x\}$.

Lemma 2.2. *Let $(E, \mathcal{M})$ be a matroid, $(E, \mathcal{A})$ an antimatroid, $X \in \mathcal{M} \wedge \mathcal{A}$ and $x \in E \setminus X$. Then $X \cup x \in \mathcal{M} \wedge \mathcal{A}$ if and only if $x \notin \sigma_{\mathcal{M}}(X)$ and $\sigma_{\mathcal{M}}(X) \in \mathscr{S}_x$.*

Proof. Assume $X \cup x \in \mathcal{M} \wedge \mathcal{A}$. Let $\gamma x \in \mathcal{L}(\mathcal{M} \wedge \mathcal{A})$ be an ordering of $X \cup x$. Let $\alpha \in \mathcal{L}(\mathcal{A})$ be a shortest word such that $\pi(\alpha) = \gamma x$. Then $\alpha = \alpha' x$. Now $\tilde{\alpha}' \subseteq \sigma_{\mathcal{M}}(X)$ which proves that $\sigma_{\mathcal{M}}(X) \in \mathscr{S}_x$.

Conversely, assume that there is a set $A \in \mathcal{A}$ such that $x \in A$ and $A \setminus x \in \sigma_{\mathcal{M}}(X)$. Let $\beta \in \mathcal{L}(\mathcal{A})$ be such that $\pi(\tilde{\beta}) = X$. Let $\beta\alpha_1 x\alpha_2 \in \mathcal{L}(\mathcal{A})$ be an ordering of $\tilde{\beta} \cup A$. Then the underlying set of $\pi(\beta\alpha_1 x)$ is $X \cup x$. $\square$

This lemma implies that feasibility in $\mathcal{M} \wedge \mathcal{A}$ can be tested in polynomial time provided that feasibility oracles for the matroid and antimatroid are given. To test if $B \in \mathcal{M} \wedge \mathcal{A}$, we build up a feasible subset $X \subseteq B$. Assume we have such an X and $X \neq B$. Let A be the (unique) $\mathcal{A}$–basis of $\sigma_{\mathcal{M}}(X)$. We check if there exists an $x \in B \setminus X$ such that $x \notin \sigma_{\mathcal{M}}(X)$ and $A \cup x \in \mathcal{A}$. If we find such an x, we add it to X. If not, $B \notin \mathcal{M} \wedge \mathcal{A}$.

As a consequence of Lemma 2.2 we obtain

Lemma 2.3. *Let $(E, \mathcal{M})$ be a matroid, $(E, \mathcal{A})$ an antimatroid and $X \in \mathcal{M} \wedge \mathcal{A}$. Then*

$$(2.2) \qquad\qquad \sigma_{\mathcal{M} \wedge \mathcal{A}}(X) = \sigma_{\mathcal{A}}(\sigma_{\mathcal{M}}(X)) .$$

Proof. By Lemma 2.2, $x \in \sigma_{\mathcal{M} \wedge \mathcal{A}}(X)$ implies that either $x \in \sigma_{\mathcal{M}}(X)$ or $x \in \sigma_{\mathcal{A}}(A)$ for all $A \in \mathcal{A}$ with $A \subseteq \sigma_{\mathcal{M}}(X)$. Hence $x \in \sigma_{\mathcal{A}}(\sigma_{\mathcal{M}}(X))$.

Conversely, if $x \notin \sigma_{\mathcal{M} \wedge \mathcal{A}}(X)$, then by Lemma 2.2, $x \notin \sigma_{\mathcal{A}}(\sigma_{\mathcal{M}}(X))$. $\square$

Recall that matroids are the interval greedoids without lower bounds and that antimatroids are the interval greedoids without upper bounds. The next result shows that the meet operation preserves the interval property.

Lemma 2.4. *$(E, \mathcal{M} \wedge \mathcal{A})$ is an interval greedoid.*

Proof. Lemma 2.3 implies that $\sigma_{\mathcal{M} \wedge \mathcal{A}}$ satisfies the conditions of V.5.2. $\square$

Lemma 2.5.

(i) $r_{\mathcal{M} \wedge \mathcal{A}}(X) \leq \beta_{\mathcal{M} \wedge \mathcal{A}}(X) \leq r_{\mathcal{M}}(X)$ *for all* $X \subseteq E$,

(ii) $r_{\mathcal{M} \wedge \mathcal{A}}(X) = \beta_{\mathcal{M} \wedge \mathcal{A}}(X) = r_{\mathcal{M}}(X)$ *for all* $X \in \mathcal{A}$.

Proof. (i) Let A be a basis of $(E, \mathcal{M} \wedge \mathcal{A})$ such that $\beta_{\mathcal{M} \wedge \mathcal{A}}(X) = |X \cap A|$. Since by construction $A \in \mathcal{M}$, we have

$$\beta_{\mathcal{M} \wedge \mathcal{A}}(X) \leq r_{\mathcal{M}}(X) \ .$$

(ii) If $\alpha \in \mathcal{A}$ is a feasible ordering of X with respect to the antimatroid $(E, \mathcal{A})$, then by definition $|\pi(\alpha)| = r_{\mathcal{M}}(X)$. Hence $r_{\mathcal{M} \wedge \mathcal{A}}(X) \geq r_{\mathcal{M}}(X)$. Since the reverse inequality obviously holds, we get the desired result. $\qquad \square$

It follows that $\mathcal{A} \subseteq \mathcal{R}_{\mathcal{M} \wedge \mathcal{A}}$. We also obtain a strengthening of V.3.3 for the meet of matroids and antimatroids.

Corollary 2.6. *Let* $X, Y \in \mathcal{A}$. *Then*

$$\beta_{\mathcal{M} \wedge \mathcal{A}}(X \cap Y) + \beta_{\mathcal{M} \wedge \mathcal{A}}(X \cup Y) \leq r_{\mathcal{M} \wedge \mathcal{A}}(X) + r_{\mathcal{M} \wedge \mathcal{A}}(Y) \ .$$

Proof. In view of Lemma 2.5,

$$\begin{aligned}
\beta_{\mathcal{M} \wedge \mathcal{A}}(X \cap Y) + \beta_{\mathcal{M} \wedge \mathcal{A}}(X \cup Y) &\leq r_{\mathcal{M}}(X \cap Y) + r_{\mathcal{M}}(X \cup Y) \\
&\leq r_{\mathcal{M}}(X) + r_{\mathcal{M}}(Y) \\
&= r_{\mathcal{M} \wedge \mathcal{A}}(X) + r_{\mathcal{M} \wedge \mathcal{A}}(Y) \ .
\end{aligned}$$
$\qquad \square$

We now describe $\mathcal{M} \wedge \mathcal{A}$ in terms of the closure of $(E, \mathcal{M})$ and the circuits of $(E, \mathcal{A})$.

Lemma 2.7. *Let* $\mathcal{K}$ *be the collection of circuits of the antimatroid* $(E, \mathcal{A})$. *Then*

$$(2.3) \qquad \mathcal{L}(\mathcal{M} \wedge \mathcal{A}) = \{x_1 \ldots x_k : \{x_1, \ldots, x_k\} \in \mathcal{M} \text{ and for } 1 \leq i \leq k$$
$$(C, x_i) \in \mathcal{K} \text{ implies } C \cap \sigma_{\mathcal{M}}(\{x_1, \ldots, x_{i-1}\}) \neq \emptyset\} \ .$$

Proof. We denote by $\mathcal{L}'$ the right hand side of the above equation and first show $\mathcal{L}(\mathcal{M} \wedge \mathcal{A}) \subseteq \mathcal{L}'$.

Let $x_1 \ldots x_k \in \mathcal{L}$ be a feasible word in the meet and assume by induction that $x_1 \ldots x_{k-1} \in \mathcal{L}'$. Then there exists a word $\beta \in \mathcal{L}(\mathcal{A})$ with $\pi(\beta) = x_1 \ldots x_{k-1}$ and $\beta x_k \in \mathcal{L}(\mathcal{A})$. Consider a circuit $(C, x_k) \in \mathcal{K}$. By Lemma III.3.6, $C \setminus x_k$ must intersect $\tilde{\beta} \subseteq \sigma_{\mathcal{M}}(\{x_1, \ldots, x_{k-1}\})$. Hence $x_1 \ldots x_k \in \mathcal{L}'$.

Conversely, let $x_1 \ldots x_k \in \mathcal{L}'$ and assume inductively that $x_1 \ldots x_{k-1} \in \mathcal{L}$ has already been shown. We claim $x_1 \ldots x_k \in \mathcal{L}(\mathcal{M} \wedge \mathcal{A})$. By Lemma 2.2, it suffices to show that $x_k \notin \sigma_{\mathcal{M}}(\{x_1, \ldots, x_{k-1}\})$ and $\sigma_{\mathcal{M}}(\{x_1, \ldots, x_{k-1}\}) \in \mathcal{S}_{x_k}$. The first is trivial since $\{x_1, \ldots, x_k\} \in \mathcal{M}$. To show the second, note that $\sigma_{\mathcal{M}}(\{x_1, \ldots, x_{k-1}\})$ intersects every circuit with root x_k and hence by Theorem III.3.14 it contains $P \setminus x$ for some x–path P in $(E, \mathcal{A})$. $\qquad \square$

The following lemma shows that meets of matroids with antimatroids can be viewed as generalizations of minimal ordered geometries where the underlying poset is replaced by a general antimatroid.

Lemma 2.8. *Let* $(E, \mathcal{M})$ *be a matroid and* $(E, \mathcal{A})$ *an antimatroid. Then*

(i) $\alpha \in \mathcal{L}(\mathcal{A} : \tilde{\alpha})$ *for every* $\alpha \in \mathcal{L}(\mathcal{M} \wedge \mathcal{A})$,
(ii)   *if* $X, Y \in \mathcal{A}, X \subseteq Y, x \in X$ *and* $r(X \backslash x) = r(X)$, *then* $r(Y \backslash x) = r(Y)$,
(iii)   *for every* $X \in \mathcal{M} \wedge \mathcal{A}$, $\Gamma_{\mathcal{M} \wedge \mathcal{A}}(X)$ *is free in* $\mathcal{A}$.

Proof. (i) is trivial. To see (ii), let B be a basis in $\mathcal{M} \wedge \mathcal{A}$ of X avoiding x and $\beta \in \mathcal{L}(\mathcal{M} \wedge \mathcal{A})$ an ordering of B, and let $\delta \in \mathcal{L}(\mathcal{A})$ be such that $\pi(\delta) = \beta$. Extend δ to a feasible ordering $\delta\epsilon$ of $Y \cup \tilde{\delta}$. Now $\pi(\delta\epsilon) = \beta\psi$, where ψ consists of elements of Y. Hence this is a basic word of Y. Moreover, $x \in \sigma_{\mathcal{M}}(B)$ and therefore x does not occur in ψ. This shows that $r_{\mathcal{M} \wedge \mathcal{A}}(Y \backslash x) = r_{\mathcal{M} \wedge \mathcal{A}}(Y)$.

(iii) Assume that $\Gamma_{\mathcal{M} \wedge \mathcal{A}}(X)$ is not free. Then it contains a circuit (C, y). Let $\alpha \in \mathcal{L}(\mathcal{M} \wedge \mathcal{A})$ be an ordering of X. Since $\alpha y \in \mathcal{L}(\mathcal{M} \wedge \mathcal{A})$, there exists a word $\beta y \in \mathcal{L}(\mathcal{A})$ such that $\pi(\beta y) = \alpha y$. Since obviously $\sigma_{\mathcal{M}}(X) \cap \Gamma_{\mathcal{M} \wedge \mathcal{A}}(X) = \emptyset$, it follows that β cannot contain any element of $\Gamma_{\mathcal{M} \wedge \mathcal{A}}(X)$. So $(\tilde{\beta} \cup y) \cap \Gamma_{\mathcal{M} \wedge \mathcal{A}}(X) = \{y\}$. This contradicts Lemma III.3.6.   □

Corollary 2.9. *A greedoid is the underlying greedoid of a minimal ordered geometry if and only if it is the meet of a matroid and a poset antimatroid.*

Proof. The "if"-part follows from the previous lemma. Conversely, let $(E, \mathcal{F})$ be a minimal ordered geometry. Then the 1–truncation

$$r(X) = \min_{Y \subseteq X}\{\rho(Y) + |X \setminus Y|\}$$

is a matroid rank function. Letting $(E, \mathcal{M})$ be the matroid induced by r and $(E, \mathcal{A})$ the poset antimatroid on the poset underlying $(E, \mathcal{F})$, it is straightforward to verify that $\mathcal{F} = \mathcal{M} \wedge \mathcal{A}$.   □

In the remaining part of this section we give two different characterizations of the meet of matroids with antimatroids. We have seen that the intersection of a matroid and an antimatroid need not always yield a greedoid. One of our goals is to show that if the intersection is a greedoid, then it is also the meet of a matroid and an antimatroid.

Our first observation in this direction is that in some cases it is enough to consider the antimatroid of partial alphabets.

Lemma 2.10. *Let* $(E, \mathcal{F})$ *be a greedoid with* $\mathcal{F} = \mathcal{M} \cap \mathcal{G}$ *for some matroid* $(E, \mathcal{M})$ *and some antimatroid* $(E, \mathcal{G})$. *Let* $\mathcal{A}$ *be the collection of partial alphabets of* $(E, \mathcal{F})$. *Then* $\mathcal{F} = \mathcal{M} \cap \mathcal{A}$.

Proof. Clearly, $\mathcal{F} \subseteq \mathcal{M} \cap \mathcal{A}$. Conversely, since $\mathcal{F} \subseteq \mathcal{G}$ and $\mathcal{G}$ is closed under union, we have that $\mathcal{A} \subseteq \mathcal{G}$ and so $\mathcal{M} \cap \mathcal{A} \subseteq \mathcal{M} \cap \mathcal{G} = \mathcal{F}$.   □

Lemma 2.11. *Let* $(E, \mathcal{M})$ *be a matroid,* $(E, \mathcal{F})$ *a greedoid, and* $\mathcal{A}$ *the collection of partial alphabets of* $(E, \mathcal{F})$. *If* $\mathcal{F} = \mathcal{M} \cap \mathcal{A}$, *then* $\mathcal{F} = \mathcal{M} \wedge \mathcal{A}$.

Proof. Clearly, $\mathcal{M} \cap \mathcal{A} \subseteq \mathcal{M} \wedge \mathcal{A}$ by the definition of the projection operation. Let $X \in \mathcal{M} \wedge \mathcal{A}$ be a feasible set in the meet. Then there exists an element $x \in X$

so that $X \setminus x \in \mathcal{M} \wedge \mathcal{A}$. We may assume by induction that $X \setminus x \in \mathcal{M} \cap \mathcal{A}$. By Lemma 2.2, there exists a set $A \in \mathcal{F}$ such that $x \in A$ and $A \setminus x \subseteq \sigma_{\mathcal{M}}(X \setminus x)$. We may again assume that $A \setminus x \in \mathcal{F}$ by choosing A as small as possible.

Augment $A \setminus x$ to a basis B of $\sigma_{\mathcal{M}}(X \setminus x)$ with respect to $\mathcal{F} = \mathcal{M} \cap \mathcal{A}$. Then, in particular, $B \in \mathcal{M}$. Since $x \notin \sigma_{\mathcal{M}}(X \setminus x) = \sigma_{\mathcal{M}}(B)$, we have $B \cup x \in \mathcal{M}$. Moreover, $B \cup x = B \cup A \in \mathcal{A}$, since $\mathcal{A}$ is closed under union. Hence $B \cup x \in \mathcal{M} \cap \mathcal{A} = \mathcal{F}$.

We now augment $X \setminus x$ from $B \cup x$ in $\mathcal{M} \wedge \mathcal{A}$. Since B is a basis of $\sigma_{\mathcal{M}}(X \setminus x)$, we get $X \setminus x \cup x = X \in \mathcal{F}$. □

Before we show that the intersection and the meet of matroids with antimatroids are equivalent, recall from Section 1 that for interval greedoids the notion of trimming and strong trimming are the same. Moreover, the trimming of an interval greedoid is again an interval greedoid.

Theorem 2.12. *For a greedoid $(E, \mathcal{F})$ the following statements are equivalent:*

(i) $\mathcal{F} = \mathcal{M} \cap \mathcal{A}$ *for some matroid $(E, \mathcal{M})$ and some antimatroid $(E, \mathcal{A})$.*
(ii) $\mathcal{F} = \mathcal{M} \wedge \mathcal{A}$ *for some matroid $(E, \mathcal{M})$ and some antimatroid $(E, \mathcal{A})$.*
(iii) $(E, \mathcal{F})$ *is a trimmed matroid.*

Proof. (i) $\Rightarrow$ (ii): By Lemma 2.10 we may assume that $\mathcal{A}$ is the collection of partial alphabets. The claim then follows from Lemma 2.11.

(ii) $\Rightarrow$ (iii): For each $x \in E$ let $T_x = \mathcal{P}_x$ be the set of paths in $\mathcal{A}$ with endpoint x. By Theorem III.3.14, $\mathcal{P}_x$ is the blocker of the circuits rooted at x. Hence

$$\mathcal{L}(\mathcal{T}) = \{x_1 \ldots x_k \in \mathcal{L}(\mathcal{M}) : \text{for all } 1 \leq i \leq k \text{ there exists a path}$$
$$(P, x) \text{ with } P \subseteq \sigma_{\mathcal{M}}(\{x_1, \ldots, x_{i-1}\})\}$$
$$= \{x_1 \ldots x_k \in \mathcal{M} : \text{for all } 1 \leq i \leq k, \ (C, x_i) \in \mathcal{K}_{x_i}$$
$$\text{implies } C \cap \sigma_{\mathcal{M}}(\{x_1, \ldots, x_{i-1}\}) \neq \emptyset\}$$
$$= \mathcal{M} \wedge \mathcal{A},$$

by Lemma 2.7.

(iii) $\Rightarrow$ (i): Assume that $\mathcal{L}(\mathcal{F}) = \mathcal{L}(\mathcal{T})$. Let $\mathcal{A}$ be the collection of partial alphabets of $\mathcal{F}$. Clearly $\mathcal{F} \subseteq \mathcal{M} \cap \mathcal{A}$. To see the converse, we consider a set $X \in \mathcal{M} \cap \mathcal{A}$ and argue as in the proof of Lemma 2.11 that there exists a set $A \in \mathcal{F}$ with $A \subseteq X$, $x \in A$ and $A \setminus x \in \mathcal{F}$. By the definition of $\mathcal{L}(\mathcal{T})$, there exists a set $T \in \mathcal{T}_x$ such that $T \subseteq \sigma_{\mathcal{M}}(A \setminus x) \subseteq \sigma_{\mathcal{M}}(X \setminus x)$. Hence $X \in \mathcal{F}$. □

We close this section by relating some subclasses of greedoids to intersections of matroids with antimatroids. We first deal with a class obtainable as intersections of matroids with poset antimatroids. Note that this is not the same question as in Lemma 2.9 since in the equivalence conditions in Theorem 2.12 the antimatroid may change.

A greedoid $(E, \mathcal{F})$ is called **intersection-closed** if $A, B \in \mathcal{F}$ implies $A \cap B \in \mathcal{F}$.

Theorem 2.13. *For an interval greedoid $(E, \mathcal{F})$ the following statements are equivalent:*

(i) $(E, \mathscr{F})$ is intersection-closed.
(ii)   For every $x \in E$ there exists a unique x-path in $\mathscr{F}$ (unless x is a loop).
(iii) $\mathscr{F}$ is the intersection of a matroid and a poset antimatroid.
(iv) $(E, \mathscr{F})$ is the underlying greedoid of a distributive supermatroid.

Proof. (iv) $\Rightarrow$ (i): Since the intersection of feasible ideals in a supermatroid is a feasible ideal, $\mathscr{F}$ is closed under intersection.

(i) $\Rightarrow$ (ii): Suppose there are two x-paths $A, B \in \mathscr{F}$. The $C = A \cap B \in \mathscr{F}$ and contains an x-path, in contradiction to A, B being x-paths.

(ii) $\Rightarrow$ (iii): Define a partial order on E by letting $x \leq y$ if the x-path is contained in the y-path. Then every feasible set is an ideal in the poset $(E, \leq)$. In fact, $A \in \mathscr{F}, y \in A$ and $x \leq y$ implies $x \in A$. Conversely, the principal ideals $I(x)$ are the paths P_x. For if $y \in I(x)$, then $y \in P_x$. Conversely if $y \in P_x$, then in particular $P_y \subseteq P_x$ and hence $y \in I(x)$.

Since every principal ideal is feasible, every ideal is a partial alphabet and thus rank feasible by the interval property. Hence if an ideal is feasible, then all of its subideals are feasible. It also follows that the rank function is submodular on ideals, and hence $\rho(X) = r(I(X))$ is a submodular function.

Let $(E, \mathscr{M})$ be the matroid induced by the 1-truncation of ρ and let $(E, \mathscr{I})$ be the poset antimatroid of $(E, \leq)$. We claim that $\mathscr{F} = \mathscr{M} \cap \mathscr{I}$. For if $A \in \mathscr{F}$, then A is an ideal, i.e. $A \in \mathscr{I}$. Also $A \in \mathscr{M}$ since for each $X \subseteq A$ we have

$$\rho(X) = r(I(X)) \geq |A \cap I(X)| \geq |X| \ .$$

Conversely, let $A \in \mathscr{M} \cap \mathscr{I}$. Then $A \in \mathscr{M}$ implies $\rho(A) \geq |A|$ and $A \in \mathscr{I}$ yields $\rho(A) = r(I(A)) = r(A)$, i.e. $A \in \mathscr{F}$.

(iii) $\Rightarrow$ (iv): $\mathscr{F} = \mathscr{M} \cap \mathscr{I}$ is a collection of ideals closed under taking subideals. Since the augmentation property holds, $(E, \mathscr{F})$ is distributive supermatroid. $\square$

Observe that we have not used the interval property to show that in intersection-closed greedoids paths are unique. The converse, however, only holds for interval greedoids as the example in Figure 14 shows.

Note, moreover, that not every intersection-closed greedoid has the interval property. A counterexample is given by the greedoid $(E, \mathscr{F})$ with $E = \{x, y, z\}$ and $\mathscr{F} = 2^E \setminus \{\{x, y\}\}$. Moreover, not every intersection of a matroid with a poset antimatroid is again a greedoid.

We have just seen that the intersection-closed greedoids with the interval property are precisely the distributive supermatroids. We will come back to the more general class of ordered geometries in a moment after relating polymatroid branchings to intersections of matroids with antimatroids.

Theorem 2.14. *Every polymatroid branching greedoid is a trimmed matroid.*

Proof. Let $(E, \mathscr{F})$ be a polymatroid branching greedoid, and let $(E', \mathscr{F}')$ be the polymatroid greedoid induced by the submodular function f as in the definition. Denote by $(E, \mathscr{M})$ the matroid induced by f restricted to E. By Lemma II.4.4, this matroid is given by

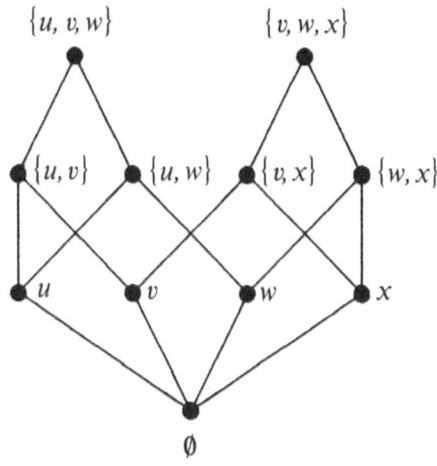

$\{u, v, w\}$ $\{v, w, x\}$

$\{u, v\}$ $\{u, w\}$ $\{v, x\}$ $\{w, x\}$

u v w x

$\emptyset$

Fig. 14

$$\mathcal{M} = \{X \subseteq E : |Y| \leq f(Y) \text{ for all } Y \subseteq X\} \, .$$

Let $\mathcal{L}'$ consist of all words $x_1 \ldots x_k$ over E such that

$$f(x_1, \ldots, x_i) \leq f(x_1, \ldots, x_{i-1}) + 1 \text{ and } f(x_1, \ldots, x_{i-1}, x_i') = f(x_1, \ldots, x_{i-1})$$

for all $1 \leq i \leq k$. Then $(E, \mathcal{L}')$ is an antimatroid by Lemma II.4.1. Let $\mathcal{A} = \mathcal{F}(\mathcal{L}')$. By Theorem 2.12 it suffices to show that $\mathcal{F} = \mathcal{M} \cap \mathcal{A}$. It is trivial that $\mathcal{F} \subseteq \mathcal{M} \cap \mathcal{A}$. Consider a set $X \in \mathcal{M} \cap \mathcal{A}$. Let $x_1 \ldots x_k \in \mathcal{L}'$ be an ordering of X. Then we have $f(x_1, \ldots, x_i) \leq i$ by the definition of $\mathcal{L}'$, but also $f(x_1, \ldots, x_i) \geq i$ by the description of $\mathcal{M}$. Hence $f(x_1, \ldots, x_i) = i$ and so $X \in \mathcal{F}$. ☐

In particular, minimal and maximal ordered geometries are intersections of matroids with antimatroids, since they are a subclass of polymatroid branching greedoids (cf. Lemma VIII.4.10). Although this argument does not apply to arbitrary ordered geometries, they also can be regarded as intersections of matroids with antimatroids.

Theorem 2.15. *Every ordered geometry $(E, \mathcal{F})$ is a trimmed matroid.*

Proof. We know from Theorem VIII.4.1 that the ideal rank function ρ induces a matroid $(E, \mathcal{M})$. In view of Theorem 2.12, it suffices to show that

$$\mathcal{F} = \mathcal{M} \cap \mathcal{A} \, ,$$

where $(E, \mathcal{A})$ is the antimatroid of partial alphabets in $\mathcal{F}$.

Clearly $\mathcal{F} \subseteq \mathcal{M} \cap \mathcal{A}$. To see the converse, consider $X \cup y \in \mathcal{A} \cap \mathcal{M}$, where we assume by induction that $X \in \mathcal{F}$ also holds for $X \in \mathcal{A} \cap \mathcal{M}$.

Since $X \cup y \in \mathcal{A}$, there exists a set $Y \in \mathcal{F}$ with

$$X \cup y = X \cup Y \, .$$

Augmenting Y from X, we obtain a set $X \setminus x \cup y \in \mathcal{F}$. Since $r(I(X) \cup I(y)) \geq r_{\mathcal{M}}(X \cup y) = |X| + 1$, there exists an element $z \leq y$ such that $X \cup z \in \mathcal{F}$.

If we augment $X \setminus x \cup y$ from $X \cup z$, we cannot add z by axiom (VIII.2.1) and hence $X \cup y \in \mathcal{F}$.   □

We close this section with construction principles which are of particular interest with respect to optimization problems. We will see in Chapter XI that the worst-out greedy algorithm is optimal for every linear objective function if and only if the bases of the greedoid are the bases of a matroid. We call such a greedoid a **slimmed matroid** and mention constructions for obtaining a slimmed matroid as the intersection of matroids with antimatroids.

Suppose that for a given matroid $(E, \mathcal{M})$ all bases are feasible in the anti-matroid $(E, \mathcal{A})$. Then clearly the meet $\mathcal{F} = \mathcal{M} \wedge \mathcal{A}$ is a slimmed matroid. For example, meet of a graphic matroid and the line-search antimatroid of the corresponding graph with respect to some root gives an undirected branching greedoid.

This construction in fact gives all slimmed matroids that are interval greedoids by taking $\mathcal{A} = \mathcal{R}$. Not every slimmed matroid, however, arises as the intersection of a matroid with an antimatroid. For example, in Section 4 we will investigate exchange systems, which are slimmed matroids but which do not necessarily have the interval property.

We can also obtain such an antimatroid directly from the matroid. Consider a matroid $(E, \mathcal{M})$ with rank function $r_{\mathcal{M}}$ and assume $r_{\mathcal{M}}(E) = k$. Let $A_1 \subseteq A_2 \subseteq \ldots \subseteq A_{k-1} \subseteq A_k = E$ be a sequence of sets with $r_{\mathcal{M}}(A_i) \geq i$.

Lemma 2.16. *The system* $(E, \mathcal{A})$ *with*

$$\mathcal{A} = \{X \subseteq E : |X \cap A_i| \geq i \text{ for } 1 \leq i \leq |X|\}$$

is an antimatroid.

Proof. We first prove that $\mathcal{A}$ is accessible. Let $X \in \mathcal{A}$. If $X \nsubseteq A_{k-1}$, then deleting any element of $X \setminus A_{k-1}$ preserves the conditions. So suppose that $X \subseteq A_{k-1}$ and let j be the smallest index such that $X \subseteq A_j$. Consider $x \in X \cap (A_j \setminus A_{j-1})$. Then clearly $|(X \setminus x) \cap A_i| \geq i$ for all $1 \leq i \leq |X| - 1$, i.e. $(E, \mathcal{A})$ is accessible.

We now verify that $(E, \mathcal{A})$ is closed under union. For this purpose assume that $X, X \cup x, X \cup y \in \mathcal{A}$, i.e. $X \subseteq A_{|X|}$, $X \cup x \subseteq A_{|X|+1}$ and $X \cup y \subseteq A_{|X|+1}$. Then clearly,

$$X \cup \{x, y\} \subseteq A_{|X|+1} \subseteq A_{|X|+2}$$

and

$$\left| (X \cup \{x, y\}) \cap A_{|X|+2} \right| \geq |X| + 2 \ .$$

Lemma III.1.2 now implies that $(E, \mathcal{A})$ is an antimatroid.   □

Lemma 2.17. $\mathcal{F} = \mathcal{M} \wedge \mathcal{A}$ *is a slimmed matroid.*

Proof. It remains to show that every basis $B \in \mathcal{M}$ is also a basis of $\mathcal{F}$. Consider a basis B of $\mathcal{M}$. Then

$$|B \cap A_i| = k - |B \cap (E \setminus A_i)|$$
$$\geq k - r_{\mathcal{M}}(E \setminus A_i)$$
$$\geq i .$$

Hence $B \in \mathcal{A}$ and $\pi(B) = B$. □

3. Balanced Interval Greedoids

In this section we concentrate on meets of antimatroids with partition matroids.

We call a greedoid $(E, \mathcal{F})$ **balanced** if there exists a partition $E_1, \ldots, E_k$ of the ground set E such that

$$|B \cap E_i| = 1$$

for all bases $B \in \mathcal{F}$ and all $1 \leq i \leq k$. Balanced matroids are usually called **partition matroids**. An important class of balanced greedoids is the class of directed branching greedoids for which $E_1, \ldots, E_k$ is the partition of the edges into in-stars. Another class is the class of simplicial clique elimination greedoids (cf. Section IV.2, Example 8). Let $x_1 \ldots x_k$ be a basic word and let Q_i be the clique deleted together with x_i. We show that for every other basic word $y_1 \ldots y_k$, each Q_i contains exactly one y_j. Since $y_1 \ldots y_k$ are independent vertices, no Q_i contains more than one of them. So it suffices to show that each y_j is contained in one of the Q_i's. Suppose not, then augment $y_1 \ldots yj - 1$ from $x_1 \ldots x_k$ to a basic word $y_1 \ldots y_{j-1} z_j \ldots z_k$. Then y_j is not adjacent to any vertex in this basic word, so it is a vertex of the graph remaining after the elimination of $\{y_1, \ldots, y_{j-1}, z_j, \ldots, z_k\}$ and their neighbors. On the other hand, y_j is simplicial after the elimination of $y_1, \ldots, y_{j-1}$ and their neighbors, so it stays simplicial. But then $y_1 \ldots y_{j-1} z_j \ldots z_k y_j$ is feasible, a contradiction.

Clearly for any partition matroid $(E, \mathcal{M})$ and any antimatroid $(E, \mathcal{A})$ the resulting interval greedoid $\mathcal{M} \wedge \mathcal{A}$ is balanced. Our first lemma shows that the converse is also true.

Lemma 3.1. *Let $(E, \mathcal{F})$ be a balanced interval greedoid. Then $\mathcal{F} = \mathcal{M} \wedge \mathcal{A}$ for some antimatroid $(E, \mathcal{A})$ and a partition matroid $(E, \mathcal{M})$.*

Proof. Let $E_1, \ldots, E_k$ be a partition of E such that $|B \cap E_i| = 1$ for all bases B and all $1 \leq i \leq k$. Let $(E, \mathcal{M})$ be the corresponding partition matroid and $(E, \mathcal{A})$ the antimatroid of the partial alphabets of $\mathcal{F}$.

Consider a set $X \in \mathcal{M} \cap \mathcal{A}$. Since in particular $X \in \mathcal{A}$, we can write $X = X_1 \cup X_2$, where $X_1 \in \mathcal{A}$, $X_2 \in \mathcal{F}$ and $|X_1| < |X|$. By induction on the cardinality of X, we assume $X_1 \in \mathcal{F}$. Augment X_1 in $\mathcal{F}$ to a basis B_1. Since $(E, \mathcal{F})$ is balanced, the augmentation of X_2 from B_1 in $\mathcal{F}$ yields a basis $B_2 = X_2 \cup (B_1 \setminus \sigma_{\mathcal{M}}(X_2))$. Moreover, since $X_1 \cup X_2 \in \mathcal{M}$, we have $X_1 \subseteq B_1 \setminus \sigma_{\mathcal{M}}(X_2)$ and so both X_1 and X_2 are subsets of B_2. The local union property of interval greedoids then implies $X_1 \cup X_2 \in \mathcal{F}$.

Hence $\mathcal{M} \cap \mathcal{A} \subseteq \mathcal{F}$, and since $\mathcal{F} \subseteq \mathcal{M} \cap \mathcal{A}$ holds trivially, we get $\mathcal{M} \cap \mathcal{A} = \mathcal{F}$. The assertion now follows from Lemma 2.11. □

Our next results indicate that balanced interval greedoids have many similarities with their defining matroids and antimatroids. Recall that the monotonicity of the matroidal closure operator $\sigma_{\mathcal{M}}$ implies

$$\sigma_{\mathcal{M}}(X) \cup \sigma_{\mathcal{M}}(Y) \subseteq \sigma_{\mathcal{M}}(X \cup Y)$$

for all $X, Y \subseteq E$. The partition matroids are precisely those matroids for which also the converse relation

$$\sigma_{\mathcal{M}}(X \cup Y) \subseteq \sigma_{\mathcal{M}}(X) \cup \sigma_{\mathcal{M}}(Y)$$

holds. We will show in a moment that this inclusion carries over to balanced interval greedoids.

Moreover, feasible sets of balanced interval greedoids are almost closed under union, where "almost" is made precise in the next result.

Lemma 3.2. Let $(E, \mathcal{M})$ be a partition matroid, $(E, \mathcal{A})$ an antimatroid and $\mathcal{F} = \mathcal{M} \wedge \mathcal{A}$. Then $A, B \in \mathcal{F}$ implies $A \cup (B \setminus \sigma_{\mathcal{M}}(A)) \in \mathcal{F}$.

Proof. We may assume that $\mathcal{A}$ is the collection of partial alphabets of $(E, \mathcal{F})$. Let $\alpha \in \mathcal{L}(\mathcal{F})$ be a feasible ordering of A. Extend α to an ordering $\alpha\gamma \in \mathcal{L}(\mathcal{A})$ of $A \cup B$. Then $\pi(\alpha\gamma) = \alpha\gamma'$, where $\tilde{\gamma}' = B \setminus \sigma_{\mathcal{M}}(A)$. □

We are now ready to prove the above-mentioned characterization of balanced interval greedoids in terms of their closure operators.

Theorem 3.3. Let $(E, \mathcal{F})$ be an interval greedoid. Then $(E, \mathcal{F})$ is balanced if and only if $\sigma(A \cup B) \subseteq \sigma(A) \cup \sigma(B)$ for all $A, B \in \mathcal{F}$.

Proof. We may assume that the greedoid is normal. The necessity of the condition follows from Lemma 3.2.

Conversely, for each $x \in E$ consider a largest set $A \in \mathcal{F}$ with $x \in \Gamma(A)$. First we show that for each $y \in \Gamma(A)$, A is also a maximal feasible set such that $y \in \Gamma(A)$. Suppose not. Then there exists a $B \in \mathcal{F}$, $|B| > |A|$ such that $y \in \Gamma(B)$. Then $y \notin \sigma(A) \cup \sigma(B)$, and hence by hypothesis, $y \notin \sigma(A \cup B)$. Let C be a basis of $A \cup B$ containing A, then $y \in \Gamma(C)$. Augment $A \cup x$ to a basis of $C \cup x \cup y$. Clearly this basis is not $C \cup x$ by the maximality of A, hence it contains y. But then $A \cup y \cup x \in \mathcal{F}$ by the interval property, a contradiction.

Next we show that if A and B are two largest feasible sets such that $x \in \Gamma(A) \cap \Gamma(B)$, then $\Gamma(A) = \Gamma(B)$. In fact, $x \notin \sigma(A) \cup \sigma(B)$, so $x \notin \sigma(A \cup B)$. Let C be a basis of $A \cup B$. Then $x \in \Gamma(C)$, so by the maximality of A and B, $|A| = |B| = |C|$ and thus A and B are cospanning. Hence $\Gamma(A) = \Gamma(B)$.

This implies that the sets $\Gamma(A)$, where A is a maximal feasible set such that $x \in \Gamma(A)$ for some $x \in E$, partition E. We show that for every basis B, $|B \cap \Gamma(A)| = 1$ for each of these A's. It is trivial from the augmentation property that $|B \cap \Gamma(A)| \geq 1$. Suppose that $|B \cap \Gamma(A)| > 1$. Then B has a feasible subset C such that $|C \cap \Gamma(A)| \geq 1$ and $|\Gamma(C) \cap \Gamma(A)| \geq 1$. Let $x \in C \cap \Gamma(A)$ and

$y \in \Gamma(A) \cap \Gamma(C)$. Then $y \notin \sigma(A) \cup \sigma(C)$, and thus by hypothesis $y \notin \sigma(A \cup C)$. So let D be a basis of $A \cup C$. Then $|D| \geq |A \cup x| > |A|$ and $y \in \Gamma(D)$, which contradicts the maximality of A.   □

Like antimatroids, balanced interval greedoids may be characterized in terms of critical circuits. Let (C, x) be a rooted set with $(C \backslash x) \cap \sigma_{\mathscr{M}}(x) = \emptyset$. (C, x) is a **critical circuit**, if for some basis B of $E \backslash (C \cup \sigma_{\mathscr{M}}(x))$, $B \cup x \notin \mathscr{F}$ but $B \cup c \in \mathscr{F}$ and $B \cup c \cup x \in \mathscr{F}$ for all $c \in C \backslash x$. Observe that a critical circuit of a balanced interval greedoid does not depend on the particular choice of the basis B.

Lemma 3.4 *Let $(E, \mathscr{F})$ be a balanced interval greedoid with partition $\{E_1, \ldots, E_k\}$. Then $x_1 \ldots x_n \in \mathscr{L}$ if and only if for each critical circuit (C, x_i), $C \cap \{x_1, \ldots, x_{i-1}\} \neq \emptyset$.*

Proof. Let $x_1 \ldots x_n \in \mathscr{L}$ and (C, x_i) be a critical circuit. Let $B \in \mathscr{F}$ be a basis of $E \backslash (C \cup \sigma_{\mathscr{M}}(x_i))$. Consider the smallest index $j \leq i$ such that $x_j \notin \sigma_{\mathscr{M}}(B)$. Then $\{x_1, \ldots, x_j\} \in \mathscr{F}, \{x_1, \ldots, x_{j-1}\} \in \sigma_{\mathscr{M}}(B)$, and hence $B \cup x_j \in \mathscr{F}$ by Lemma 3.2. Since $B \cup x_i \notin \mathscr{F}$, we have $j < i$ and thus $C \cap \{x_1, \ldots, x_{i-1}\} \neq \emptyset$.

Conversely, let $x_1 \ldots x_n$ be a feasible word in the language determined by the critical circuits. Assume by induction that $x_1, \ldots, x_{n-1} \in \mathscr{L}$. Suppose $x_1 \ldots x_n \notin \mathscr{L}$. Augment $x_1 \ldots x_{n-1}$ to a largest feasible word $\alpha \in E \backslash \sigma_{\mathscr{M}}(x_n)$ such that $\alpha x_n \notin \mathscr{L}$. Let $C = \{y \in E \backslash \sigma_{\mathscr{M}}(x_n) : \alpha y \in \mathscr{L}\} \cup \{x_k\}$. Then (C, x_n) is a critical circuit and $(C \backslash x_n) \cap \{x_1, \ldots, x_{n-1}\} = \emptyset$, a contradiction.   □

In Chapter X we will investigate the relationship between decomposition processes and greedoids. It will turn out that we can characterize balanced interval greedoids as those arising from certain elimination procedures.

We have seen that directed branching greedoids are an important class of balanced interval greedoids. We close this section with a characterization of directed branching greedoids given by Schmidt [1985a, 1988] and extended by Goecke [1986].

Theorem 3.5. *Let $(E, \mathscr{F})$ be a balanced interval greedoid with partition $\{E_1, \ldots, E_k\}$. Let $(E, \mathscr{M})$ be the corresponding partition matroid. Then $(E, \mathscr{F})$ is a directed branching greedoid if and only every $x \in E$ is the root of at most one critical circuit and every critical circuit (C, x) has $r_{\mathscr{M}}(C \backslash x) = 1$.*

Proof. For a directed branching greedoid, a critical circuit with root $x = (u, v)$ consists of all arcs (w, u) such that w is covered by a maximal branching in $V \backslash \{u, v\}$. Hence every arc $x = (u, v)$ is the root of at most one critical circuit (C, x) and all non-root arcs enter u. Thus $r_{\mathscr{M}}(C \backslash x) = 1$.

For the converse, we construct a directed graph D as follows. For each block E_i define a node v_i and let $V(D) = \{v_1, \ldots, v_k\} \cup \{r\}$. Identify each element $e \in E_i$, $1 \leq i \leq k$, with an arc a_e by

$$a_e = \begin{cases} (v_j, v_i) & \text{if } (C, e) \text{ is a critical circuit for some } C \text{ with } C \backslash e \subseteq E_j, \\ (r, v_i) & \text{otherwise.} \end{cases}$$

We claim that $(E, \mathcal{L})$ is the directed branching greedoid $(E, \mathcal{L}')$ on D.

By construction, the blocks E_i correspond to the in-stars of non-root nodes. Consider $x_1 \ldots x_m \in \mathcal{L}$. We assume by induction that $x_1 \ldots x_{m-1} \in \mathcal{L}'$. Then the head v_m of x_m is different from the heads of the arcs x_i, $1 \leq i \leq m-1$, since otherwise $x_1 \ldots x_m$ has two elements in the block E_m. The initial vertex v of x_m is either the root r or some other vertex v_j. In the latter case there exists a critical circuit (C, x_m) with $C \backslash x_m \subseteq E_j$. Since by Lemma 3.4 $C \cap \{x_1, \ldots, x_{m-1}\}) \neq \emptyset$, there exists an arc x_i with endpoint v_j. Hence $x_1 \ldots x_m$ is a rooted arborescence. So $\mathcal{L} \subseteq \mathcal{L}'$.

Conversely, let $x_1 \ldots x_m \in \mathcal{L}'$ and assume by induction that $x_1 \ldots x_{m-1} \in \mathcal{L}$. Let $x_m = (u, v)$. Then v is not covered by the arcs $x_1, \ldots, x_{m-1}$, and hence $\{x_1, \ldots, x_m\} \in \mathcal{M}$. If $\{x_m\} \in \mathcal{F}$, then $x_1 \ldots x_m \in \mathcal{L}$ by Lemma 3.2. If not, let (C, x_m) be the unique critical circuit with root x_m. By the construction of D, every arc in $C \backslash x_m$ enters u. Let $U \in \mathcal{F}$, $x_m \in U$. By Lemma 3.2, $U' = x_1 \ldots x_{m-1} \cup (U \backslash \sigma_{\mathcal{M}}(\{x_1 \ldots x_{m-1}\})) \in \mathcal{F}$. Since $x_m \in U'$, we have $U' \cap (C \backslash x_m) \neq \emptyset$ by Lemma 3.4. But every arc in U' enters a different node, and some x_j $(1 \leq j \leq m-1)$ enters u, so $x_j \in C \backslash x_m$. Hence by Lemma 3.4, $x_1 \ldots x_m \in \mathcal{L}$. □

Corollary 3.6. *A local poset greedoid $(E, \mathcal{F})$ is a directed branching greedoid if and only if for all $A, B \in \mathcal{F}$ the following holds:*

$$(3.1) \qquad\qquad \sigma(A) \cap \sigma(B) \subseteq \sigma(A \cup B) \subseteq \sigma(A) \cup \sigma(B) .$$

Proof. Clearly every directed branching greedoid is a local forest greedoid. So the necessity follows from Lemma VII.4.5 and Theorem 3.3.

Conversely, let $(E, \mathcal{F})$ be a local poset greedoid satisfying (3.1). Since every local poset greedoid has the interval property (Lemma VII.2.1), $(E, \mathcal{F})$ is a balanced interval greedoid by Theorem 3.3. Let $E_1, E_2, \ldots, E_k$ be the blocks of the underlying partition.

Suppose the non-root part $C \backslash x$ of a critical circuit (C, x) contains some elements x_i, x_j of two different blocks E_i and E_j, say.

Let B be a basis of $E \backslash (C \cup \sigma_{\mathcal{M}}(x))$. Then $B \cup x_i, B \cup x_j, B \cup \{x_i, x\}, B \cup \{x_j, x\} \in \mathcal{F}$. Lemma 3.2 now implies that $B \cup \{x_i, x_j, x\} \in \mathcal{F}$. The local intersection property then yields $B \cup x \in \mathcal{F}$, contradicting the circuit properties. So $C \backslash x \subseteq E_i$.

Next, consider two critical circuits (C_1, x) and (C_2, x) with the same root. Suppose $C_1 \backslash x \subseteq E_1$ and $C_2 \backslash x \subseteq E_2$. For $i = 1, 2$, let B_i be the bases of $E \backslash (C_i \cup \sigma_{\mathcal{M}}(x))$. If $\sigma_{\mathcal{M}}(B_1) \neq \sigma_{\mathcal{M}}(B_2)$, then Lemma 3.2 implies $B_1 \cup (B_2 \backslash \sigma_{\mathcal{M}}(B_1)) \cup x \in \mathcal{F}$. Thus $x \notin \sigma(B_1 \cup B_2)$ but $x \in \sigma(B_1) \cap \sigma(B_2)$, contradicting (3.1).

Hence $\sigma_{\mathcal{M}}(B_1) = \sigma_{\mathcal{M}}(B_2)$ and, again by Lemma 3.2, we conclude that $C_1 = C_2$.

Hence every $x \in E$ is the root of at most one critical circuit and every critical circuit (C, x) has $r_{\mathcal{M}}((C \backslash x) = 1$. So the result follows from Theorem 3.5. □

4. Exchange Systems and Gauss Greedoids

In order to characterize classes of greedoids which are closed under deletion, Brylawski and Dieter [1986] introduced exchange systems as special greedoids.

A greedoid $(E, \mathscr{F})$ is an **exchange system** if for any subset $E' \subseteq E$ the restriction $(E', \mathscr{F}_{E'})$ is a slimmed matroid.

Lemma 4.1. *A greedoid $(E, \mathscr{F})$ is an exchange system if and only if it satisfies:*

(4.1)   *For all $X, Y \in \mathscr{F}$ with $|X| = |Y|$ and all $x \in X \setminus Y$ there exists an element $y \in Y \setminus X$ such that either $X \cup y \in \mathscr{F}$ or $X \setminus x \cup y \in \mathscr{F}$.*

Proof. Let $(E, \mathscr{F})$ be a greedoid with property (4.1). Consider any subset $E' \subseteq E$. Then all maximal feasible subsets of E' have the same cardinality $r(E')$ and (4.1) assures the basis exchange property for matroids. Hence $(E, \mathscr{F})$ is an exchange system.

Conversely, let $(E, \mathscr{F})$ be an exchange system. Then, by definition, $(E, \mathscr{F})$ is a greedoid. Consider two feasible sets $X, Y \in \mathscr{F}$ with $|X| = |Y|$ and set $E' = X \cup Y$. If $|X| < r(E')$, we can augment X by some $y \in Y \setminus X$. Otherwise, X is a basis of E' and the basis exchange property of $(E', \mathscr{F}_{E'})$ guarantees property (4.1). $\square$

Antimatroids are exchange systems since property (4.1) holds trivially. As another example, consider an undirected branching greedoid $(E, \mathscr{F})$. The maximal feasible subsets of the restriction to some set $E' \subseteq E$ are the spanning trees of the connected component of E' containing the root. Hence every restriction is a slimming of a graphic matroid.

Let $(E, \mathscr{F})$ be an ear decomposition greedoid with specified edge e_0. As mentioned in Chapter IV, for any subset $A \subseteq E$ of edges the bases of A are of the form $B \cup T$, where B is the block of $A \cup e_0$ and T is a spanning tree of the connected component of $A \cup e_0$ containing e_0 with B shrunken to a single vertex. Since these sets are the bases of a matroid, we again have an exchange system.

A second characterization will be given in terms of operations on greedoids. For a given greedoid $(E, \mathscr{F})$ and an element $e \in E$ define two set systems:

$$\mathscr{F} * e = \{X \subseteq E : X \in \mathscr{F} \text{ or } X \cup e \in \mathscr{F}\},$$

$$\mathscr{F} \cup e = \{X \subseteq E : X \in \mathscr{F}\} \cup \{X \cup e : X \in \mathscr{F}\}.$$

Like the trace operation, these constructions will not preserve the greedoid properties. For exchange systems, however, the situation is different.

Lemma 4.2. *For a greedoid $(E, \mathscr{F})$, the following statements are equivalent:*

(i)   *$(E, \mathscr{F})$ is an exchange system.*
(ii)   *$\mathscr{F} \cup e$ is a greedoid for all $e \in E$.*
(iii) *$\mathscr{F} * e$ is a greedoid for all $e \in E$.*
(iv)   *$\mathscr{F} : (E \setminus e)$ is a greedoid for all $e \in E$.*

Proof. (i) $\Rightarrow$ (ii): Consider two feasible sets $X, Y \in \mathscr{F} \cup e$ with $|Y| < |X|$. A case analysis shows that we can either augment Y by an element $x \in X \setminus Y$ or by e unless $X = X' \cup e$ with $X' \in \mathscr{F}$, $e \in Y$ and $|X'| = |Y|$. But in this case, (4.1) implies that either $Y \cup x \in \mathscr{F}$ or $Y \setminus e \cup x \in \mathscr{F}$ for some $x \in X \setminus Y$. So $Y \cup x \in \mathscr{F} \cup e$.

(ii) $\Rightarrow$ (iii): Observe that $X \in \mathscr{F} * e$ implies $X \cup e \in \mathscr{F} \cup e$, and $X \in \mathscr{F} \cup e$ implies $X \setminus e \in \mathscr{F} * e$. Now consider two sets $X, Y \in \mathscr{F} * e$ with $|Y| < |X|$. If $|Y \cup e| < |X \cup e|$, then we augment $Y \cup e$ from $X \cup e$ in $\mathscr{F} \cup e$ and obtain an augmented set in $\mathscr{F} * e$. Otherwise, $e \in X \setminus Y$. Then either $X \in \mathscr{F}$ and we can augment Y from X in $\mathscr{F}$, or $X \cup e \in \mathscr{F}$ and hence $Y \cup e \in \mathscr{F}$.

(iii) $\Rightarrow$ (iv): Clear, since $\mathscr{F} : (E \setminus e)$ is the restriction of $\mathscr{F} * e$ to $E \setminus e$.

(iv) $\Rightarrow$ (i): Consider two sets $X, Y \in \mathscr{F}$ with $|X| = |Y|$. For any $x \in X \setminus Y$ we have $X \setminus x, Y \in \mathscr{F} : (E \setminus x)$. Since $|X \setminus x| < |Y|$, there exists an element $y \in Y \setminus X$ such that $X \setminus x \cup y \in \mathscr{F} : (E \setminus x)$. Thus either $X \setminus x \cup y \in \mathscr{F}$ or $X \cup y \in \mathscr{F}$.   □

In Chapter III we remarked that unions of antimatroids are antimatroids. This property carries over to the more general class of exchange systems.

Lemma 4.3. *Exchange systems are closed under unions.*

Proof. Consider two exchange systems $(E_1, \mathscr{F})$ and $(E_2, \mathscr{G})$ and some subset $E \subseteq E_1 \cup E_2$. It is straightforward to verify that the maximal feasible subsets in $(E, (\mathscr{F} \cup \mathscr{G})_E)$ are precisely the unions of maximal feasible subsets in $(E, \mathscr{F}_E)$ and $(E, \mathscr{G}_E)$. By assumption, these maximal feasible sets are the bases of two matroids $(E, \mathscr{M}_1)$ and $(E, \mathscr{M}_2)$. Now matroids are closed under unions (cf. Welsh [1976]). Since the union $(E_1 \cup E_2, \mathscr{F} \cup \mathscr{G})$ is obviously accessible, the restriction to any $E \subseteq E_1 \cup E_2$ is a slimmed matroid.   □

Further constructions like series and parallel connection, the two-sum and erections may be generalized from matroid theory to greedoids. Some of them turn out to be well behaved with respect to exchange systems. For more details, see Brylawski and Dieter [1986].

A particular subclass of exchange systems was investigated by Goecke [1986]. Recall that a matroid $(E, \mathscr{M}_1)$ is a **strong map** of another matroid $(E, \mathscr{M}_2)$ if every closed set in $\mathscr{M}_1$ is also closed in $\mathscr{M}_2$. Equivalently, $\mathscr{M}_1$ is a strong map of $\mathscr{M}_2$ if

$$r_1(X) + r_2(Y) \geq r_1(X \cup Y) + r_2(X \cap Y) \quad \text{for all } X, Y \subseteq E ,$$

where r_i is the rank function of the matroid $\mathscr{M}_i$.

Let $(E, \mathscr{M}_i)$, $i = 0, \ldots, m$, be a family of matroids on a common ground set so that $r_i(E) = i$ for $0 \leq i \leq m$ and $\mathscr{M}_i$ is a strong map of $\mathscr{M}_{i+1}$ for $1 \leq i \leq m - 1$. We call the hereditary language $(E, \mathscr{L})$ with

$$\mathscr{L} = \{x_1 \ldots x_k : \{x_1, \ldots, x_i\} \in \mathscr{M}_i \text{ for } 0 \leq i \leq k\} ,$$

a **Gauss greedoid** .

Lemma 4.4. *Every Gauss greedoid is an exchange system.*

Proof. We first show that it is a greedoid. Let $x_1 \ldots x_k \in \mathscr{L}$ and $y_1 \ldots y_{k+1} \in \mathscr{L}$. Then we can augment $\{x_1, \ldots, x_k\}$ from $\{y_1, \ldots, y_{k+1}\}$ by some y_j in $\mathscr{M}_{k+1}$. Hence $x_1 \ldots x_k y_j \in \mathscr{L}$.

We show that $(E, \mathcal{L})$ is a slimming of $(E, \mathcal{M}_m)$. Let B be a basis of $\mathcal{M}_m$. The rank of B in $\mathcal{M}_{m-1}$ is $m-1$ and hence there exists an element $x_m \in B$ such that $B \backslash x_m \in \mathcal{M}_{m-1}$. By induction, $B \backslash x_m$ has an ordering $x_1 \ldots x_{m-1} \in \mathcal{L}$. Then $x_1 \ldots x_m \in \mathcal{L}$. So $(E, \mathcal{L})$ is a slimmed matroid. Since this follows similarly for its restrictions, it is an exchange system.   □

It can also be seen from the proof that if $(E, \mathcal{F})$ is the Gauss greedoid defined by a sequence of matroids $(E, \mathcal{M}_i)$ and $\mathcal{B}_i$ is the set of bases of $(E, \mathcal{M}_i)$, then $\mathcal{F} = \mathcal{B}_0 \cup \mathcal{B}_1 \cup \ldots \cup \mathcal{B}_m$.

The name Gauss greedoid is motivated by the following connection with Example IV.2.3. Let A be an (m, n)–matrix with full row rank and let E be the set of column indices. If we perform Gaussian elimination on A, we obtain an assignment of row indices i to column indices j_i such that for every $k \leq m$, the submatrix $(a_{ij_r})_{1 \leq i \leq k, 1 \leq r \leq k}$ is nonsingular.

Denoting by $(E, \mathcal{M}_k)$ the linear matroid represented by the first k rows, $1 \leq k \leq m$, one readily verifies that $(1, j_1), (2, j_2), \ldots, (k, j_k)$ are a pivot sequence of a Gaussian elimination procedure if and only if $\{j_1, j_2, \ldots, j_k\}$ is a basis of $\mathcal{M}_k$. Hence the Gaussian elimination greedoid is a Gauss greedoid.

Let (S, T, E) be a graph and $s_1, \ldots, s_m$ an ordering of the nodes. Let $(E, \mathcal{F})$ be the bipartite matching greedoid, i.e. $\mathcal{F}$ consists of all subsets $X \subseteq T$ such that the graph induced by X and the first $|X|$ nodes of S has a perfect matching. A closer look at Edmonds' [1967] proof that every transversal matroid is linear reveals this example as a special case of Gaussian elimination for matrices. Hence the bipartite matching greedoid is a Gauss greedoid.

We give a further example of Gauss greedoids. Let $G = (V, E)$ be a directed graph and $W \subseteq V$ a subset of nodes. Given a fixed ordering $w_1, \ldots, w_m$ of the elements of w, the linking greedoid $(E, \mathcal{F})$ consists of all subsets $X \subseteq V$ which can be linked by node disjoint paths to the first $|X|$ elements of W. The set systems $\mathcal{M}_k = \{X \subseteq V : X \text{ can be linked to } \{w_1, \ldots, w_k\}\}$ is a matroid (cf. Welsh [1976]). The fact that they are in strong map relation follows, for example, from Menger's Theorem. Hence the linking greedoids are Gauss greedoids.

We point out here that Gauss greedoids need not always have the interval property. Consider the bipartite matching greedoid on the graph in Figure 15. Here $x, y, \{y, z\}, \{x, y, z\} \in \mathcal{F}$ but $\{x, y\} \notin \mathcal{F}$.

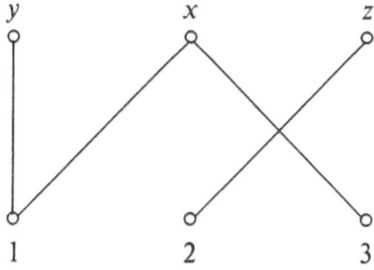

Fig. 15

After these examples we proceed to characterize Gauss greedoids. For this we need an equivalent formulation of strong maps.

Lemma 4.5. *Let $(E, \mathcal{M}_1)$ and $(E, \mathcal{M}_2)$ be two matroids with $\mathcal{M}_1 \subseteq \mathcal{M}_2$ and $r_1(E) = r_2(E) - 1$. Then $\mathcal{M}_1$ is a strong map of $\mathcal{M}_2$ if and only if the following holds:*

(4.2) *If B is a basis of $\mathcal{M}_1, z \in B, x, y, \notin B$, and $B \setminus z \cup \{x, y\}$ is a basis of $\mathcal{M}_2$, then $B \cup y$ is a basis of $\mathcal{M}_2$ or $B \setminus z \cup y$ is a basis of $\mathcal{M}_1$.*

Proof. Let $\mathcal{M}_1$ be a strong map of $\mathcal{M}_2$, then

$$r_1(B \setminus z \cup y) + r_2(B \cup y) \geq r_1(B \cup y) + r_2(B \setminus z \cup y) = 2|B| .$$

Hence if $B \cup y$ is not a basis of $\mathcal{M}_2$, then $B \setminus z \cup y$ is a basis of $\mathcal{M}_1$.

Conversely, assume that (4.2) holds. We have to show that every set closed in $\mathcal{M}_1$ is also closed in $\mathcal{M}_2$. Since in a matroid every closed set is the intersection of hyperplanes, it suffices to show that hyperplanes of $\mathcal{M}_1$ are closed in $\mathcal{M}_2$.

Suppose the hyperplane H of $\mathcal{M}_1$ is not closed in $\mathcal{M}_2$. Then there exists an element $z \in E \setminus H$ with $r_2(H \cup z) = r_2(H)$ and a basis B of $\mathcal{M}_1$ with $z \in B \subseteq H \cup z$. Hence

$$r_1(E) = r_1(H \cup z) \leq r_2(H \cup z) = r_2(H) \leq r_1(H) + 1 = r_1(E) ,$$

i.e. $r_2(H) = r_1(E) = r_2(H \cup z)$.

Since $B \setminus z \in \mathcal{M}_1 \subseteq \mathcal{M}_2$, we have $r_2(B \setminus z) = r_1(E) - 1$. Thus there exists an element $y \in H \setminus (B \setminus z) = H \setminus B$ with $B \setminus z \cup y \in \mathcal{M}_2$, and an element $x \neq z$ such that $B \setminus z \cup \{x, y\}$ is a basis of $\mathcal{M}_2$.

Now $r_2(B \cup y) \leq r_2(H \cup z) = r_1(E) < r_2(E)$ and $r_1(B \setminus z \cup y) \leq r_1(H) < r_1(E)$ yield a contradiction to (4.2). Consequently $\mathcal{M}_1$ is a strong map of $\mathcal{M}_2$. □

With the help of this preparatory lemma we can now formulate

Theorem 4.6. *A greedoid $(E, \mathcal{F})$ is a Gauss greedoid if and only if for any $B \in \mathcal{F}, z \in B$ and $x, y \notin B$*

(4.3) $B \setminus z \cup \{x, y\} \in \mathcal{F}$ *implies* $B \cup y \in \mathcal{F}$ *or* $B \setminus z \cup y \in \mathcal{F}$.

Proof. The necessity is immediate from Lemma 4.5. To see that (4.3) is also sufficient, consider for each $k, 1 \leq k \leq r(E)$, the set system

$$\mathcal{B}_k = \{X \in \mathcal{F} : |X| = k\} .$$

We show that $\mathcal{B}_k$ is the set of bases of a matroid. Consider two sets $B_1, B_2 \in \mathcal{B}_k$ and an element $x \in B_1 \setminus B_2$. Since $\mathcal{F}$ is accessible, there exists a largest subset $X \subseteq B_1$ with $x \in X$ such that $X \in \mathcal{F}$ and $X \setminus x \in \mathcal{F}$.

If $X = B_1$, then we augment $B_1 \setminus x$ from B_2 and we are done. So we may assume that X is a proper subset of B_1. Let $Y \subseteq B_2$ be such that $Y \in \mathcal{F}$ and $|Y| = |X|$. By induction on k there exists an element $y \in Y \setminus X$ with $X \setminus x \cup y \in \mathcal{F}$.

Let $B_1 = X \cup \{x_1, \ldots, x_m\}$, where $X \cup \{x_1, \ldots, x_i\} \in \mathscr{F}$ for $1 \le i \le m$. Successively augment $X \setminus x \cup y$ from B_1. Suppose at some step i we have to add the element x, i.e. $B = X \setminus x \cup y \cup \{x_1, \ldots, x_{i-1}\} \in \mathscr{F}, B \cup x_i \notin \mathscr{F}, B \cup x \in \mathscr{F}$.

Then $B \in \mathscr{F}$ and $B \setminus y \cup \{x, x_i\} \in \mathscr{F}$. Hence by (4.3), either $B \cup x_i \in \mathscr{F}$ or $B \setminus y \cup x_i \in \mathscr{F}$. The first case violates our assumptions on x_i and the second contradicts the choice of X, since augmentation of $B \setminus y \cup x_i$ from B_1 yields a larger set $X' \subseteq B_1$ with $X' \in \mathscr{F}$ and $X' \setminus x \in \mathscr{F}$. Hence $B_1 \setminus x \cup y \in \mathscr{F}$ and $y \in B_2 \setminus B_1$ since otherwise we would again be able to construct a larger set X'.

Thus for $0 \le k \le r(E)$, $\mathscr{B}_k$ is the set of bases of a matroid $\mathscr{M}_k$, which, by Lemma 4.5, is a strong map of $\mathscr{M}_{k+1}$. Hence $(E, \mathscr{F})$ is a Gauss greedoid.   □

We close this Section with the remark that in contrast to the more general class of exchange systems, Gauss greedoids are not closed under unions and trace operations.

On the other hand, full Gauss greedoids are closed under a "duality" operator. Let $(E, \mathscr{F})$ be a full greedoid. We define its **dual** by $\mathscr{F}^* = \{E \setminus X : X \in \mathscr{F}\}$. In general the dual of a greedoid is not a greedoid.

Theorem 4.7. *The dual of a full Gauss gredoid is a full Gauss greedoid.*

Proof. Consider the matroids $(E, \mathscr{M}_m^*), (E, \mathscr{M}_{m-1}^*), \ldots, (E, \mathscr{M}_0^*)$. It is easy to see that $(E, \mathscr{M}_k^*)$ is a strong map of $(E, \mathscr{M}_{k-1}^*)$ and that the rank of $(E, \mathscr{M}_k^*)$ is $m - k$. Let $\mathscr{B}_i$ be the set of bases of $(E, \mathscr{M}_i^*)$. Then $\mathscr{F}^* = \mathscr{B}_m^* \cup \mathscr{B}_{m-1}^* \cup \ldots \cup \mathscr{B}_0^*$ is a Gauss greedoid.   □

Matroid duality may be viewed as a special case of the previous theorem. Given a matroid $(E, \mathscr{M})$, define a sequence of matroids $(E, \mathscr{M}_i), 0 \le i \le |E|$, by $\mathscr{M}_i = \mathscr{M}^{(i)}$ if $i \le r(E)$ and $\mathscr{M}_i = ((\mathscr{M}^*)^{(|E|-i)})^*$ if $i > r(E)$. Then it is easy to check that $r_i(E) = i$ and $\mathscr{M}_i$ is a strong map of $\mathscr{M}_{i+1}$. The Gauss greedoid $(E, \mathscr{F})$ defined by this sequence is a full greedoid whose feasible sets are the independent and spanning sets of $\mathscr{M}$. Moreover, $(E, \mathscr{F}^*)$ arises from the dual matroid by the same construction.

There are other classes of full greedoids whose duals are greedoids, for example poset antimatroids. Another class can be obtained from twisted antimatroids (cf. Goecke [1986]).

Chapter X. Transposition Greedoids

Many decomposition procedures (series-parallel decomposition, dismantling etc.) lead to languages that are not locally free, but possess a property close to this. Local freeness says that if at a certain point both x and y are feasible choices in the process, then so is x followed by y. The transposition property requires the same except when x and y are "twins" (i.e. whenever x occurs in the continuation of the process, it can be replaced by y and vice versa). While this property does not imply that the language is an antimatroid, it does imply that it is a greedoid. In fact, transposition greedoids form a proper superclass of interval greedoids.

In Section 1 we introduce the transposition property, and in Section 2 we give some examples of transposition greedoids based on Korte and Lovász [1986b]. In Section 3 we present a general framework for simplicial decomposition schemes of combinatorial structures proposed by Faigle, Goecke and Schrader [1986], which covers many of the examples listed in Chapter IV. This setting may be regarded as a special instance of Church-Rosser systems (cf. Newman [1942] and Huet [1980]), where the decomposition schemes satisfy a Jordan-Dedekind chain condition. Finally we describe two principles for deriving greedoids from simplicial elimination schemes and show that one leads to transposition greedoids and the other to interval greedoids.

1. The Transposition Property

To illustrate the property we shall be dealing with in this chapter, recall the process of series-parallel reductions in graphs without loops. Here we may delete an edge if it is either pending, in series or parallel with another edge. Consider two parallel edges x and y. If at some stage we may delete one, x say, we may also delete y. After deleting x, y may no longer be reducible but may again become reducible after we have deleted a set B of other edges. Then the same will happen if we delete y instead of $x : x$ can be reduced after we have deleted the edges in B. So as soon as x and y become reducible, they behave like twins. This observation is the basis for the so-called transposition property. It is shared by a substantially large class of elimination processes which can be used to inductively verify certain properties of combinatorial objects. These elimination processes very often induce greedoids. However, a direct proof of the greedoid properties is often tedious while the transposition property is easily verified.

Given a set system $(E, \mathcal{F})$, we say that two sets $X, Y \subseteq E$ are **twins** if for every $B \subseteq E \setminus (X \cup Y)$, $X \cup B \in \mathcal{F}$ if and only if $Y \cup B \in \mathcal{F}$. (Note that this property is slightly weaker than the equivalence defined in Section V.5).

We say that a set system $(E, \mathcal{F})$ has the **transposition property** if the following holds:

(1.1)   If $A \subseteq E$, $x, y \in E \setminus A$ such that $A \cup x \in \mathcal{F}$, $A \cup y \in \mathcal{F}$ but $A \cup \{x, y\} \notin \mathcal{F}$, then $A \cup x, A \cup y$ are twins.

Since (1.1) is obviously a strengthening of the augmentation property (IV.1.7), any accessible set system with the transposition property is a greedoid by Theorem IV.1.3. Again, as in the case of condition (IV.1.7) one might ask whether (1.1) can be weakened so as to hold only for feasible sets $A \in \mathcal{F}$. Unlike that case, however, (1.1) and its weaker version turn out to be equivalent for accessible set systems.

Theorem 1.1. *Let $(E, \mathcal{F})$ be an accessible set system. Then (1.1) holds for every $A \subseteq E$ if and only if (1.1) holds for every $A \in \mathcal{F}$.*

Proof. Suppose the transposition property holds for every $A \in \mathcal{F}$. We assume by induction on $|E|$ that the statement is true for the restrictions of $\mathcal{F}$ to any proper subset. Then, in particular, the restrictions form a greedoid.

Let $A \subseteq E$, $x, y \in E \setminus A$ and $B \subseteq E \setminus (A \cup x \cup y)$ be such that $A \cup x, A \cup y, A \cup x \cup B \in \mathcal{F}$, $A \cup \{x, y\} \notin \mathcal{F}$. Suppose $A \cup y \cup B \notin \mathcal{F}$. Among all elements $v \in A' = A \cup \{x, y\}$ with $A' \setminus v \in \mathcal{F}$ and $A' \setminus v \cup B \in \mathcal{F}$, choose one for which $A' \setminus \{x, v\}$ contains a subset $C \in \mathcal{F}$ with $|C|$ maximum.

Then C can be augmented from $A \cup y = A' \setminus x$ by some element. Since C is maximally feasible in $A' \setminus \{x, v\}$, this element must be v. Hence $C \cup v \in \mathcal{F}$.

Now let $C' \subseteq A' \setminus v$ be a maximal subset with $C \subseteq C', C' \in \mathcal{F}$ and $C' \cup v \in \mathcal{F}$. Since $A' \notin \mathcal{F}, C'$ is a proper subset of $A' \setminus v \in \mathcal{F}$. So $C' \cup w \in \mathcal{F}$ for some $w \in A' \setminus v$. Furthermore, $C' \cup \{v, w\} \notin \mathcal{F}$ since otherwise $C' \cup w$ would contradict the maximality of C'. Finally,

$$C' \cup w \cup (A' \setminus (C' \cup \{v, w\})) \cup B = A' \setminus v \cup B \in \mathcal{F}.$$

The transposition property (1.1) applied to C' then yields

$$C' \cup v \cup (A' \setminus (C' \cup \{v, w\})) \cup B = A' \setminus w \cup B \in \mathcal{F}.$$

Similarly, the transposition property implies $A' \setminus w \in \mathcal{F}$. But then $C \cup v \subseteq A \setminus \{x, w\}$, contradicting the choice of C.   □

Corollary 1.2. *Interval greedoids have the transposition property.*

Proof. To see (1.1), augment $A \cup y$ from $A \cup x \cup B$ to some set $A \cup y \cup B'$. By the interval property and since $A \cup \{x, y\} \notin \mathcal{F}$, $x \notin B'$. Hence $A \cup y \cup B \in \mathcal{F}$.   □

We will now reformulate the transposition property in a language framework. The ordered version consists of two conditions, where the first one is a very special exchange property.

(1.2)   If $\alpha xy\beta \in \mathcal{L}$ and $\alpha y \in \mathcal{L}$, then $\alpha yx\beta \in \mathcal{L}$.

(1.3)   If $\alpha x, \alpha y \in \mathcal{L}$ and $\alpha xy \notin \mathcal{L}$, then $\alpha x\beta \in \mathcal{L}$ and $y \notin \beta$ imply $\alpha y\beta \in \mathcal{L}$, and $\alpha x\beta yy \in \mathcal{L}$ implies $\alpha y\beta xy \in \mathcal{L}$.

Our first observation shows that with the help of (1.3) the exchange property in (1.2) can be generalized.

Lemma 1.3. *Let $(E, \mathcal{L})$ be a hereditary language satisfying properties (1.2) and (1.3), and suppose $\alpha\beta xy$, $\alpha x \in \mathcal{L}$. Then there exists a word β' with $\tilde{\beta}' = \tilde{\beta}$ and $\alpha x\beta'\gamma \in \mathcal{L}$.*

Proof. Let $\beta = \beta_1\beta_2$ be such that $\alpha\beta_1 x\beta_2'\gamma \in \mathcal{L}$ for some word β_2' with $\tilde{\beta}_2' = \tilde{\beta}_2$ and $|\beta_1|$ minimal. We want to show that $|\beta_1| = 0$. Suppose not, then we can write $\beta_1 = \beta_3 y$ and distinguish two cases.

  (i) $\alpha\beta_3 x \in \mathcal{L}$. Then from $\alpha\beta_3 yx\beta_2'\gamma \in \mathcal{L}$ we obtain by (1.2) $\alpha\beta_3 xy\beta_2'\gamma \in \mathcal{L}$, contradicting the minimality of $|\beta_1|$.

  (ii) $\alpha\beta_3 x \notin \mathcal{L}$. Let δ be the smallest beginning section of β_3 such that $\alpha\delta x \notin \mathcal{L}$. Clearly $|\delta| > 0$ and we can write $\delta = \delta_1 z$ and $\beta_3 = \delta\beta_4$. Then $\alpha\delta_1 x \in \mathcal{L}$, $\alpha\delta_1 z \in \mathcal{L}$ but $\alpha\delta_1 zx \notin \mathcal{L}$. Moreover, $\alpha\delta_1 z\beta_4 yx\beta_2'\gamma \in \mathcal{L}$. Hence by (1.3) $\alpha\delta_1 x\beta_4 yz\beta_2'\gamma \in \mathcal{L}$, again contradicting the minimality of $|\beta_1|$. $\square$

Theorem 1.4. *Let $(E, \mathcal{L})$ be a hereditary language satisfying properties (1.2) and (1.3). Then $(E, \mathcal{L})$ is a transposition greedoid.*

Proof. We show that $(E, \mathcal{L})$ satisfies the special exchange property I.1.5.

  Let $\alpha, \beta \in \mathcal{L}$ be feasible words with $\tilde{\beta} \subseteq \tilde{\alpha}$ and $|\alpha| = |\beta| + 1$. Let γ be the longest common beginning section of α and β. We prove by induction on $|\beta| - |\gamma|$ that there exists a letter $x \in \alpha$ such that $\beta x \in \mathcal{L}$.

  Since this obviously holds if $\beta = \gamma$, we may assume that β is of the form $\beta = \gamma y\beta_1$. As $\tilde{\beta} \subseteq \tilde{\alpha}$, we may also partition α into $\alpha = \gamma\alpha_1 y\alpha_2$. Lemma 1.3 then implies that there exists a word α_1' with $\tilde{\alpha}_1' = \tilde{\alpha}_1$ such that $\alpha' = \gamma y\alpha_1'\alpha_2 \in \mathcal{L}$.

  By the induction hypothesis applied to α' and β, there exist an $x \in \tilde{\alpha}' = \tilde{\alpha}$ such that $\beta x \in \mathcal{L}$. So $(E, \mathcal{L})$ has property I.1.1.

  Now Lemma I.1.1 implies that $\mathcal{L} = \mathcal{L}(\mathcal{F})$ for some accessible family $\mathcal{F}$. Then (1.3) implies that $\mathcal{F}$ satisfies (1.1) for every $A \in \mathcal{F}$. Hence by Theorem 1.1 $(E, \mathcal{F})$ is a transposition greedoid. $\square$

In Chapter V we remarked that the free erection of a greedoid is again a greedoid. The interval property, however, may be lost under this operation. To see this, consider the free erection of the matroid with bases $\{x, y\}, \{x, z\}$ on $E = \{x, y, z\}$. In contrast to this, the transposition property is preserved.

Lemma 1.5. *If $(E, \mathcal{F})$ is a transposition greedoid, then its free erection $(E, \mathcal{F}')$ is also a transposition greedoid.*

Proof. Suppose that $A \cup x, A \cup y, A \cup x \cup B \in \mathcal{F}'$, $A \cup x \cup y \notin \mathcal{F}'$. Then $A \cup x \in \mathcal{F}$ and $A \cup y \in \mathcal{F}$. If $A \cup x \cup B \in \mathcal{F}$, then by the transposition property of $(E, \mathcal{F})$,

$A \cup y \cup B \in \mathscr{F}$. Otherwise B can be written as $B = B' \cup z$ such that $A \cup x \cup B'$ is a basis of $(E, \mathscr{F})$. By the transposition property of $(E, \mathscr{F})$, $A \cup y \cup B' \in \mathscr{F}$ and so $A \cup y \cup B \in \mathscr{F}'$. $\square$

2. Applications of the Transposition Property

We now show that the tranposition property can be applied to give simple proofs that certain hereditary languages are greedoids. Some of these have been mentioned among the examples in Section IV.2 but their proofs were postponed until this chapter.

a. Series-Parallel Reduction in Graphs. Suppose that having eliminated a set A we can eliminate either edge x or edge y, but not both. Then trivially x and y are parallel or in series. Hence eliminating x results in the same graph as eliminating y. This proves the transposition property.

b. Retraction in Directed Graphs. Suppose that having eliminated a set X we can eliminate either x or y, but not both. This can only happen if x is retractable to y and y is retractable to x. But then the graphs $G \backslash (A \cup x)$ and $G \backslash (A \cup y)$ are isomorphic and hence $A \cup x$ and $A \cup y$ are twins. Hence retract elimination sequences form a transposition greedoid.

c. Dismantling in directed graphs. Dismantling sequences form a transposition greedoid by a similar argument.

d. Bisimplicial elimination. Suppose that having eliminated edges $(s_1, t_1), \ldots, (s_k, t_k)$ from a bipartite graph $G = (S, T, E)$, either (x, u) or (y, v) can be eliminated, but not both (where $x \neq y$). Then we must have $u = v$ and hence by the definition of a bisimplicial edge, x and y must have the same neighbors. Hence eliminating (x, u) or (y, v) results in the same graph up to exchanging x and y.

Sometimes even for interval greedoids the transposition property provides an easy way to prove that they are greedoids. Consider simplicial clique elimination. If at some point either x or y can be eliminated, but not both, then they must be adjacent and have the same neighbors. So deleting x together with its neighbors yields the same graph as deleting y together with its neighbors.

3. Simplicial Elimination

The examples in the previous section indicate that there may be a general phenomenon behind the transposition property. In this section we introduce a general framework to explain this.

Consider an abstract class $\mathscr{K}$ of combinatorial objects, e.g. the class of graphs, directed graphs or ordered sets. We assume that $\mathscr{K}$ is closed under taking subobjects of objects and that there is a notion of isomorphism between objects taking subobjects into subobjects. We furthermore assume that for each subobject A of an object $K \in \mathcal{K}$, denoted by $A \leq K$, a **factor object** K/A exists and is a subobject of K. To exclude trivial cases, we will always assume that K/A is not isomorphic to K. If A_2 is a subobject of K/A_1, then we write $K/(A_1, A_2)$ instead of $(K/A_1)/A_2$.

Let $\mathscr{E}$ be a property of pairs (K, A) where $A \leq K$ (abstracting "eligible for elimination"). We call $\mathscr{E}$ **simplicial** if it is invariant under isomorphisms and if for all $K \in \mathcal{K}$ and $A, B \leq K$ at least one of the following possibilities occurs:

(3.1) $K/A \simeq K/B$,

(3.2) $K/(A, B') \simeq K/(B, A')$ for some $A' \leq K/B$ and $B' \leq K/A$ with property $\mathscr{E}$.

For the remainder of this section we will always assume that $\mathscr{E}$ is simplicial. We then call a subobject $A \leq K$ **simplicial** if (K, A) has property $\mathscr{E}$.

Consider an object $K \in \mathcal{K}$ and a collection $A_1, \ldots, A_n$ of objects such that, for $i = 1, 2, \ldots, n, A_i$ is a simplicial subobject of $K/(A_1, \ldots, A_{i-1})$. If $K' = K/(A_1, \ldots, A_n)$ has no simplicial subobject, we say that the **elimination sequence** $(A_1, A_2, \ldots, A_n)$ for K is **maximal** and call K' a (simplicially) **irreducible kernel** of K.

A fundamental observation is the following

Lemma 3.1. *Let the property $\mathscr{E}$ be simplicial on $\mathcal{K}$ and $K \in \mathcal{K}$ an object. Then any two maximal elimination sequences $(A_1, \ldots, A_n)$ and $(B_1, \ldots, B_m)$ of K have the same length, i.e. $n = m$, and the irreducible kernels $K/(A_1, \ldots, A_n)$ and $K/(B_1, \ldots, B_m)$ are isomorphic.*

Proof. Proceeding by induction on the length of the elimination sequence we consider two cases.

(i) A_1 and B_1 satisfy condition (3.1), i.e. K/A_1 and K/B_1 are isomorphic. Since $(A_2, \ldots, A_n)$ and $(B_2, \ldots, B_m)$ are maximal elimination sequences for K/A_1 and K/B_2, the statement follows from the induction hypothesis.

(ii) A_1 and B_1 satisfy condition (3.2). Then we can find maximal elimination sequences $(A_1, B', C_3, \ldots, C_k)$ and $(B_1, A', D_3, \ldots, D_\ell)$ for suitable $B' \leq A_1$ and $A' \leq B_1$. Since $K/(A_1, B') \simeq K/(B_1, A')$ we conclude, as in case (i), that $k = n = m = \ell$, and similarly,

$$K/(A_1, A_2, \ldots, A_n) \simeq K/(A_1, B', C_3, \ldots, C_k)$$
$$\simeq K/(B_1, A', D_3, \ldots, D_\ell)$$
$$\simeq K/(B_1, B_2, \ldots, B_n) .   \qquad \square$$

Let $\mathcal{K}$ be the class of **semimodular ordered sets**, i.e. the class of all ordered sets P having a unique minimal element and such that every pair of elements covering a common element is also covered by a common element.

For $P \in \mathcal{K}$, the simplicial subobjects may be taken to be the principal ideals $I(x)$, where $x \in P$ covers the unique minimal element 0 in P. The associated quotient intervals P/x are the principal filters $F(x)$.

In this case, (3.2) is always satisfied and maximal elimination sequences correspond to maximal chains in P. Proposition 3.1 thus reduces to the Jordan-Dedekind-chain condition: all maximal chains of a semimodular order have the same length.

Whereas this example yields a complete decomposition of the combinatorial object, the following example shows that a simplicial decomposition may leave irreducible kernels. A subgraph M of a graph G is a **module** of G if for every node $m \in M$ and $x \in G \setminus M$, x is adjacent to all $y \in M$ whenever x is adjacent to m. G/M then denotes the graph obtained from G by contracting M to a single node. Let $\mathcal{K}$ be the class of graphs G such that both G and its complement $\overline{G}$ are connected.

Contraction of maximal proper modules now gives rise to a simplicial elimination scheme for $\mathcal{K}$. This follows from the fact that for every $G \in \mathcal{K}$ and maximal proper modules $A, B \le G$, either $A = B$ or $A \cap B = \emptyset$. Thus Lemma 3.1 assures uniqueness of this **substitution decomposition**. This is the key result for the investigation of comparability invariants of ordered sets (cf. Faigle and Schrader [1984]).

For the rest of this section we assume that each object $K \in \mathcal{K}$ is a finite ground set endowed with some structure, and we will often identify the ground set with the structure. We will focus on elimination schemes where the quotient objects are constructed in a straightforward way: only the simplicial set is removed, i.e. K/A is the structure induced by K on the difference set $K \setminus A$. Many recognition algorithms for optimization structures are based on such elimination principles. We will now use "$K \setminus A$" instead of "K/A".

We start with an example that is a slight generalization of simplicial elimination (cf. IV.2.8). Let $\mathcal{K}$ be the class of all finite graphs $G = (V, E)$ and $i \in \mathbb{N}$ a fixed natural number. We say that a subset $A \subseteq V$ of nodes is **simplicial** if the following holds:

(3.3) A is a clique of size i and the set of all proper neighbors of A induces a complete subgraph,

where a node $v \in V$ is a **proper neighbor** of A if $v \notin A$ but is adjacent to some node $w \in A$.

Note that the case $i = 1$ is the classical simplicial shelling, which results in a complete decomposition of the graph G if and only if G is chordal (cf. Fulkerson and Gross [1965]).

To see that (3.3) is a simplicial property, consider two simplicial sets A and B with $|A| = |B| = i$. If $A \cap B = \emptyset$, then clearly B remains simplicial in $G \setminus A$ and conversely.

If not, there exists a node $w \in A \cap B$. Consider three arbitrary nodes $a \in A \setminus B$, $b \in B \setminus A$ and $x \in V \setminus (A \cup B)$. We claim that x is a neighbor of a if and only if x is a neighbor of b. Indeed, if x is a neighbor of a, both x and b are proper neighbors of A since b is a neighbor of w. Hence (3.3) implies that x and b are neighbors.

Thus $G \setminus A$ and $G \setminus B$ must be isomorphic.

In a related elimination scheme we call a subset $A \subseteq V$ **clique-simplicial** if for some fixed $i \in \mathbb{N}$

(3.4) A is a clique and consists of some i-clique K_i together with its proper neighbors.

An argument similar to the one above shows that (3.4) is a simplicial property. For $i = 1$, we get the simplicial clique elimination greedoid.

The notion of **simple** nodes in a graph (cf. Section I.2) gives rise to three related but different elimination schemes: firstly we may just take simple nodes as simplicial sets, secondly we may take the closed neighborhood $\overline{N}$ of simple nodes, and thirdly sets of the form $N_2(v) = \bigcup\{\overline{N}(z) : z \in \overline{N}(v)\}$ where v is a simple node. In all three cases property (3.2), and hence simpliciality, is easily verified making use of the following observation.

Lemma 3.2. *Let v and w be two simple nodes of the graph G and assume $v \in N_2(w)$. Then $N_2(v) = N_2(w)$.*

Proof. Assume $v \in N_2(w)$. Then there exists a node $z \in \overline{N}(w)$. Now consider an arbitrary node y. If $y \in N_2(w)$, i.e. $y \in \overline{N}(x)$ for some $x \in \overline{N}(w)$, then by hypothesis $v \in \overline{N}(z)$ and w is simple, hence either v is a neighbor of x or y is a neighbor of z . In both cases $y \in N_2(v)$. Similarly one shows $N_2(v) \subseteq N_2(w)$. $\square$

The class of graphs that allow complete elimination by simple nodes is exactly the class of **strongly chordal** graphs (cf. Anstee and Farber [1984]) and hence is essentially equivalent to the class of totally balanced matrices or hypergraphs (see also Lovász [1979]).

A decomposition scheme related to the clique elimination is the cover simplicial elimination. Here we let $\mathscr{K}$ be the class of all pairs $(E, \mathscr{H})$, where E is a finite ground set and $\mathscr{H}$ is a family of subsets of E. We call a member $S \in \mathscr{H}$ **cover simplicial** if

(3.5)   there exists an element $x \in S$ such that for all $S' \in \mathscr{H}$, $x \in S'$ implies $S' \subseteq S$.

To avoid ambiguities, let us point out that elimination of a cover simplicial set A means restricting $(E, \mathscr{H})$ to $(E \setminus A, \mathscr{H}/(E \setminus A))$, where

$$\mathscr{H}/(E \setminus A) = \{S \setminus A : S \in \mathscr{H}\}.$$

It is easy to see that (3.5) is indeed a simplicial property. For let $A \neq B$ be two cover simplicial members of $\mathscr{H}$. Then there exist $x \in A$ and $y \in B$, each satisfying (3.5). Because $A \neq B$, we must have $x \in A \setminus B$ and $y \in B \setminus A$. Thus $A \setminus B$ and $B \setminus A$ are cover simplicial with respect to the induced structures.

In most of the preceding examples we have eliminated sets of nodes of a graph. Bisimplicial elimination leads to two different elimination processes, one on the edges, and one on the vertices (cf. IV.2.4).

Lemma 3.3. *Removing bisimplicial edges defines a simplicial elimination scheme.*

Proof. Let $e = (u, v)$ and $f = (x, y)$ be two bisimplicial edges of G. Then clearly f remains bisimplicial in $G \setminus e$ unless (u, y) and (v, x) are edges of G. In this case, we claim that the complete bipartite graphs $B(e)$ and $B(f)$ induced by e and f, respectively, are identical.

Let z be a neighbor of x. Since u is a neighbor of y and (x, y) is a bisimplicial edge, z is also neighbor of u. Thus the interchange of x with u and of y with v yields an isomorphism between $G \setminus e$ and $G \setminus f$.   □

The class of bipartite graphs which can be decomposed by repeatedly eliminating bisimplicial edges without leaving an irreducible kernel is the class of **chordal bipartite graphs,** i.e. bipartite graphs with no chordless cycle of length greater than four (cf. Faigle et al. [1986]). We thereby obtain the bisimplicial elimination antimatroid.

Golumbic and Goss [1978] investigated a related decomposition where not only the bisimplicial edge e but also the set $N(e)$ of adjacent edges is eliminated. Note that this is equivalent to removing the endnodes of the bisimplicial edge from the graph.

Lemma 3.4. *Removing bisimplicial edges together with their adjacent edges defines a simplicial elimination scheme.*

Proof. We distinguish two cases.

(i) The bisimplicial edges e and f are not incident. Then clearly property (3.2) applies trivially.

(ii) $e = (x, y)$ and $f = (x, z)$ for suitable vertices x, y, z. Then the map which interchanges y and z and leaves the other vertices of G fixed is an automorphism, i.e. property (3.1) applies. Thus obviously condition (3.2) is satisfied with the induced isomorphism.   □

A further elimination process for which the simpliciality is readily verified is the **retract operation** in directed graphs (cf. Example IV.2.5). This implies several results of Duffus and Rival [1976].

As a last example, we mention the upper bipartite elimination in posets (see Example IV.2.19). Here property (3.1) is always satisfied, showing that this elimination process is simplicial. This decomposition is of particular interest for the setup optimization problem since at each step the setup number can be updated (cf. Faigle et al. [1985] and Faigle and Schrader [1986]).

This list of examples suggests that simplicial elimination schemes are closely related to greedoids. We will now present two construction principles for the derivation of greedoids from simplicial elimination schemes. It should be noted however, that simplicial elimination, even when restricted to the elimination of singletons, does not always give rise to a greedoid. Consider the elimination process of removing bisimplicial edges in the bipartite graph of Figure 16. Here the string (x, x') cannot be augmented from the string $(z, z')(y, z')$.

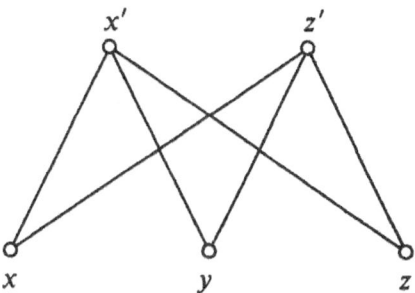

Fig. 16

For the first construction principle let $\mathcal{E}$ be a simplicial property and assume that for each $K \in \mathcal{K}$ the set underlying K has been partitioned into "red" and "green" points such that for all substructures $K' \subseteq K$ induced by simplicial elimination and all simplicial subsets $A, B \subseteq K'$ containing exactly one red point each, the following condition is satisfied:

(3.6) A is simplicial in $K' \setminus B$ and B is simplicial in $K' \setminus A$ or there exists a color-preserving isomorphism between $K' \setminus A$ and $K' \setminus B$ which is the identity on $K' \setminus (A \cup B)$.

This coloring naturally induces a language $\mathcal{L} = \mathcal{L}(E, K)$, where $x_1 \ldots x_k \in \mathcal{L}$ if there exists an elimination sequence $(A_1, \ldots, A_k)$ for K such that x_i is the unique red point of A_i.

Lemma 3.5. *Let $(A_1, \ldots, A_k)$ be an elimination sequence yielding the word $x_1 \ldots x_k \in \mathcal{L}$ and let $x_1 \ldots x_k x_{k+1} \ldots x_\ell \in \mathcal{L}$. Then there exist sets $A_{k+1}, \ldots, A_\ell$ such that $(A_1, \ldots, A_k, A_{k+1}, \ldots, A_\ell)$ is an elimination sequence for the word $x_1 \ldots x_k x_{k+1} \ldots x_\ell$.*

Proof. We proceed by induction on k. Let $x_1 \ldots x_{k+1} \ldots x_\ell \in \mathcal{L}$ and $(A_1, \ldots, A_{k+1})$ be an elimination sequence for $x_1 \ldots x_{k+1}$. By induction there exist sets $B_{k+1}, \ldots, B_\ell$ such that $(A_1, \ldots, A_k, B_{k+1}, \ldots, B_\ell)$ yields $x_1 \ldots x_k x_{k+1} \ldots x_\ell$. Since $A_{k+1} \cap B_{k+1} \neq \emptyset$, A_{k+1} cannot be simplicial in $K' \setminus B_{k+1}$, where $K' := K \setminus (A_1 \cup \ldots \cup A_k)$. Hence condition (3.6) guarantees the existence of an isomorphism

$$\phi : K' \setminus B_{k+1} \to K' \setminus A_{k+1} ,$$

which acts as the identity on $K \setminus (A_{k+1} \cup B_{k+1})$. For $j = k + 2, \ldots, \ell$ we set $A_j = \phi(B_j)$ and claim that $(A_1, \ldots, A_k, A_{k+1}, \ldots, A_\ell)$ represents $x_1 \ldots x_\ell$. It suffices to show that x_j is the only red point in $A_j = \phi(B_j)$. But this follows immediately from the observation $\phi(B_j) \subseteq (B_{k+1} \setminus A_{k+1}) \cup B_j$. $\square$

Theorem 3.6. *$\mathcal{L} = \mathcal{L}(E, K)$ is a transposition greedoid.*

Proof. Clearly $\mathcal{L}$ is a hereditary language. It remains to verify (1.2) and (1.3).

For (1.2), let $\alpha x y \beta$, $\alpha y \in \mathcal{L}$. In view of Lemma 3.5, we may assume that α is the empty word and that (A_1, A_2) and B_1 are elimination sequences for xy and y.

If A_1 is simplicial in $K \setminus B_1$ and B_1 is simplicial in $K \setminus A_1$, then (A_1, B_1) also generates xy and (B_1, A_1) generates $yx \in \mathcal{L}$. Hence by Lemma 3.5, $\beta \in \mathcal{L}(K \setminus (A_1 \cup B_1))$, i.e. $yx\beta \in \mathcal{L}(K)$.

If $K \setminus A_1$ and $K \setminus B_1$ are isomorphic as in the second part of (3.6), then the same argument as in the proof of Lemma 3.5 yields an elimination sequence for $yx\beta$.

To verify (1.3), consider $\alpha x, \alpha y \in \mathcal{L}$ with $x \neq y$ such that $\alpha x y \notin \mathcal{L}$. Again, by Lemma 3.5 we may assume $\alpha = \emptyset$.

If A and B are simplicial sets yielding x and y respectively, B cannot be simplicial in $K \setminus A$ because $xy \notin \mathcal{L}$. Thus there is a color-preserving isomorphism $\phi : K \setminus A \to K \setminus B$ as in (3.6). In particular, $\phi(y) = x$. Hence ϕ induces an isomorphism between the two languages $\mathcal{L}_x = \{\beta : x\beta \in \mathcal{L}\}$ and $\mathcal{L}_y = \{\beta : y\beta \in \mathcal{L}\}$, thereby proving (1.3). □

We illustrate this result by three examples: First, let $G = (S, T, E)$ be a bipartite graph. We color the vertices of the graph red and green and define a language $\mathcal{L}$ of sequences $x_1 \ldots x_k$ such that there exists a bisimplicial elimination scheme $(s_1, t_1), \ldots, (s_k, t_k)$ with x_i being the only red point in $\{s_i, t_i\}$. The automorphism given in the proof of Lemma 3.3 clearly satisfies (3.6). Hence the language $(S \cup T, \mathcal{L})$ defines a transposition greedoid.

In the special case where S is red and T is green, we obtain the bisimplicial elimination greedoid (cf. IV.2.4).

Secondly, consider the series-parallel reduction in a graph G (cf. IV.2.7). One easily verifies that sets consisting of a single reducible edge are simplicial. Hence if we color all vertices red, we see that the series-parallel reduction greedoid has the transposition property.

Thirdly, the retract operation in directed graphs (IV.2.5) has a natural isomorphism which makes it a simplicial elimination.

We now turn to the second construction principle. We call the simplicial property $\mathcal{S}$ on $\mathcal{K}$ **locally free** if for all $K \in \mathcal{K}$ and all simplicial subsets $A, B \subseteq K$ with $A \neq B$

(3.7) $B \setminus A$ is simplicial in $K \setminus A$.

The element $x \in K$ is **relatively simplicial** if for all (possibly empty and not necessarily complete) elimination sequences $(A_1, \ldots, A_k)$ of K with $x \notin A_1 \cup \ldots \cup A_k$ there is exactly one simplicial subset A of $K \setminus (A_1 \cup \ldots \cup A_k)$ containing x.

We then associate with K a language $\mathcal{L} = \mathcal{L}(K, \mathcal{S})$ in the following manner:

(3.8) $x_1 \ldots x_k \in \mathcal{L}$ if there exists an elimination sequence $(A_1, \ldots A_k)$ of K such that for $i = 1, \ldots, k$, $x_i \in A_i$ and x_i is relatively simplicial in $K \setminus (A_1 \cup \ldots \cup A_{i-1})$.

Theorem 3.7. *If $\mathcal{S}$ is locally free on K, then $(K, \mathcal{L})$ is a balanced interval greedoid.*

Proof. Observing that for any word $x_1 \ldots x_k \in \mathscr{L}$ the corresponding sequence $(A_1, \ldots, A_k)$ is uniquely determined, we prove the conditions (V.5.3.1) and (V.5.3.2).

To verify (V.5.3.1), let $\alpha x y \beta$, $\alpha y \in \mathscr{L}$. Without loss of generality we may assume $\alpha = \emptyset$. If A and B are simplicial sets for x and y respectively, we have $y \notin A$ and $x \notin B$ since x and y are relatively simplicial. By (3.7), $(A, B \setminus A)$ and $(B, A \setminus B)$ are simplicial elimination sequences, implying $y x \beta \in \mathscr{L}$.

To prove (V.5.3.2), we again assume $\alpha = \emptyset$. $xy \notin \mathscr{L}$ now implies $x \in B$ and $y \in A$, and thus $A = B$ because x and y are relatively simplicial. Therefore $x \beta \in \mathscr{L}$ yields $y \notin B$ and $y \beta \in \mathscr{L}$. The claim now follows from Lemma V.5.3.

To prove that $(K, \mathscr{L})$ is balanced we use induction on $|K|$. Assume without loss of generality that $(K, \mathscr{L})$ is normal. Let x_1 be relatively simplicial in K with simplicial set A_1, and $(A_1, A_2, \ldots, A_k)$ a complete elimination sequence. We claim that $(A_1, A_2, \ldots, A_k)$ is the desired partition.

Let $y_1 y_2 \ldots y_k$ be a basic word of $\mathscr{L}$ with associated elimination sequence $(B_1, \ldots, B_k)$. Then $x_1 \in B_i$ for some $1 \leq i \leq k$, because otherwise by (3.7) $A_1 \setminus (B_1 \cup \ldots \cup B_k)$ is simplicial in $K \setminus (B_1 \cup \ldots \cup B_k)$, and so $y_1 \ldots y_k x_1 \in \mathscr{L}$.

Since x_1 is relatively simplicial in K and both $A_1 \setminus (B_1 \cup \ldots \cup B_{i-1})$ and B_i contain x_1 and are simplicial, we get $A_1 \setminus (B_1 \cup \ldots \cup B_{i-1}) = B_i$. Therefore $y_1 \ldots y_{i-1} x_1 y_{i+1} \ldots y_k \in \mathscr{L}$ and, using (3.7) again, $x_1 y_1 \ldots y_{i-1} y_{i+1} \ldots y_k \in \mathscr{L}$.

Consider $K' = K \setminus A_1$. Then $\mathscr{L}' = \mathscr{L}(K', \mathscr{E})$ is the contraction of x_1 in $\mathscr{L}$. Hence by induction $|\{y_1, \ldots y_{i-1}, y_{i+1}, \ldots, y_k\} \cap A_j| = 1$ for $j = 2, \ldots k$ and, by the choice of the index i, $\{y_1, \ldots, y_{i-1}, y_{i+1}, \ldots, y_k\} \cap A_1 = \emptyset$. This together with $y_1 \in B_i \subseteq A_1$ proves the result. □

As a converse to this result, we show:

Theorem 3.8. *Every balanced interval greedoid is induced by a locally free elimination process.*

Proof. Let $E_1, \ldots, E_k$ be the partition associated with the balanced interval greedoid $(E, \mathscr{F})$ and $(E, \mathscr{M})$, the corresponding partition matroid. We define inductively a set A_i to be relatively simplicial in $E \setminus (A_1, \ldots, A_{i-1})$ if for some block E_j with $(A_1 \cup \ldots \cup A_{i-1}) \cap E_j = \emptyset$, $E_i \subseteq E_j$ and if for all $x \in A_i$ with $(C, x) \in \mathscr{C}$

$$C \cap (\sigma_{\mathscr{M}}(A_1) \cup \ldots \cup \sigma_{\mathscr{M}}(A_{i-1})) \neq \emptyset.$$

It is straightforward to check that this definition yields a locally free simplicial elimination process for which every $x \in A_i$ is relatively simplicial. Hence the assertion follows using Lemma IX.2.7. □

Chapter XI. Optimization in Greedoids

Matroids and related structures (polymatroids, submodular functions) are used in many different ways in combinatorial optimization; some of these have been sketched in Chapter II. Most notable are the optimality of the greedy algorithm, nice polyhedral descriptions, and intersection results. In this chapter we explore the extension of some of these approaches to greedoids.

Section 1 describes a class of objective functions for which the greedy algorithm works optimally. In analogy to the algorithmic characterization of matroids, we show how greedoids can be characterized as those hereditary languages for which the greedy algorithm solves optimization problems with a generalized bottleneck objective function to optimality.

Unfortunately this class does not include linear functions. Section 2 therefore treats the question under what conditions a linear weight function may be optimized over a greedoid by the greedy strategy. For arbitrary greedoids a sufficient condition is that the level sets of the objective function are rank feasible. On the other hand, the greedoids for which the greedy algorithm optimizes any linear function are precisely those whose hereditary closure is a matroid. The section closes with an algorithmic characterization of Gauss greedoids (Goecke [1988]).

It is in general exponentially hard to optimize a linear objective function over the feasible sets of a greedoid, and consequently we cannot expect to obtain a nice polyhedral characterization of the convex hull of feasible sets. In Section 3 we present polyhedral results for some special classes of greedoids, mostly antimatroids.

The last two sections contain extensions of the matroid intersection theorem to greedoids under appropriate conditions. A classical equivalent version of the matroid intersection theorem is the Rado-Hall theorem on the existence of a set which is transversal to a given set system and at the same time independent in a given matroid. In Section 4 we show that this result has a rather natural extension to greedoids.

Section 5 gives a min-max characterization of the maximum size of a set which is feasible in two supermatroids defined on a distributive lattice, due to Tardos [1988].

1. General Objective Functions

Let $(E, \mathscr{L})$ be a greedoid and $w : \mathscr{L} \to \mathbb{R}$ an objective function. The problem we consider in this section is

(1.1) $\max\{w(\alpha) : \alpha$ is a basic word in $\mathscr{L}\}$.

The greedy algorithm for this problem successively augments feasible words by a new element until it reaches a basis. If there is a choice among several elements, it selects one which is locally best. The procedure can be described as follows:

Greedy Algorithm

(1)   Set $\alpha = \emptyset$.
(2)   If $\Gamma(\alpha) = \emptyset$, stop.
(3)   Choose $x \in \Gamma(\alpha)$ such that $w(\alpha x) \geq w(\alpha y)$ for all $y \in \Gamma(\alpha)$.
(4)   Set $\alpha = \alpha x$ and go to 2.

It is intuitively clear that this simple algorithm will in general not produce an optimal solution when w is an arbitrary objective function. We therefore need some extra conditions which link the objective function to the structure of the greedoid.

Given a greedoid $(E, \mathscr{L})$, we call a function $w : \mathscr{L} \to \mathbb{R}$ $\mathscr{L}$-**admissible**, if for every $\alpha \in \mathscr{L}$ and $x \in \Gamma(\alpha)$ such that

$$w(\alpha x) \geq w(\alpha y) \text{ for all } y \in \Gamma(\alpha)$$

the following two conditions hold:

(1.2) $w(\alpha\beta x\gamma) \geq w(\alpha\beta z\gamma)$ for all $z \in E$ and $\beta, \gamma \in E^*$ such that $\alpha\beta x\gamma, \alpha\beta z\gamma \in \mathscr{L}$,

and

(1.3) $w(\alpha x \beta z\gamma) \geq w(\alpha z \beta x\gamma)$ for all $z \in E$ and $\beta, \gamma \in E^*$ such that $\alpha x \beta z\gamma$, $\alpha z \beta x\gamma \in \mathscr{L}$.

The intuitive meaning of the two conditions may be stated as follows: (1.2) says that if x is the best possible choice at a certain point, then it is also the best candidate in any later iteration and, moreover, independent of the particular completion of the string. Secondly, (1.3) says that it is always better to choose first x and later some letter z than the other way round.

Our goal is to show that for any greedoid, the greedy algorithm will produce a basic word of maximal weight if the weight function is $\mathscr{L}$-admissible. In order to do so, we need two preliminary lemmas.

Lemma 1.1. *Let $(E, \mathscr{L})$ be a greedoid and $\alpha\beta \in \mathscr{L}$, $\alpha x \in \mathscr{L}$ for some $x \in E$. Then β can be partitioned into a sequence of letters y_i and (possibly empty) substrings β_i of the form*

$$\beta = y_1\beta_1 y_2\beta_2 \ldots y_r\beta_r$$

such that $\alpha\beta' \in \mathcal{L}$, where β' is given by

$$\beta' = x\beta_1 y_1 \beta_2 y_2 \ldots y_{r-1}\beta_r \ .$$

Proof. We perform induction on $|\beta|$, where for $|\beta| = 1$ we have $\beta' = x$ and there is nothing to prove. Suppose now that $\alpha\beta z \in \mathcal{L}$. By induction, β can be split into s, say, letters and substrings such that for

$$\beta' = x\beta_1 y_1 \beta_2 \ldots y_{s-1}\beta_s$$

we have $\alpha\beta' \in \mathcal{L}$. If we augment $\alpha\beta'$ from $\alpha\beta$, we get a $v \in \alpha\beta$ such that $\alpha\beta'v \in \mathcal{L}$. Then either $v = y_s$ and we set $r = s+1$ and $\beta_r = \emptyset$, or $v = z$ in which case we set $r = s$ and $\beta_r = \beta_s z$. $\qquad\square$

Lemma 1.2. *Let $(E, \mathcal{L})$ be a greedoid. Let $\alpha\beta \in \mathcal{L}$, $\alpha x \in \mathcal{L}$ and suppose β is partitioned as above. Then for $t = 1, \ldots, r$, $\alpha\beta^t \in \mathcal{L}$ where*

$$\beta^t = y_1\beta_1 y_2\beta_2 \ldots \beta_{t-1}x\beta_t y_t\beta_{t+1} \ldots y_{r-1}\beta_r \ .$$

Proof. By assumption we have $\gamma = \alpha y_1\beta_1 y_2 \ldots y_{t-1}\beta_{t-1} \in \mathcal{L}$ and by Lemma 1.1, $\delta = \alpha x\beta_1 y_1 \ldots \beta_{t-1}y_{t-1} \in \mathcal{L}$.

Augmenting γ from δ we obtain $\gamma x \in \mathcal{L}$, and if we augment γx from $\delta\beta_t y_t \ldots y_{r-1}\beta_r$, we get $\alpha\beta_t = \gamma x\beta_t y_t \ldots y_{r-1}\beta_r \in \mathcal{L}$. $\qquad\square$

We are now ready for

Theorem 1.3. *Let $(E, \mathcal{L})$ be a greedoid. Then the greedy algorithm gives the optimal solution for any $\mathcal{L}$-admissible objective function.*

Proof. Let α be a solution obtained by the greedy algorithm. Among all optimal solutions β, choose one such that the common initial string γ of α and β is maximal, and set $\alpha = \gamma\alpha'$ and $\beta = \gamma\beta'$. If $\alpha \neq \beta$, then α' can be written as

$$\alpha' = x\alpha'' \ .$$

By Lemmas 1.1 and 1.2, β' can be partitioned into

$$\beta' = y_1\beta_1 y_2\beta_2 \ldots y_r\beta_r$$

such that for every $t = 1, \ldots, r$

$$\gamma\beta^t = \gamma y_1\beta_1 \ldots y_{t-1}\beta_{t-1}x\beta_t y_t \ldots y_{r-1}\beta_r \in \mathcal{L} \ .$$

According to the greedy algorithm, x was chosen such that for every $y \in E$ with $\gamma y \in \mathcal{L}$ we have

$$w(\gamma x) \geq w(\gamma y) \ .$$

If we use property (1.3) repeatedly, we get

$$w(\gamma\beta^1) = w(\gamma x\beta_1 y_1 \ldots \beta_{r-1} y_{r-1}\beta_r) \geq w(\gamma\beta^2) \geq \ldots \geq w(\gamma\beta^r)$$
$$= w(\gamma y_1\beta_1 \ldots y_{r-1}\beta_{r-1} x\beta_r) \geq w(\gamma y_1\beta_1 \ldots y_{r-1}\beta_{r-1} y_r\beta_r)$$
$$= w(\beta) ,$$

where the last inequality follows from property (1.2). Since β is an optimal solution, $\gamma\beta^1$ must also be optimal. But $\gamma\beta^1$ has a longer initial string in common with α, contradicting the choice of β. Hence $\alpha = \beta$.   □

Many objective functions of combinatorial interest do not depend on the ordering and therefore they may be viewed as functions defined on $\mathscr{F}$ and not on $\mathscr{L}$. For such a function, condition (1.3) is trivially satisfied and condition (1.2) can be written as follows:

(1.4)   Let $A \subseteq B, x \in E\backslash B$, and assume that $A, B, A \cup x, B \cup x \in \mathscr{F}$. If for all $y \in \Gamma(A), w(A \cup y) \leq w(A \cup x)$, then for all $z \in \Gamma(B), w(B \cup z) \leq w(B \cup x)$.

Let us give some examples of $\mathscr{L}$-admissible functions. If $(E, \mathscr{F})$ is a matroid, then linear objective functions are $\mathscr{L}$-admissible. So Theorem 1.3 implies the Edmonds-Rado Theorem. Another trivial example is the cardinality function $A \rightarrow |A|$ for any greedoid. Given $f : E \rightarrow \mathbb{R}$ define $w(X) = \min\{f(x) : x \in X\}$. Then this **bottleneck function** satisfies (1.4).

An order-dependent generalization of bottleneck functions can be obtained as follows:

Let $f : E \times \mathbb{N} \rightarrow \mathbb{R}$ be monotone in the second component, i.e.

$$f(x, k) \leq f(x, k+1) \text{ for every } x \in E, k \in \mathbb{N} .$$

For $\alpha = x_1 x_2 \ldots x_n$ let

$$w(\alpha) = \min\{f(x_i, i) : 1 \leq i \leq n\} .$$

The reader readily verifies that this **generalized bottleneck function** is $\mathscr{L}$-admissible for any greedoid.

Our next example shows that Dijkstra's shortest path algorithm may also be viewed in this framework. Let $(E, \mathscr{F})$ be a local poset greedoid and $c : E \rightarrow \mathbb{R}_+$ a weighting of its elements. For each $X \in \mathscr{F}$ and $x \in X$ consider the unique x-path P_x contained in X and define $\bar{c}(X) = \sum_{x \in X} c(P_x)$. Then $-\bar{c}(X)$ satisfies (1.4). Let $A \subseteq B, x \in E\backslash B$ such that $A, B, A \cup x, B \cup x \in \mathscr{F}$ and for all $y \in \Gamma(A), \bar{c}(A \cup y) \geq \bar{c}(A \cup x)$. Let $z \in \Gamma(B)$. Let P_z be the unique z-path in $B \cup z$. Let y be the first element of P_z not in A (in some feasible ordering), then P_z contains a y-path $P_y \subseteq A \cup y$. Let P_x be the unique x-path in $A \cup x$. Then $\bar{c}(A \cup x) = \bar{c}(A) + c(P_x), \bar{c}(A \cup y) = \bar{c}(A) + c(P_y)$, and hence by the choice of x, we have $c(P_x) \leq c(P_y) \leq c(P_z)$. Hence $\bar{c}(B \cup z) = \bar{c}(B) + c(P_z) \geq \bar{c}(B) + c(P_x) = \bar{c}(B \cup x)$.

So it follows that a basis B minimizing $\bar{c}(B)$ can be found by the greedy algorithm. This example is motivated by Dijkstra's algorithm for finding a shortest path in a graph with edge-weights c_e from a root to any other vertex. In fact it is easy to show that there is a spanning tree containing a shortest path from a root to every vertex. Clearly this tree minimizes $\bar{c}$ in the undirected branching greedoid.

Conversely, every spanning tree minimizing $\bar{c}$ contains a shortest path from the root to each vertex. The greedy algorithm minimizing $\bar{c}$ over the branching greedoid translates to Dijkstra's algorithm. If we try to maximize $\bar{c}$ by the greedy algorithm, it results in depth-first search. However, the greedy algorithm does not give the maximum of $\bar{c}$. In fact the problem of maximizing $\bar{c}$ is NP-hard (it includes the problem of finding a Hamilton path.)

As for matroids, the greedy algorithm may serve as an algorithmic characterization of greedoids among all hereditary languages.

Theorem 1.4. *Let $(E, \mathcal{L})$ be a simple hereditary language. Then $(E, \mathcal{L})$ is a greedoid if and only if the greedy algorithm maximizes the cardinality function and all generalized bottleneck functions over $(E, \mathcal{L})$.*

Proof. We have just seen that if $(E, \mathcal{L})$ is a greedoid, then the greedy algorithm will maximize these functions. Conversely, observe that the greedy algorithm maximizes the cardinality function if and only if all basic words have the same length. Let $\alpha, \beta \in \mathcal{L}, \alpha = x_1 \dots x_k$ and $\beta = y_1 \dots y_\ell$ with $k = \ell + 1$. We must show that we can augment β from α. Clearly we may assume $\beta \not\subseteq \alpha$. Define a function f by

$$f(z, i) = \begin{cases} 0 & \text{for } i \le k \text{ and } z \notin \tilde{\alpha} \cup \tilde{\beta} \\ k+1 & \text{for all } i \in \mathbb{N} \text{ and } z \in \tilde{\beta} \\ k & \text{otherwise} \end{cases}$$

and let w be the induced generalized bottleneck function. Then for any basic word γ starting with α we have $w(\gamma) = k$. On the other hand, a greedy solution starts with the string β. After l steps it adds a new letter z which is either in $\tilde{\alpha} \backslash \tilde{\beta}$ or in $E \backslash (\tilde{\alpha} \cup \tilde{\beta})$. But if $z \in E \backslash (\tilde{\alpha} \cup \tilde{\beta})$, then the solution produced by the greedy algorithm will have value 0. Hence the greedy algorithm can only be optimal if $\beta z \in \mathcal{L}$ for some $z \in \tilde{\alpha}$. □

Generalizing a result of Lawler [1973] on a scheduling problem over a poset antimatroid, we relax generalized bottleneck functions to **nested bottleneck functions**

$$w(x_1 \dots x_n) = \min\{f(x_i, \{x_1, \dots, x_i\}), 1 \le i \le n\},$$

where $f : E \times 2^E \to \mathbb{R}$ is monotone increasing in the second component. This class of functions yields an algorithmic characterization of antimatroids (cf. Boyd and Faigle [1990]).

Theorem 1.5. *Let $(E, \mathcal{L})$ be a hereditary language. Then the greedy algorithm maximizes the cardinality function and all nested bottleneck functions over $(E, \mathcal{L})$ if and only if $(E, \mathcal{L})$ is a truncated antimatroid.*

Proof. Let $(E, \mathcal{L})$ be a truncated antimatroid of rank $k + 1$. We may assume by induction on k that the greedy algorithm is optimal for all truncations of smaller rank. Let $\alpha x_{k+1} = x_1 \dots x_{k+1}$ be the sequence generated by the greedy algorithm and suppose there is a word $\beta y_{k+1} = y_1 \ldots y_{k+1}$ with $w(\beta y_{k+1}) > w(\alpha x_{k+1})$. Then $f(y_i, \{y_1, \ldots, y_i\}) > f(x_{k+1}, \{x_1, \ldots, x_{k+1}\})$ for $1 \leq i \leq k+1$.

If $\alpha = \beta$, then $\alpha y_{k+1} \in \mathcal{L}$ contradicts the choice of x_{k+1} by the greedy algorithm. So let j be the first index in β such that $y_j \notin \tilde{\alpha}$. Then $\alpha y_j \in \mathcal{L}$ as $(E, \mathcal{L})$ is a truncated antimatroid. Since $f(y_j, \{x_1, \ldots, x_k, y_j\}) \geq f(y_j, \{y_1, \ldots y_j\})$, y_j should have been chosen in step $k+1$, a contradiction. So the greedy solution must be optimal.

Conversely, assume that the greedy algorithm optimizes every nested bottleneck function. Then in particular $(E, \mathcal{L})$ is a greedoid. Suppose it is not a truncated antimatroid. Then there exists a word $\alpha x = a_1 \ldots a_m x$ and a shortest word $\alpha y_1 \ldots y_k$ such that $x \notin \{y_1, \ldots, y_k\}, m+k < r(E)$ and $\alpha x y_1 \ldots y_k \notin \mathcal{L}$. Define a function f as follows:

$$f(z, B) = \begin{cases} 1 & \text{if } z \in \{a_1, \ldots, a_m, x, y_1, \ldots, y_k\} \text{ or if } y_k \in B \\ 0 & \text{otherwise .} \end{cases}$$

Clearly f is monotone in the second component. The greedy algorithm will generate a solution β starting with $\alpha x y_1 \ldots y_{k-1}$ such that $w(\beta) = 0$, while any word starting with $\gamma = \alpha y_1 \ldots y_k$ has value $w(\gamma) \geq 1$, a contradiction. $\qquad \square$

2. Linear Functions

We now turn to the problem of maximizing linear objective functions over greedoids, i.e. for a weighting $c : E \to \mathbb{R}$ of the ground set we set $c(X) = \sum_{x \in X} c(x)$ and consider the problems

$$\max \{c(X) : x \in \mathcal{F}\}$$

and

$$\max \{c(X) : X \in \mathcal{B}\} .$$

Note that if $c \geq 0$, then these two problems are equivalent. Furthermore, in the second problem we can assume without loss of generality that $c \geq 0$, e.g. by adding a large constant to each $c(e)$.

In general, linear functions are not $\mathcal{L}$-admissible. Hence we cannot expect the greedy algorithm to be optimal. A natural question to ask is under what conditions a linear objective function may be optimized over a greedoid by the greedy algorithm. We approach this problem from two sides: First, we derive a worst-case bound for the maximization of special classes of linear functions over hereditary languages. In particular, for greedoids we describe classes of linear functions for which the greedy algorithm is optimal. Secondly, we characterize those greedoids for which the greedy algorithm always yields an optimal solution when applied to linear functions.

Let $(E, \mathcal{F})$ be an accessible set system. We define three versions of the concept of rank. For any $X \subseteq E$ let $\ell r : 2^E \to \mathbb{N}$ be the **lower rank** of X given by

(2.1) $$\ell r(X) = \min \{|A| : A \text{ basis of } X\} .$$

The **(upper) rank** is $r(X) = \max\{|A| : A \text{ basis of } X\}$. Finally, the **basis rank** is defined by $\beta(X) = \max\{|X \cap A| : A \in \mathcal{F}\}$. Let $\mathcal{R} = \{X \subseteq E : \beta(X) = r(X)\}$ be the system of rank feasible sets.

The **rank quotient** $q(\mathcal{F})$ is defined as

$$q(\mathcal{F}) = \min_{X \subseteq E} \frac{\ell r(X)}{r(X)} .$$

Obviously $q(\mathcal{F}) = 1$ if and only if $(E, \mathcal{F})$ is a greedoid.

Let $c : E \to \mathbb{R}_+$ be a weight function. A **level set** of c is a set of the form $\{x \in E : c(x) \geq z\}$. We define

$$q(c, \mathcal{F}) = \min \{\frac{lr(X)}{\beta(X)} : X \text{ level set of } c\} .$$

We say that c is $\mathcal{R}$-**compatible** if every level set of c is rank feasible. In this case $q(c, \mathcal{F}) \geq q(\mathcal{F})$.

We consider the following optimization problem:

(2.2)   maximize $\{c(X)\} : X \in \mathcal{F}\}$.

Theorem 2.1. *Let $(E, \mathcal{F})$ be an accessible set system and $c : E \to \mathbb{R}_+$ a weight function. Let $O \in \mathcal{F}$ be an optimal solution of problem (2.2), and $G \in \mathcal{F}$ a solution produced by the greedy algorithm. Then*

$$q(c, \mathcal{F}) \, c(O) \leq c(G) \leq c(O) .$$

If c is $\mathcal{R}$-compatible, then

$$q(\mathcal{F}) \, c(O) \leq c(G) \leq c(O) .$$

In particular, if $(E, \mathcal{F})$ is a greedoid and c is $\mathcal{R}$-compatible, then the greedy solution is optimal.

Proof. Let $E = E^1 \cup E^2 \cup \ldots \cup E^k$ be a partition of the ground set such that for any $x \in E^i$ and $y \in E^j$, $c(x) = c(y)$ if $i = j$ and $c(x) < c(y)$ if $i > j$. Hence for any block E^i there is a unique function value denoted by c_i.

Set $E_0 = \emptyset$ and $E_j = E^1 \cup \ldots \cup E^j$, $G_j = G \cap E_j$, $O_j = O \cap E_j$ and $d_j = c_j - c_{j+1}$ with $c_{k+1} = 0$. Since G is obtained by the greedy algorithm, G_j is a basis of E_j, i.e.

$$|G_j| \geq \ell r(E_j) \geq q(c, \mathcal{F})\beta(E_j) .$$

Moreover, $|O_j| \leq \beta(E_j)$. Thus

$$c(G) = \sum_{j=1}^{k} (|G_j| - |G_{j-1}|)c_j = \sum_{j=1}^{k} |G_j| (c_j - c_{j+1})$$

$$\geq \sum_{j=1}^{k} \ell r(E_j)d_j \geq q(c, \mathcal{F}) \sum_{j=1}^{k} \beta(E_j)d_j \geq q(c, \mathcal{F}) \sum_{j=1}^{k} |O_j| d_j$$

$$= q(c, \mathcal{F}) \sum_{j=1}^{k} (|O_j| - |O_{j-1}|)c_j = q(c, \mathcal{F})c(O) .$$

The other two assertions follow trivially. $\square$

Note that if in particular the accessible set system is an independence system, then $\mathcal{R} = 2^E$ and the above theorem reduces to the well-known worst-case analysis of the greedy algorithm for independence structures.

We now turn to the question under what condition on the greedoid every linear function can be optimized by the greedy algorithm. The algorithm of Prim [1957] for optimum spanning trees shows that the undirected branching greedoid has this property. We will see that Gauss greedoids also have this property. We say that a greedoid $(E, \mathcal{F})$ has the **strong exchange property** if the following holds:

(2.3)   For $A \in \mathcal{F}, B \in \mathcal{B}, A \subseteq B$ and $x \in E \setminus B$ with $A \cup x \in \mathcal{F}$ there exists a $y \in B \setminus A$ such that $A \cup y \in \mathcal{F}$ and $B \setminus y \cup x \in \mathcal{F}$.

Theorem 2.2. *Let $(E, \mathcal{F})$ be a greedoid. Then the following statements are equivalent:*

(i)   *For every linear objective function $c \geq 0$ the greedy algorithm is optimal.*
(ii) $\mathcal{M} = \{X \subseteq B : B$ *is a basis of* $\mathcal{F}\}$ *is a matroid and every closed set in $(E, \mathcal{F})$ is also closed in $(E, \mathcal{M})$.*
(iii) $(E, \mathcal{F})$ *has the strong exchange property.*

Proof. (i) $\Rightarrow$ (ii) : Consider two sets $X, Y \in \mathcal{M}$ with $|Y| > |X|$, let $0 < t < 1$ and

$$c_t(x) = \begin{cases} 1 & \text{for } x \in X, \\ t & \text{for } x \in Y \setminus X, \\ 0 & \text{otherwise} . \end{cases}$$

The set of greedy bases is independent of t. If t is very small, then every optimum basis must contain X and hence every greedy basis must contain X. If t is close to 1, then $c_t(Y) > c_t(X)$ and hence every optimum basis must contain at least one element of $Y \setminus X$. So every greedy basis must contain some $y \in Y \setminus X$. Hence $X \cup y \in \mathcal{M}$ and $(E, \mathcal{M})$ is a matroid.

Now let X be an $\mathcal{F}$-closed set, A an $\mathcal{F}$-basis of X, and A' an $\mathcal{M}$-basis of X with $A \subseteq A'$. Define the linear function

$$c(x) = \begin{cases} 1 & \text{if } x \in A' \\ 0 & \text{otherwise} . \end{cases}$$

A' is contained in some $\mathcal{F}$-basis, thus every optimum $\mathcal{F}$-basis must contain A' and so every greedy $\mathcal{F}$-basis must contain A'. On the other hand, for any $y \in E \setminus X$ there is a greedy solution B starting with $A \cup y$. So

$$A \cup y \subseteq (A \cup y) \cup A' = A' \cup y \subseteq B .$$

Thus $A' \cup y \in \mathcal{M}$, i.e. $y \notin \sigma_{\mathcal{M}}(A') = \sigma_{\mathcal{M}}(X)$.

(ii) $\Rightarrow$ (iii): The set $B \cup x$ contains a fundamental circuit C in the matroid $(E, \mathcal{M})$. Clearly $x \in C \setminus \sigma_{\mathcal{F}}(A)$. Moreover, $(C \setminus x) \setminus \sigma_{\mathcal{F}}(A) \neq \emptyset$ since otherwise

$$\sigma_{\mathcal{M}}(C \setminus x) \subseteq \sigma_{\mathcal{M}}(\sigma_{\mathcal{F}}(A)) = \sigma_{\mathcal{F}}(A) ,$$

and hence $x \in \sigma_{\mathscr{F}}(A)$. So for any $y \in (C \backslash x) \backslash \sigma_{\mathscr{F}}(A)$ we have $A \cup y \in \mathscr{F}$ and $B \cup x \backslash y \in \mathscr{M}$, but $B \cup x \backslash y$ is a basis of $\mathscr{M}$ and hence it is in $\mathscr{F}$.

(iii) $\Rightarrow$ (i) : Let c be any objective function and $A = \{a_1, a_2, \ldots, a_m\}$ a greedy solution. Among all optimal bases choose a basis B such that $\{a_1, \ldots, a_k\} \subseteq B$, $a_{k+1} \notin B$ and k is maximal. By the strong exchange property there exists an element $y \in B \backslash A$ such that $A \cup y \in \mathscr{F}$ and $B \cup a_{k+1} \backslash y \in \mathscr{F}$. Since a_{k+1} was chosen by the greedy principle, we have $w(y) \leq w(a_{k+1})$ and hence $w(B \cup a_{k+1} \backslash y) = w(B)$, contradicting the choice of B. □

For linear functions, we may also think of applying a greedy-type algorithm which is dual to the previous one. Instead of starting with the empty set and augmenting it step by step by a best possible element, we now start with the whole set and delete singletons with the smallest possible weight as long as the current set contains a basis.

It is immediately clear that this **worst-out greedy algorithm** for greedoids $(E, \mathscr{F})$ is equivalent to applying the **(best-in)** greedy to the independence system whose bases are the complements of the bases of $(E, \mathscr{F})$. Hence we have the following

Proposition 2.3. *Let $(E, \mathscr{F})$ be a greedoid. Then for every linear objective function the worst-out greedy algorithm is optimal if and only if $(E, \mathscr{F})$ is a slimmed matroid.*

□

Next we consider the problem of maximizing a (not necessarily non-negative) objective function over the feasible sets of a greedoid.

If we want to attack this problem by a greedy strategy we can try the following: For every $k = 0, \ldots, r(E)$ let $(E, \mathscr{F}^{(k)})$ be the k-truncation of $(E, \mathscr{F})$. Let $X^{(k)}$ be a greedy solution for the problem $\max\{c(X) : X$ basis in $\mathscr{F}^{(k)}\}$. Among all these solutions choose one with maximum value. Note that to obtain $X^{(k)}$ it suffices to take $X^{(k-1)}$ and (if possible) augment it greedily. This leads to the following algorithm.

Modified Greedy Algorithm

(1) Set $\alpha^* = \alpha = \emptyset$
(2) If $\Gamma(\alpha) = \emptyset$, Stop.
(3) Choose $x \in \Gamma(\alpha)$ such that $c(x) \geq c(y)$ for all $y \in \Gamma(\alpha)$.
(4) Set $\alpha = \alpha x$. If $c(\alpha) > c(\alpha^*)$, set $\alpha^* = \alpha$. Go to 2.

This algorithm gives rise to the following algorithmic characterization of Gauss greedoids, due to Goecke [1988]:

Theorem 2.4. *Let $(E, \mathscr{F})$ be a set system with $\emptyset \in \mathscr{F}$. Then the following statements are equivalent:*

(i)   *For every weight function c, the modified greedy algorithm finds a feasible set of maximal weight.*

(ii) $(E, \mathcal{F}^{(k)})$ *has the strong exchange property (2.3) for every* $k = 0, \ldots, r(E)$.

(iii) $(E, \mathcal{F})$ *is a Gauss greedoid.*

Proof. The fact that (ii) implies (i) follows from Theorem 2.2.

(i) $\Rightarrow$ (iii): For $A \subseteq E$ let

$$c_A(e) = \begin{cases} 1 & \text{if } e \in A \\ -|E| & \text{if } e \notin A . \end{cases}$$

If $A \in \mathcal{F}$, then A is the unique solution for the problem

$$\max\{c_A(X) : X \in \mathcal{F}\} .$$

The modified greedy algorithm must produce A, and therefore there exists an ordering $a_1 \ldots a_k$ of the elements of A such that $a_1 a_2 \ldots a_i \in \mathcal{F}$ for $0 \le i \le k$. Hence $(E, \mathcal{F})$ is accessible.

For any $A \subseteq E$ the greedy solutions correspond to bases of A with respect to $(E, \mathcal{F})$. Lemma IV.1.4 then implies that $(E, \mathcal{F})$ is a greedoid.

Now let $B \in \mathcal{F}$, $B \backslash z \cup \{x, y\} \in \mathcal{F}$ with $z \in B$, $x, y \notin B$, $x \ne y$, and suppose $B \cup y \notin \mathcal{F}$ and $B \backslash z \cup y \notin \mathcal{F}$. Let

$$c(e) = \begin{cases} 1 & \text{for } e \in B \backslash z \\ c_z & \text{for } e = z \\ c_y & \text{for } e = y \\ c_x & \text{for } e = x \\ -|E| & \text{otherwise}, \end{cases}$$

where

$$-|E| < c_x < c_z < c_y < 1 .$$

Observe that the algorithm will always generate the same basic word α, no matter what values for c_x, c_z and c_y are chosen, as long as the above inequalities hold. Moreover,

$$\alpha = \beta\gamma \text{ for some } \beta \text{ with } \tilde{\beta} = B .$$

To see this, let $c_x = 1 - |E|$, $c_z = 1/3$ and $c_y = 1/2$. Then B is the unique optimal solution for $\max\{c(X) : X \in \mathcal{F}\}$.

Now let $c_x = -1/2$, $c_z = -1/3$ and $c_y = 2/3$. Since $B \cup y$, $B \backslash z \cup y \notin \mathcal{F}$, $B \backslash z \cup \{x, y\}$ is the unique optimal solution for this objective function. However, $B \not\subseteq B \backslash z \cup \{x, y\}$, contradicting the optimality of the greedy algorithm. Hence either $B \cup y \in \mathcal{F}$ or $B \backslash z \cup y \in \mathcal{F}$, proving that $(E, \mathcal{F})$ is a Gauss greedoid.

(iii) $\Rightarrow$ (ii): Let $(E, \mathcal{M}_0), (E, \mathcal{M}_1), \ldots, (E, \mathcal{M}_{r(E)})$ be the matroids in the definition of Gauss greedoids. Let B be a basis of $\mathcal{F}^{(k)}$, $A \subseteq B$, $A \in \mathcal{F}^{(k)}$ and $x \in E \backslash B$. Assume that $A \cup x \in \mathcal{F}^{(k)}$. Let C be the fundamental circuit of x with respect to B in the matroid $(E, \mathcal{M}_k)$. Let $|A| = t$. Suppose $C \backslash x \subseteq \sigma_{\mathcal{M}_{t+1}}(A)$. $\sigma_{\mathcal{M}_{t+1}}(A)$ is closed in $\mathcal{M}_{t+1}$ and hence, by the properties of strong maps, $\sigma_{\mathcal{M}_{t+1}}(A)$ is closed in $\mathcal{M}_k$. Since C is a circuit in $\mathcal{M}_k$, it follows that $x \in \sigma_{\mathcal{M}_{t+1}}(A)$. But then $A \cup x \notin \mathcal{F}$, a contradiction. So $C \backslash x \not\subseteq \sigma_{\mathcal{M}_{t+1}}(A)$, and hence there exists a $y \in C \backslash x$ such that $A \cup y \in \mathcal{F}$. Moreover, $B \backslash y \cup x$ is a basis of $\mathcal{M}_k$ and hence it belongs to $\mathcal{F}$.   □

We close this section with the remark that the problem of finding the maximum of a linear objective function over quite simply presented greedoids is NP-complete, as we can reduce the Steiner problem to it. Let G be a graph and $v_1, \ldots, v_t \in V(G)$. The Steiner problem, which is NP-complete, asks for a subtree of G with at most k edges containg $v_1, \ldots, v_t$. Consider v_1 as the root and add edges $(v_2, u_2), \ldots, (v_t, u_t)$ to the graph where $u_2, \ldots, u_t$ are new vertices. Consider the $(k + t - 1)$-truncation of the branching greedoid of the resulting graph and weights 1 on $(v_2, u_2), \ldots, (v_t, u_t)$ and 0 elsewhere. Then there exists a basis of weight $t - 1$ if and only if the Steiner problem has a solution.

There are also other classes of greedoids for which the problem of maximizing a linear objective function over bases can be solved in polynomial time, e.g. directed branchings (Edmonds [1967]). But unfortunately no general characterization of such classes is known.

3. Polyhedral Descriptions

In the first part of this section we study polyhedra associated with antimatroids. Let $(E, \mathcal{A})$ be an antimatroid and define

$$\text{conv}(\mathcal{A}) = \text{conv}\{\chi^A : A \in \mathcal{A}\} = \{\sum_A \lambda_A \chi^A : \lambda_A \geq 0, \sum_A \lambda_A = 1\},$$

where χ^A is the incidence vector of the set A and **conv** is the convex hull operator in Euclidean space. The convex hull of incidence vectors of convex sets, which is another natural polytope to consider, can be obtained by reflecting conv(A) in the center of the unit cube. Sometimes we will have to be content with finding a linear description of

$$\text{cone}(\mathcal{A}) = \text{cone}\{\chi^A : A \in \mathcal{A}\} = \{\sum_A \lambda_A \chi^A : \lambda_A \geq 0\}.$$

Of course a linear description of conv($\mathcal{A}$) yields a linear description of cone($\mathcal{A}$). For a detailed treatment of algorithmic aspects of polyhedra, see Grötschel, Lovász and Schrijver [1988].

The accessibility of $\mathcal{A}$ implies that no linear equation $ax = b$ is satisfied by all $x \in \text{conv}(\mathcal{A})$. Hence conv($\mathcal{A}$) has full dimension. We are interested in a description of this polytope in terms of linear inequalities. In particular, we look for facets, i.e. maximal subpolytopes of conv($\mathcal{A}$) satisfying one of the defining inequalities by equation.

The reader easily verifies that the box constraints

$$0 \leq x_e \leq 1 \text{ for all } e \in E$$

are valid inequalities for conv($\mathcal{A}$). Our first result characterizes the facets (essential inequalities) among these.

Lemma 3.1.

(i)   The inequality $x_e \geq 0$ is a facet of conv($\mathcal{A}$) if and only if $\{e\}$ is convex.
(ii) The inequality $x_e \leq 1$ is a facet of conv($\mathcal{A}$) if and only if there is no circuit $(C, e) \in \mathcal{C}$ with root e and $|C| = 2$.

Proof. (i) If $\{e\}$ is convex, then $E \backslash e \in \mathcal{A}$. Hence there is a chain of feasible sets connecting the empty set to $E \backslash e$. Their incidence vectors are affinely independent and satisfy $x_e = 0$, i.e. $x_e \geq 0$ is a facet. Conversely, if $x_e \geq 0$ is a facet, then there exist $n - 1$ feasible subsets of $E \backslash e$ with affinely independent incidence vectors. The union of these sets is $E \backslash e$. Hence $E \backslash e \in \mathcal{A}$.

(ii) Clearly, if there is a two-element circuit $(\{e, f\}, e)$ with root e, then $x_e = 1$ implies $x_f = 1$ and hence $x_e \leq 1$ cannot be a facet. Conversely, assume that no such circuit exists. Then for any $a \in E \backslash e$ there exists a subset $A \in \mathcal{A}$ with $e \in A$, $a \notin A$ but $A \cup a \in \mathcal{A}$. Hence the unit vector $\chi^a = \chi^{A \cup a} - \chi^A$ is in the linear span of incidence vectors satisfying $x_e = 1$. Since

$$\chi^e = \chi^A - \sum_{a \in A \backslash e} \chi^a$$

is also in this span, these vectors generate $\mathbb{R}^{|E|}$, i.e. $x_e \leq 1$ is a facet.   □

A second class of inequalities is given by the **circuit constraints:**

$$x_r \leq x(C \backslash r),$$

where (C, r) is a circuit rooted at r. Clearly the circuit constraints define valid inequalities for conv($\mathcal{A}$) by Lemma III.3.6. In fact, this lemma implies that the 0-1 vectors satisfying the circuit constraints are exactly the incidence vectors of feasible sets. We will see later on that they do not always suffice to describe conv($\mathcal{A}$). In our first example (which is a reformulation of a well-known fact in greedoid terms), however, they give, together with the box constraints, a complete description of the polytope.

Theorem 3.2. Let $(E, \mathcal{A})$ be a poset antimatroid on the ordered set $(E, \leq)$. Then

$$\mathrm{conv}(\mathcal{A}) = \{x \in \mathbb{R}^{|E|} : 0 \leq x_e \leq 1, \; x_e \leq x_f \text{ for all } e > f \text{ in } E\} \,.$$

Proof. We must show that the vertices of the above polytope P on the right hand side are precisely the ideals of $(E, \leq)$. Clearly P contains these vectors.

Let y be the incidence vector of some extreme point of P. Suppose y is not the incidence vector of an ideal. Clearly y is not a 0-1 vector. Then let $A = \{e \in E : 0 < y_e < 1\}$ and $\varepsilon = \min\{y_e, 1 - y_e : e \in A\}$. Define

$$y_e^1 = \begin{cases} y_e^1 + \varepsilon & \text{for } e \in A \\ y_e^1 & \text{for } e \notin A \end{cases}$$

and

$$y_e^2 = \begin{cases} y_e^2 - \varepsilon & \text{for } e \in A \\ y_e^2 & \text{for } e \notin A \end{cases}.$$

Then $y^1, y^2 \in P$ and $y = \frac{1}{2}(y^1 + y^2)$, in contradiction to the assumption on y.   □

We continue with some further examples of classes of antimatroids for which the convex hull of feasible sets can be described. In some cases it will be more convenient to work with convex sets rather than feasible sets. Let N be the family of convex sets. We consider the polytope

$$\text{conv}(N) = \text{conv}\{\chi^U : U \in N\}.$$

Since the mapping $x \mapsto 1-x$ maps $\text{conv}(A)$ onto $\text{conv}(N)$, our previous discussions can be applied to describe facets of $\text{conv}(N)$ of the form $x_e \geq 0$ or $x_e \leq 1$. Furthermore, the circuit constraints for $\text{conv}(N)$ look like this: if (C, r) is a circuit, then

$$x(C \setminus r) - x(r) \leq |C| - 2.$$

The following lemma is useful for eliminating some inequalities from the list of possible facets:

Lemma 3.3. *Assume that every element of E is convex as a singleton set. Let $b^T x \leq 0$ define a facet of* $\text{conv}(N)$*. Then it is equivalent to a non-negativity constraint.*

Proof. Since every singleton $\{e\}$ is convex, it follows that $b_e \leq 0$ and so the inequality $b^T x \leq 0$ is implied by the non-negativity constraints. Since it is a facet, it must be equivalent to one of them.   □

Next we characterize the convex hull of convex sets in a double shelling antimatroid on a poset $(E, \leq)$. Let us introduce the following constraints. Let $e_1 < e_2 < \ldots < e_{2k+1}$ be a chain of odd length ($k \geq 0$). Then

$$x_{e_1} - x_{e_2} + \ldots - x_{2k} + x_{2k+1} \leq 1$$

is trivially valid for every convex set. We call these constraints **alternating chain constraints**. The elements $e_1, e_3, \ldots$ are called **odd**, the elements $e_2, e_4, \ldots$ are called **even** elements in the chain. Note that in particular the constraints $x_e \leq 1$ as well as the circuit constraints are special alternating chain constraints (with $k = 0$ and $k = 1$, respectively). The following theorem is a consequence of a result of Gröflin [1984] on path-systems; here we give a direct proof.

Theorem 3.4. *The convex hull of convex sets of the double shelling antimatroid of a poset is given by the non-negativity constraints and the alternating chain constraints.*

Proof. The proof depends on the main theorem on total dual integrality, due to Hoffman [1974] and Edmonds and Giles [1977]. A system $Ax \leq b$ of linear inequalities is called **total primal integral** if every face of the polyhedron it defines contains an integer point. In other words, for every vector c the linear program

$$\max \{cx : \ Ax \le b\}$$

has an integral optimum solution if it has an optimal solution at all. (In the case of a bounded polyhedron, this is equivalent to saying that every vertex is integral.) The system is **total dual integral** if for every integer vector c the dual linear program

$$\min \{yb : \ yA = c, y \ge 0\}$$

has an integer optimal solution if it has an optimal solution at all. Now the main theorem on total dual integrality states that *every totally dual integral system of linear inequalities is totally primal integral.* (For a detailed exposition of total dual integrality see Schrijver [1986].)

Consider the system of linear inequalities consisting of the non-negativity constraints and the alternating chain constraints. As we remarked earlier, the only integral solutions of this system are the incidence vectors of convex sets. Therefore it suffices to show that this system is totally primal integral. By the theorem quoted above it suffices to show that it is totally dual integral.

So consider an integral vector c and an optimum dual solution. In this dual program we will have variables y_A associated with the alternating chains A and variables associated with the non-negativity constraints. The latter can be eliminated easily, and so we get for each $e \in E$ the dual constraint

$$\sum \{y_A : \ e \text{ odd in } A\} - \sum \{y_A : \ e \text{ even in } A\} \le c(e).$$

Let us denote by $w(e)$ the left hand side of this equation. The objective function is $\sum_A y_A$. We call a chain A **active** if $y_A > 0$.

Next we want to simplify the dual solution so that each element e is either odd in all active chains A containing it, or it is always even. Let A_1 and A_2 be two active chains, and assume that e is an even element of A_1 but an odd element of A_2. Let A_1' be the chain formed by the elements of A_1 above e and the elements of A_2 below e; let A_2' be the chain formed by the elements of A_2 above e and the elements of A_1 below e (so e is not contained in these new chains). Let $\delta = \min\{y_{A_1}, y_{A_2}\}$. Add δ to $y_{A_1'}$ and to $y_{A_2'}$, and subtract δ from y_{A_1} and y_{A_2}. Then clearly we obtain a new optimal dual solution, and $\sum_A y_A|A|$ has decreased. So if among all optimal solutions we choose one for which $\sum_A y_A|A|$ is minimal, then every element e will be either odd in every active chain through it, or even in every such chain. We call the element e **odd** and **even** accordingly. It is clear that $w(e) \le 0$ for the even elements and $w(e) \ge 0$ for the odd elements.

We now construct an auxiliary minimum cost circulation problem. Consider two elements e' and e'' for every $e \in E$, and two new elements s and t. Define a directed graph G on $\{e', e'' : \ e \in E\} \cup \{s, t\}$ as follows: Connect s to all points e' where e is odd, and also connect all points e'' to t where e is odd. Connect e' to e'' for all e. Connect e'' to f' if one of e and f is odd, the other is even, and $e < f$. Finally, connect t to s.

Define capacities on the edges as follows: if e is an even element, then give a lower bound of $-c(e)$ to the edge $e'e''$ and a lower bound of 0 to every other edge. If e is an odd element, then give an upper bound of $c(e)$ to $e'e''$ and infinity to the other edges.

Define the cost of ts to be 1 and the cost of the other edges to be 0.

Observe that each active chain $e_1 < \ldots < e_{2k+1}$ gives rise to a cycle through ts: take the edges se'_1, $e'_1 e''_1$, $e''_1 e'_2$, …, $e''_{2k} e'_{2k+1}$, $e'_{2k+1} e''_{2k+1}$, $e''_{2k+1} t$ and ts. Conversely, every cycle in G passes through ts and corresponds to a chain in $(E, \leq)$ with odd length.

If we add up the cycles corresponding to active chains A with the corresponding coefficients y_A, we get a circulation that satisfies the capacity constraints and whose cost is $\sum_A y_A$. By the Integrality Theorem in flow theory, there is an integral circulation whose cost is at most this large. Now this circulation can be written as a linear combination of cycles with positive integral coefficients. Each of these cycles corresponds to a chain with odd length, and the corresponding coefficients give an integral optimum dual solution.

This proves the total dual integrality of the system and completes the proof.

$\square$

Next we study the polytope associated with the convex sets of the edge shelling antimatroid of a tree. To this end we introduce some constraints, which are analogous to the alternating chain constraints for the double shelling antimatroids of posets. For a tree $T = (V, E)$, a $\{0, +1, -1\}$-vector $x \in \mathbb{R}^E$ is called **alternating** if the edges corresponding to its non-negative entries cover all nodes, and every connected component of the graph formed by these edges contains exactly one edge corresponding to an entry 1. For each alternating vector a, the inequality

$$a^T x \leq 1$$

is called an **alternating vector constraint**. Note that in particular the constraints $x_e \leq 1$ as well as the circuit constraints are special alternating vector constraints. The following result is a special case of a theorem of Gröflin and Liebling [1981].

Theorem 3.5. *The convex hull of convex sets of the edge shelling antimatroid of a tree is given by the non-negativity constraints and the alternating vector constraints.*

Proof. Let $b^T x \leq \beta$ define a facet of $\text{conv}(N)$. Since $\emptyset$ is convex, we must have $\beta \geq 0$. If $\beta = 0$, then by Lemma 3.3 we have a non-negativity constraint. Assume that $\beta > 0$, then we may assume that $\beta = 1$.

Call a convex set U **tight** if χ^U lies on the facet $b^T x = 1$, i.e. if $b(U) = 1$. Observe that

(3.1)   if $U, V \in N$ are tight and $U \cup V \in N$, then $U \cup V$ and $U \cap V$ are also tight.

Moreover, we may observe that

(3.2)   if e and e' are two edges, then there must exist a tight convex set containing exactly one of them.

(Otherwise the equation $x_e = x_{e'}$ would be satisfied by all vertices of $\text{conv}(N)$ on the facet $b^T x = 1$, and hence these two linear equations would be equivalent, which is not the case.)

Consider the partition $E = E_0 \cup E_+ \cup E_-$, where

$$E_0 = \{e \in E : b_e = 0\} ,$$

$$E_+ = \{e \in E : b_e > 0\} ,$$

$$E_- = \{e \in E : b_e < 0\} .$$

Let W be the edge set of any connected component of $E_0 \cup E_+$. Note that if U is any tight subtree such that $V(U) \cap V(W) \neq \emptyset$, then $U \cup W \in N$ and

$$b(U \cup W) = b(U) + b(W \setminus U) \geq b(U) = 1 .$$

Since $b^T x \leq 1$ is valid for conv(N), we must have equality here. This implies that $U \cup W$ is tight and U must contain every $e \in W \cap E_+$. By (3.2), this implies that $|W \cap E_+| \leq 1$.

For any node v, let U_v denote the intersection of all tight subtrees containing v. Then by (3.1), U_v is also tight. This implies that U_v contains at least one edge incident with v; by (3.2), it must contain exactly one such edge e. Then we must have $b_e \geq 0$, otherwise deleting e from U_v would result in a subtree violating the inequality $b^T x \leq 1$.

Hence if $v \in V(W)$, then $W \cup U_v$ is a tight subtree containing no edge of E_- incident with v. So $W = \cap_{v \in V(W)}(W \cup U_v)$ is a tight subtree. Hence W must contain a unique edge $e_W \in E_+$ and $b_{e_W} = 1$.

To conclude the proof, it suffices to show that for $e \in E_-$ we have $b_e = -1$. Let $e = uv$. Then $Q = U_u \cup U_v \cup \{e\}$ is a subtree and so

$$2 + b_e = b(U_u) + b(U_v) + b_e = b(Q) \leq 1 .$$

This implies that $b_e \leq -1$.

Every edge e must be contained in a tight subtree, otherwise $x_e = 0$ would be valid for all subtrees on the facet $b^T x = 1$, which is impossible. Hence by (3.1), E is tight. Now $E_0 \cup E_+ = E \setminus E_-$ has exactly $|E_-| + 1$ connected components, and hence

$$1 = b(E) = b(E_-) + b(E_+ \cup E_0) \leq -|E_-| + (|E_-| + 1) = 1 .$$

We must have equality here, and hence $b_e = -1$ for every $e \in E_-$. $\square$

Now consider the vertex shelling of a tree. A vector $d \in \mathbb{R}^V$ is an **indegree vector** if for some orientation of T the indegree of each vertex v is $d(v)$. Equivalently, a nonnegative vector $d \in \mathbb{R}^V$ is an indegree vector if and only if $d(V) = |V| - 1$ and for each edge $e \in E$ we have

(3.3) $$d(V') \leq |V'| ,$$

where $V' \subseteq V$ is a connected component of $T \setminus e$.

Theorem 3.6. *Let $(V, \mathscr{A})$ be the vertex shelling greedoid of a tree T. Then*

$$\mathrm{conv}(\mathscr{A}) = \{x \in \mathbb{R}^E : 0 \leq x \leq 1 , \sum_{i \in V}(1 - d_i)x_i \geq 0 \text{ for all indegree vectors } d\} .$$

Proof. We prefer to work with convex sets rather than their complements. So our claim is equivalent to

$$\text{conv}(\mathcal{N}) = \{x \in \mathbb{R}^V : x \geq 0, \sum (1 - d_i)x_i \leq 1\} .$$

To see that the indegree constraints are valid, consider an arbitrary degree vector d and an incidence vector x of a convex set $K \in \mathcal{N}$. Then

$$\sum_{i \in V} (1 - d_i)x_i = \sum_{i \in K} (1 - d_i) = |K| - d(K) \leq 1 ,$$

since the edges in the subtree induced by K already give a contribution of $|K| - 1$ to $d(K)$.

Conversely, let $a^T x \leq b$ be some facet of $\text{conv}(\mathcal{N})$. Since $\emptyset \in \mathcal{N}$, we have $b \geq 0$. If $b = 0$, then by Lemma 3.3 the facet is equivalent to a non-negativity constraint. So we may assume that $b = 1$.

Call a convex set U tight if χ^U lies on the facet $a^T x = 1$. For every two nodes u and v, there exists a tight set containing exactly one of them. For suppose not, then $x_u = x_v$ for all vertices on the hyperplane $a^T x = 1$. So this must be the hyperplane $x_u = x_v$, which is impossible.

Let $uv \in E(T)$. There exists a tight set $K \in \mathcal{N}$ with $u \in K$ and $v \notin K$, say. Suppose that there also exists a tight set $K' \in \mathcal{N}$ with $u \notin K$, $v \in K$. Then $K \cup K' \in \mathcal{N}$ and $K \cap K' = \emptyset$. Hence $a(K \cup K') = 2 > 1$, contradicting the validity of $a^T x \leq 1$.

Orient the edge uv from u to v if there exists a tight set K with $u \in K$, $v \notin K$. Let d be the indegree vector of this orientation. Consider any tight set K. By definition, all edges between K and $V \setminus K$ are oriented from K to $V \setminus K$. Therefore

$$\sum_{v \in K} (1 - d_v) = |K| - d(K) = |K| - |K| + 1 = 1 .$$

Thus every vertex on $a^T x = 1$ lies on the hyperplane given by the above indegree inequality and hence $a^T x = 1$ is this inequality. □

We remark that every indegree inequality defines a facet. Consider a vertex $v \in V$ with predecessors $u_1, \ldots, u_k$ in that orientation. For each $1 \leq i \leq k$, let $T_i = (V_i, E_i)$ be the component of $T \setminus (u_i, v)$ containing u_i and set

$$T_0 = \bigcup_{i=1}^{k} T_i \cup \{v\} .$$

Then for any $0 \leq i \leq k$, edges connecting T_i to $T \setminus T_i$ are oriented out of T_i. Hence for the corresponding incidence vectors χ^{T_i}, the indegree inequality is tight. Since

$$\chi^v = \chi^{T_0} - \sum_{i=1}^{k} \chi^{T_i} ,$$

these vectors generate all unit vectors. Thus the indegree inequality is a facet.

For the next two classes of antimatroids we have no explicit description of conv($\mathscr{A}$), but at least from the algorithmic point of view, the polyhedra are still well behaved.

Theorem 3.7. *Let* $(E, \mathscr{A})$ *be the antimatroid of lower convex shelling in the plane. Then the validity of an inequality* $a^T x \leq b$ *for* conv($\mathscr{A}$) *can be tested in polynomial time.*

Proof. We describe the algorithm for the special case when the cone C in Definition III.2.2 of lower convex shelling is the non-negative orthant. Let $e_1, \ldots, e_n$ be an ordering of the points in E by increasing first coordinates and let $a^T x \leq b$ be an inequality which we want to test for validity. Since $a^T x \leq b$ is valid if and only if $a(K) \geq a(E) - b$ for all convex sets K, it suffices to compute

$$\min \{a(K) : K \in N\}.$$

Consider $i < j$ and let $f(i, j)$ be the minimum of $a(K)$ among all convex sets K whose two rightmost vertices are e_i and e_j. Set $H(i, j) = (\text{conv}(e_i, e_j) + \mathbb{R}^2_+) \cap E$ and

$$g(i, j) = \min\{f(k, i) : k < i , \ e_i \notin H(k, j)\} .$$

We set $g(i, j) = \infty$ if no such k exists (e.g. if $i = 1$). Then

$$f(i, j) = \min\{a(H(i, j)), \ g(i, j) + a(H(i, j) - H(i, i))\} .$$

So we can recursively compute $f(i, j)$ for all $i < j$. Hence

$$\min\{a(K) : K \in N\} = \min\{\min_{i<j} f(i, j), \ \min_i a(H(i, i))\} . \qquad \square$$

Similarly, we show

Theorem 3.8. *Let* $(E, \mathscr{A})$ *be the antimatroid of convex shelling in the plane. Then the validity problem for* conv($\mathscr{A}$) *can be solved in polynomial time.*

Proof. As before, we seek to minimize $a(C)$ among all convex sets C.

For each ordered quadruple (x, y, z, u) of points in E which forms a quadrilateral in this order, let $f(x, y, z, u)$ be the minimum of $a(C)$ over all convex sets C such that x, y, z, u are four consecutive extreme points in C. Let $P(x, y, z, u)$ be the set of all points $e \in E$ such that (x, y, z, u, e) is a convex pentagon in this order. Then

$$|P(x, y, z, u)| > |P(x, y, u, e)| \quad \text{for all } e \in P(x, y, z, u) .$$

We denote $H(A) = \text{conv}(A) \cup E$ for $A \subseteq E$. We have

$$f(x, y, z, u) = \min\{a(H(\{x, y, z, u\}), \ \min\{f(x, y, u, e) : \ e \in P(x, y, z, u)\}$$
$$+ a(H(\{y, z, u\}) - a(H(\{y, u\}))\} ,$$

and this yields a recursive formula to compute $f(x, y, z, u)$ (again, if no e exists, the second minimum is ∞). To obtain a minimum weight convex set, it suffices to find the minimum of f and compare it to the weights of convex hulls of singletons, pairs and triples. $\qquad\qquad\qquad\qquad\qquad\qquad\qquad\qquad\qquad\square$

The construction in the previous two proofs is adopted from Chvátal and Klincsek [1980]. There seems to be no obvious way of generalizing this result to higher dimensions.

Next we exhibit a class of antimatroids, namely point-line search antimatroids (cf. Definition III.2.12), for which no decent linear description of $\mathrm{conv}(\mathscr{A})$ can be expected. However, if we consider the cone generated by the incidence vectors of feasible sets, then we can derive a linear description. (Note that a description of $\mathrm{conv}(\mathscr{A})$ yields a description of $\mathrm{cone}(\mathscr{A})$, but not vice versa.)

Theorem 3.9. *For point-line search antimatroids it is NP-hard to decide if a given vector belongs to* $\mathrm{conv}(\mathscr{A})$ *(membership problem).*

Proof. Consider the face

$$F = \mathrm{conv}(\mathscr{A}) \cap \{x \in \mathbb{R}^E : x_{uv} = 1 \text{ for all } (u, v) \in E(G)\} \, .$$

If $\chi^A \in F$ for $A \in \mathscr{A}$, then $A \setminus E(G)$ is a vertex cover for G. Hence F is essentially the vertex cover polytope of G. Since the membership problem for the vertex cover polytope is NP-hard, the claim follows. $\qquad\qquad\qquad\qquad\qquad\square$

Theorem 3.10. *Let* $(E, \mathscr{A})$ *be the point-line search antimatroid on a graph G with* $E = E(G) \cup V(G)$. *Then*

$$\mathrm{cone}(\mathscr{A}) = \{x \in \mathbb{R}^E : x \geq 0, \ x_{uv} \leq x_u + x_v \text{ for all } (u, v) \in E(G)\} \, .$$

Proof. We show that every extreme ray $x \in \mathbb{R}^E$ of the cone on the right is of the form $x = \lambda \cdot \chi^A$, where χ^A is the incidence vector of some element $A \in \mathscr{A}$.

There must exist $|E| - 1$ linearly independent constraints which are satisfied by x with equality. Choose this system so that it contains as many non-negativity constraints as possible, and call its members **basic**. Let n_1 be the number of basic inequalities of type $x_k \geq 0$ and m_1 the number of type $x_{uv} \geq 0$, and let k count the basic circuit inequalities $x_{uv} \leq x_u + x_v$. Observe that if $x_{uv} \geq 0$ is a basic inequality, then $x_{uv} \leq x_u + x_v$ cannot be basic since $0 = x_{uv} = x_u + x_v$ yields $x_u = 0$ and $x_v = 0$. Hence we could replace $x_{uv} \leq x_u + x_v$ by one of $x_u \geq 0$ and $x_v \geq 0$ in the family of basic constraints. Thus $k \leq |E(G)| - m_1$ and

$$n_1 + m_1 + (E(G) - m_1) \geq |E| - 1$$
$$= |V(G)| + |E(G)| - 1 \, .$$

Hence $n_1 \geq |V(G)| - 1$. If $n_1 = |V(G)|$, then clearly $x = 0$. So let $n_1 = |V(G)| - 1$, i.e. $x_r = \lambda > 0$ for some $r \in V(G)$ and $x_v = 0$ for $v \in V(G) \setminus r$. Moreover, for every edge $(u, v) \in E(G)$, either $x_{uv} = 0$ or $x_{uv} = x_u + x_v$ holds.

Thus $x_{uv} = 0$ with the possible exception of some edges (u, r) incident with r, for which $x_{ur} = \lambda$. Then x is, up to scaling, the incidence vector of some feasible set. $\square$

Unfortunately even cone($\mathscr{A}$) can be difficult to describe.

Theorem 3.11. *It is NP-complete to test the validity of a linear inequality $a^T x \leq b$ for cone($\mathscr{A}$) of a line search antimatroid.*

Proof. We reduce the Steiner tree problem to the validity problem as follows.

Given k vertices $v_0, \ldots, v_{k-1}$ of a graph G and some integer t, the Steiner tree problem is to decide if G has a connected subgraph containing $v_0, \ldots, v_{k-1}$ with less than t edges.

Add k new vertices $r_0, \ldots, r_{k-1}$ and new edges (r_i, v_i) to obtain a new graph G'. Set $a(e) = 1$ for all old edges, $a(r_0, v_0) = t(k-1)$, and $a(r_i, v_i) = -t$ for $1 \leq i \leq k-1$. Then G has a connected subgraph with less than t edges containing $v_0, \ldots, v_{k-1}$ if and only if the line search antimatroid in G' with root r_0 has a feasible set A with $a(A) < 0$. $\square$

A further natural polytope associated with antimatroids is the convex hull of free sets. This is closely related to finding a maximum free set. For arbitrary antimatroids this problem turns out to be NP-complete (see Goecke [1986b] for a reduction of the dominating set problem to the determination of a maximal free set in a point search antimatroid). For certain classes, however, the maximum free set problem can be solved in polynomial time. For the tree shelling antimatroid, the leaves obviously form a maximum cardinality free set. For the poset antimatroid and the k-shelling of an ordered set, the problem is equivalent to determining maximal antichains and k-antichains (cf. Dilworth [1950] and Greene and Kleitman [1976]). Chvátal and Klincsek [1980] describe an algorithm to find a maximal free set in the convex shelling antimatroid in $\mathbb{R}^2$. However, no general theory has been developed.

There are two important min-max results of Edmonds [1967b], [1973], which give linear characterizations of two polyhedra associated with branchings. Since branchings form a prominent class of greedoids, it is natural to try to extend these results to wider classes of greedoids. Such extensions were found by Schmidt [1985a], [1988].

We consider interval greedoids whose closure operator σ has the following property:

(3.4) $\sigma(X) \cap \sigma(Y) \subseteq \sigma(X \cup Y)$ for all $X, Y \in \mathscr{A}$.

Observe that for local poset greedoids this property characterizes local forest greedoids. In particular, directed branchings have this property.

Lemma 3.12. *Let $(E, \mathscr{F})$ be an interval greedoid with property (3.4). Then for any two partial alphabets X, Y,*

$$|\sigma(X)| + |\sigma(Y)| \leq |\sigma(X \cup Y)| + |\sigma(\lambda(X) \cap \lambda(Y))| \ .$$

Proof. We may assume that X and Y are λ-closed. By the interval property, $\sigma(X) \subseteq \sigma(X \cap Y) \cup \sigma(X \cup Y)$. Similarly, $\sigma(Y) \subseteq \sigma(X \cap Y) \cup \sigma(X \cup Y)$. So

$$\sigma(X) \cup \sigma(Y) \subseteq \sigma(X \cup Y) \cup \sigma(X \cap Y) .$$

Moreover, by (3.4), $\sigma(X) \cap \sigma(Y) \subseteq \sigma(X \cup Y)$. We show that $\sigma(X) \cap \sigma(Y) \subseteq \sigma(X \cap Y)$. Let A be a basis of $X \cap Y$ and assume that there exists an element $a \in (\sigma(X) \cap \sigma(Y)) \setminus \sigma(X \cap Y)$. Then $A \cup a \in \mathscr{F}$ and $A \cup a \subseteq \sigma(X)$, so $A \cup a \in v(\sigma(X)) = \lambda(X) = X$. Similarly $A \cup a \subseteq Y$, but then A is not a basis of $X \cap Y$. Hence we know that $\sigma(X) \cap \sigma(Y) \subseteq \sigma(X \cup Y) \cap \sigma(X \cap Y)$.

Thus

$$\begin{aligned}
|\sigma(X)| + |\sigma(Y)| &= |\sigma(X) \cup \sigma(Y)| + |\sigma(X) \cap \sigma(Y)| \\
&\leq |\sigma(X \cup Y) \cup \sigma(X \cap Y)| + |\sigma(X \cup Y) \cap \sigma(X \cap Y)| \\
&= |\sigma(X \cup Y)| + |\sigma(X \cap Y)| .
\end{aligned}$$

$\square$

The next result generalizes Edmonds' Disjoint Branching Theorem.

Theorem 3.13. *Let $(E, \mathscr{F})$ be a balanced interval greedoid with property (3.4). Then the maximum number of disjoint bases is equal to the minimum of $|\Gamma(A)|, A \in \mathscr{F} \setminus \mathscr{B}$.*

Proof. Let $k = \min\{|\Gamma(A)| : A \in \mathscr{F} \setminus \mathscr{B}\}$. Since every set $\Gamma(A)$ intersects every basis, there are at most k disjoint bases.

For $S \subseteq E$ let σ_S and λ_S denote the closure operators for the greedoid $(E \setminus S, \mathscr{F} \setminus S)$. Consider a largest feasible set $S \in \mathscr{F}$ such that

(3.5) $|\Gamma(X) \setminus S| \geq k - 1$

for all $X \in \mathscr{F} \setminus S$ with $r(X) < r(E)$. If S is a basis, then by induction $(E \setminus S, \mathscr{F} \setminus S)$ has $k - 1$ disjoint bases which together with S prove the claim.

If $S \notin \mathscr{B}$, we claim that we can augment S and still satisfy (3.5). We distinguish two cases.

(i) $|\Gamma(X) \setminus S| > k - 1$ for all $X \in \mathscr{F} \setminus S$ such that $\sigma(X \cup S) \neq E$. Consider an arbitrary element $a \in \Gamma(S)$. Suppose $S \cup a$ violates (3.5), i.e. there exists a set $X \in \mathscr{F} \setminus (S \cup a)$ with $r(X) < r(E)$ and $|\Gamma(X) \setminus (S \cup a)| < k - 1$. Then $|\Gamma(X) \setminus S| = k - 1$ and $a \in \Gamma(X)$. Hence, using Theorem IX.3.3, we get

$$a \in \Gamma(S) \cap \Gamma(X) = E \setminus (\sigma(S) \cup \sigma(X)) \subseteq E \setminus (\sigma(S \cup X)) .$$

So $\sigma(S \cup X) \neq E$ and therefore $|\Gamma(X) \setminus S| = k - 1$, which contradicts the assumption.

(ii) Choose a maximal set $X \in \mathscr{F} \setminus S$ with $\sigma(S \cup X) \neq E$ such that $|\Gamma(X) \setminus S| = k - 1$. Since $|E \setminus \sigma(S \cup X)| \geq k$, there exists an element $a \in \sigma(X) \setminus \sigma(S \cup X)$. Then by property (3.4), $a \notin \sigma(S)$, i.e. $S \cup a \in \mathscr{F}$.

Consider $Z \in \mathscr{F} \setminus (S \cup a)$. If $a \in \sigma(Z)$, then $|\Gamma(Z) \setminus (S \cup a)| = |\Gamma(Z) \setminus S| \geq k - 1$. So let $a \notin \sigma(Z)$. Augment X to a basis Y_1 of $X \cup S$ and a basis Y_2 of

$X \cup Z$. Using Theorem IX.3.3, $a \notin \sigma(Z) \cup \sigma(Y_1) \supseteq \sigma(Z \cup Y_1)$ and hence by the interval property, $a \notin \sigma(Y_2)$. Similarly, $a \notin \sigma(S) \cup \sigma(Y_2)$ yields $a \notin \sigma(Y_2 \cup S)$.

Hence $X \subseteq Y_2 \subseteq X \cup Z \subseteq E \setminus S$, $\sigma(Y_2 \cup S) \neq E$, and X is properly contained in Y_2 since $a \in \sigma(X) \setminus \sigma(Y_2)$. By the choice of X and the preceding lemma we then get

$$
\begin{aligned}
|\Gamma(Z) \setminus S| = |E \setminus \sigma_S(Z)| &= |E| - |\sigma_S(Z)| \\
&\geq |E| - |\sigma_S(X \cup Z)| - |\sigma_S(\lambda_S(X) \cap \lambda_S(Z))| + |\sigma_S(X)| \\
&\geq |E - \sigma_S(X \cup Z)| - (|E| - (k-1)) + (|E| - (k-1)) \geq k .
\end{aligned}
$$

Hence $|\Gamma(Z) \setminus (S \cup a)| \geq k - 1$. $\square$

Theorem 3.13 can be formulated as follows: consider the pair of dual linear programs

(P) $\min \left\{ \sum_{e \in E} x_e : x(B) \geq 1 \text{ for all } B \in \mathscr{B}, x \geq 0 \right\}$,

(D) $\max \left\{ \sum_{B \in \mathscr{B}} y_B : \sum_{B \ni e} y_B \leq 1 \text{ for all } e \in E, y \geq 0 \right\}$.

Then (P) and (D) have optimal integer solutions.

By introducing parallel elements, we obtain that for all non-negative weight functions $c : E \to \mathbb{Z}_+$ the linear programs

$$\min \left\{ c^T x : x(B) \geq 1 \ (B \in \mathscr{B}), x \geq 0 \right\}$$

$$\max \left\{ \sum_B y_B : \sum_{B \ni e} y_B \leq c(e), y \geq 0 \right\}$$

have integer optimum solutions. It follows that the polyhedron $\text{conv}\{\Gamma(A) : A \in \mathscr{F} \setminus \mathscr{B}\} + \mathbb{R}_+^E$ is described by the inequalities

$$x \geq 0, \ x(B) \geq 1 \ (B \in \mathscr{B}) .$$

Using a theorem of Lehmann [1979], this implies the following

Corollary 3.14. *Let $(E, \mathscr{F})$ be a balanced interval greedoid with property (3.4). Then the incidence vectors of bases are the vertices of the polyhedron*

$$x \geq 0, \ x(\Gamma(A)) \geq 1 \ (A \in \mathscr{F} \setminus \mathscr{B}) .$$ $\square$

4. Transversals and Partial Transversals

Consider a family $A_1, \ldots, A_n$ of subsets of E. A subset $T \subseteq E$ is a **transversal** of $A_1, \ldots, A_n$ if there is a bijection $\phi : T \to \{1, \ldots, n\}$ such that

$$x \in A_{\phi(x)} \text{ for all } x \in T .$$

Similarly, T is a **partial transversal** of $A_1, \ldots, A_n$ if T is a transversal of some subfamily of $A_1, \ldots, A_n$. In this case the cardinality $|T|$ is called the **length** of the partial transversal.

It seems that Hall [1935] was the first to formulate conditions for a family of sets to possess a transversal. This result was extended by Rado [1942] to transversals which in addition are independent in a matroid.

In this section we show how the Rado-Hall Theorem carries over to greedoids and formulate necessary and sufficient conditions for a family of closure feasible sets to have a transversal which is feasible (cf. Korte and Lovász [1983] and Ding and Yue [1987]).

Given a greedoid $(E, \mathscr{F})$ we call a transversal T **feasible** if $T \in \mathscr{F}$.

Theorem 4.1. *Let $(E, \mathscr{F})$ be a greedoid with rank function r and let $A_1, \ldots, A_n \in \mathscr{C}$ be a family of closure feasible sets. Then $A_1, \ldots, A_n$ has a feasible transversal if and only if for all $1 \le i_1 < \ldots < i_k \le n$,*

$$(4.1) \qquad r(A_{i_1} \cup \ldots \cup A_{i_k}) \ge k.$$

Proof. It is easily seen that (4.1) is necessary. Let $T \in \mathscr{F}$ be a feasible transversal of $A_1, \ldots, A_n$. Then by Lemmas V.3.8 and V.3.9,

$$A_{i_1} \cup \ldots \cup A_{i_k} \in \mathscr{C} \subseteq \mathscr{R}.$$

Hence

$$r(A_{i_1} \cup \ldots \cup A_{i_k}) = \beta(A_{i_1} \cup \ldots \cup A_{i_k}) \ge T \cap (A_{i_1} \cup \ldots \cup A_{i_k}) \ge k.$$

Conversely, assume that condition (4.1) holds. We perform induction on n.

For $n = 1$ we have $r(A_1) \ge 1$. Hence there exists an element $t_1 \in A_1$ with $t_1 \in \mathscr{F}$. Clearly t_1 is a feasible transversal.

Suppose now that $n > 1$ and distinguish two cases.

(i) For all $1 \le k \le n$ and $1 \le i_1 \le \ldots \le i_k \le n$ we have $r(A_{i_1} \cup \ldots \cup A_{i_k}) \ge k+1$. In particular, there exists an element $t_1 \in A_1$ with $t_1 \in \mathscr{F}$. Set $U = \{t_1\}$ and $A_i' = A_i \setminus U$ for $2 \le i \le n$. By Corollary V.4.3, $A_i' \in \mathscr{C}/U$, i.e. the A_i' are closure feasible in the contraction. For every $2 \le i_1 \le \ldots \le i_k \le n$,

$$r_U(A_{i_1}' \cup \ldots \cup A_{i_k}') = r(A_{i_1}' \cup \ldots \cup A_{i_k}' \cup U) - r(U)$$
$$\ge r(A_{i_1} \cup \ldots \cup A_{i_k}) - 1$$
$$\ge k.$$

Hence by induction there exists a transversal $T' \in \mathscr{F}/U$ for $A_2', \ldots, A_n'$. Then $T = T' \cup \{t_1\} \in \mathscr{F}$, and T is a feasible transversal for $A_1, \ldots, A_n$.

(ii) $r(A_{i_1} \cup \ldots \cup A_{i_k}) = k$ for some $1 \le k < n$ and some $1 \le i_1 \le \ldots \le i_k \le n$. By renumbering, we may assume $i_j = j$, $1 \le j \le k$.

Set $U = A_1 \cup \ldots \cup A_k$. Then $U \in \mathscr{C} \subseteq \mathscr{R}$ and

$$A_i' = A_i \setminus U \in \mathscr{C}/U.$$

As before,

$$r_U(A'_{i_1} \cup \ldots \cup A'_{i_\ell}) = r(A'_{i_1} \cup \ldots \cup A'_{i_\ell} \cup U) - r(U)$$
$$= r(A_1 \cup \ldots \cup A_k \cup A_{i_1} \cup \ldots \cup A_{i_\ell}) - k$$
$$\geq \ell .$$

By induction, $A'_{k+1}, \ldots, A'_n$ has a transversal T' with $T' \in \mathcal{F}/U$ and $A_1, \ldots, A_k$ has a transversal $T'' \in \mathcal{F}$. Then $r(U) = k = |T''|$, i.e. T' is a basis of U. Now $T = T' \cup T'' \in \mathcal{F}$ is a feasible transversal of $A_1, \ldots, A_n$. $\square$

The following example shows that the condition (4.1) is not necessary if we only assume that $A_i \in \mathcal{R}$ or even $A_i \in \mathcal{F}$.

Example. Let $E = \{a, b, c, d\}$, $\mathcal{F} = \{\emptyset, a, b, c, \{a, d\}, \{b, d\}, \{c, d\}, \{a, b, d\}, \{a, c, d\}\}$ and $A_1 = \{a\}$, $A_2 = \{b\}$, $A_3 = \{c, d\}$. Then $\{a, b, d\}$ is a feasible transversal although

$$r(A_1 \cup A_2) = 1 < 2 .$$

It is an open question whether condition (4.1) remains sufficient if closure feasibility of the A_i is not assumed.

Corollary 4.2. *Let $(E, \mathcal{F})$ be a greedoid with rank function r, let $A_1, \ldots A_n \in \mathcal{C}$ be a family of closure feasible sets and $p_1, \ldots, p_n$ be positive integers. There exists a set $X \in \mathcal{F}$ which can be partitioned into $X_1 \cup \ldots \cup X_n$ with $X_i \cap X_j = \emptyset$, $X_i \subseteq A_i$ and $|X_i| = p_i$ for each i if and only if for all $1 \leq i_1 \leq \ldots \leq i_k \leq n$,*

$$r(A_{i_1} \cup \ldots \cup A_{i_k}) \geq \sum_{j=1}^{k} p_{i_j} .$$

Proof. It is easily seen that a set $X \in \mathcal{F}$ with the required properties exists if and only if the family consisting of p_1 copies of A_1, p_2 copies of $A_2, \ldots, p_n$ copies of A_n has a feasible transversal.

By theorem 4.1, this is the case if and only if for all $1 \leq i_1 \leq \ldots \leq i_k \leq m$ and all $1 \leq \ell_{i_j} \leq p_{i_j}$ with $j = 1, \ldots, k$

$$r(A_{i_1} \cup \ldots \cup A_{i_k}) \geq \sum_{j=1}^{k} \ell_{i_j} ,$$

i.e. if and only if

$$r(A_{i_1} \cup \ldots \cup A_{i_k}) \geq \sum_{j=1}^{k} p_{i_j} .   \square$$

Corollary 4.3. *Let $(E, \mathcal{F})$ be a greedoid with rank function r and let m be an integer with $0 \leq m \leq n$. The family $A_1, \ldots, A_n$ of closure feasible sets has a feasible partial*

transversal of length m if and only if for all $1 \le k \le n$ and indices $1 \le i_1 \le \ldots \le$
$i_k \le n$

$$r(A_{i_1} \cup \ldots \cup A_{i_k}) \ge k - n + m .$$

Proof. Let E' be a ground set with $E \cap E' = \emptyset$ and $|E'| = n - m$. Let $(E', \mathcal{M})$ be the free matroid on E', i.e. $\mathcal{M} = 2^{E'}$, and consider the direct sum $(E \cup E', \mathcal{F} + \mathcal{M})$, which is a greedoid by Lemma V.4.4. Then $A_1, \ldots, A_n$ has a feasible partial transversal in $(E, \mathcal{F})$ if and only if $A_1 \cup E', \ldots, A_n \cup E'$ has a feasible transversal in the direct sum $(E \cup E', \mathcal{F} + \mathcal{M})$. By Theorem 4.1, the latter is equivalent to the condition that for all $1 \le i_1 < \ldots < i_k \le n$,

$$
\begin{aligned}
k &\le r_{\mathcal{F}+\mathcal{M}}(A_{i_1} \cup \ldots \cup A_{i_k} \cup E') \\
&= r_{\mathcal{F}}(A_{i_1} \cup \ldots \cup A_{i_k}) + r_{\mathcal{M}}(E') \\
&= r_{\mathcal{F}}(A_{i_1} \cup \ldots \cup A_{i_k}) + n - m .
\end{aligned}
$$

$\square$

5. Intersection of Supermatroids

The results in the previous section can be viewed as intersection theorems for certain special classes of greedoids. Unfortunately there is no general greedoid intersection theorem. In this section we discuss another special case which has been settled by Tardos [1990]. For two distributive supermatroids defined on the same poset she described an algorithm for finding a set X of maximum cardinality which is feasible in both supermatroids, and also gave a min-max formula for this cardinality.

If we allow the two supermatroids to be defined over two different posets, then the intersection problem cannot be solved in polynomial time since the matroid matching problem can be reduced to it (for details see Tardos [1990]).

Let $(E, \mathcal{F}, \le)$ be a distributive supermatroid. For any subset $X \subseteq E$ let

$$i(X) = \bigcap \{A \in \mathcal{F} : A \text{ is a basis of } X\}$$

be the set of **isthmuses** of X. Clearly $i(X)$ is an ideal and

(5.1) $\qquad i(Y) \cap X \subseteq i(X)$ for any two ideals $X \subseteq Y$.

Let X be an ideal, $Y \subseteq i(X)$, and let $A \in \mathcal{F}$ be such that $|A \cap (X \setminus Y)| = \beta(X \setminus Y)$. Then $Y \subseteq A$. Since $A \cap X \in \mathcal{F}$, we may assume that A is a basis of X. Hence

(5.2) $\qquad \beta(X \setminus Y) = r(X) - |Y| .$

For $X \subseteq E$, let $F(X)$ denote the filter generated by X. For an ideal X we set $D(X) = X \cup F(X \setminus i(X))$. Then

(5.3) $\qquad D(X) \subseteq D(Y)$ for any two ideals $X \subseteq Y$.

Lemma 5.1. *Let $(E, \mathcal{F}, \leq)$ be a distributive supermatroid. Then for any ideal X, $\beta(D(X)) = r(X)$.*

Proof. Suppose $\beta(D(X)) > r(X)$ and let Y be a smallest set with $X \subseteq Y \subseteq D(X)$ and $\beta(Y) = \beta(D(X))$. Then there exist elements $y \in Y \backslash X$ and $x \in X \backslash i(X)$ with $x \leq y$. Since x is not an isthmus, there exists a basis A of X not containing x. By Lemma V.3.1 we can augment A to a feasible set $B \in \mathcal{F}$ with $|B \cap Y| = \beta(D(X))$. Then B must contain y and hence x, contradicting the rank feasibility of X. $\square$

This lemma allows us to strengthen the submodularity condition of the rank function on ideals.

Lemma 5.2. *Let $(E, \mathcal{F})$ be a distributive supermatroid and X, Y two ideals. Then*

$$r(X) + r(Y) \geq r(X \cap Y) + r(X \cup Y) + |i(X) \cap (D(Y) \backslash Y)| .$$

Proof. Since ideals are rank feasible, we can choose feasible sets $A \subseteq B \subseteq C$ with $r(X \cap Y) = |A|, r(Y) = |B|$ and $r(X \cup Y) = |C|$. By Lemma 5.1, $C \backslash B$ does not contain elements from $D(Y)$, i.e. $|A \cup (C \backslash B)| \leq r(X \backslash (D(Y) \backslash Y))$. Using (5.2), we obtain

$$
\begin{aligned}
r(X \cap Y) + r(X \cup Y) &= |A| + |C| \\
&= |B| + |A \cup (C \backslash B)| \\
&\leq r(Y) + r(X \backslash (D(Y) \backslash Y)) \\
&\leq r(Y) + r(X \backslash (i(X) \cap (D(Y) \backslash Y))) \\
&\leq r(Y) + r(X) - |i(X) \cap (D(Y) \backslash Y)| .
\end{aligned}
$$
$\square$

Given two distributive supermatroids $(E, \mathcal{F}_1)$ and $(E, \mathcal{F}_2)$ on the same poset, we use subscripts to denote their rank functions, isthmuses and so on.

We say that two ideals X_1, X_2 are **complementary** (with respect to the given supermatroids) if

(5.4) $i_1(X_1) = i_2(X_2) = X_1 \cap X_2,$
          $F(X_1 \backslash X_2) = E \backslash X_2$ and $F(X_2 \backslash X_1) = E \backslash X_1 .$

We will use complementary ideals in the statements of our min-max results. In the proofs, however, it will be convenient to work with weaker conditions.

Lemma 5.3. *Let $(E, \mathcal{F}_1), (E, \mathcal{F}_2)$ be two distributive supermatroids on the same poset. For any two ideals A_1, A_2 with $i_2(A_2) \subseteq A_1 \cap A_2 \subseteq i_1(A_1)$ and $F(A_2 \backslash A_1) = E \backslash A_1$, there exist complementary ideals X_1, X_2 with*

$$r_1(X_1) + r_2(X_2) - |X_1 \cap X_2| \leq r_1(A_1) + r_2(A_2) - |A_1 \cap A_2| .$$

Proof. Choose X_1, X_2 such that X_1 is minimal and X_2 is maximal with respect to the properties $i_2(X_2) \subseteq X_1 \cap X_2 \subseteq i_1(X_1)$, $F(X_2 \setminus X_1) = E \setminus X_1$ and $r_1(X_1) + r_2(X_2) - |X_1 \cap X_2|$ is minimal. Then we claim that X_1 and X_2 are complementary.

(i) $i_1(X_1) = X_1 \cap X_2$. It suffices to show that $i_1(X_1) \subseteq X_1 \cap X_2$. Set $B_1 = X_1$ and $B_2 = X_2 \cup i_1(X_1)$. Then by (5.1), $i_2(B_2) \subseteq i_2(X_2) \cup i_1(X_1) \subseteq B_1 \cap B_2 \subseteq i_1(B_1)$ and $F(B_2 \setminus B_1) = E \setminus B_1$.

Hence

$$r_1(B_1) + r_2(B_2) - |B_1 \cap B_2| = r_1(X_1) + r_2(X_2 \cup (i_1(X_1) \setminus X_2))$$
$$- |X_1 \cap X_2| - |i_1(X_1) \setminus X_2|$$
$$\leq r_1(X_1) + r_2(X_2) - |X_1 \cap X_2|.$$

So by the choice of X_2, $X_2 = B_2$ and $i_1(X_1) \subseteq X_1 \cap X_2$.

(ii) $F(X_1 \setminus X_2) = E \setminus X_2$. Trivially $F(X_1 \setminus X_2) \subseteq E \setminus X_2$. Consider $B_1 = X_1$ and $B_2 = X_2 \cup (E \setminus F(X_1 \setminus X_2))$. As before, B_1, B_2 satisfy the same conditions as X_1 and X_2. Moreover, Lemma 5.1 implies that $r(B_1) + r(B_2) - |B_1 \cap B_2| = r(X_1) + r(X_2) - |X_1 \cap X_2|$. Hence $B_2 = X_2$, as X_2 was chosen to be maximal. So $E \setminus F(X_1 \setminus X_2) \subseteq X_2$ or, equivalently, $E \setminus X_2 \subseteq F(X_1 \setminus X_2)$.

(iii) $i_2(X_2) = X_1 \cap X_2$. We have to show that $X_1 \cap X_2 \subseteq i_2(X_2)$. Suppose not. Let x be a maximal element in $(X_1 \cap X_2) \setminus i_2(X_2)$ and set $B_1 = X_1 \setminus F(x), B_2 = X_2$. Again $i_2(B_2) \subseteq B_1 \cap B_2 \subseteq i_1(B_1), F(B_2 \setminus B_1) = E \setminus B_1$ and, since $x \in i_1(X_1)$, we get

$$r_1(B_1) + r_2(B_2) - |B_1 \cap B_2| \leq r_1(X_1) - 1 + r_2(X_2) - (|X_1 \cap X_2| - 1),$$

which contradicts the choice of X_1. $\qquad \square$

We are now ready to state and prove the intersection theorem.

Theorem 5.4. *Let $(E, \mathscr{F}_1)$ and $(E, \mathscr{F}_2)$ be two distributive supermatroids on the same poset. Then*

$$\max\{|A| : A \in \mathscr{F}_1 \cap \mathscr{F}_2\} = \min\{r_1(X_1) + r_2(X_2) - |X_1 \cap X_2|\},$$

where the minimum is taken over all complementary ideals.

Proof. Let A be any common feasible set and X_1, X_2 complementary ideals. Using Lemma 5.1, (5.2) and (5.4), we get

$$r_1(X_1) + r_2(X_2) - |X_1 \cap X_2| = \beta_1(D_1(X_1)) + \beta_2(X_2 \setminus X_1)$$
$$= \beta_1(E \setminus (X_2 \setminus X_1)) + \beta_2(X_2 \setminus X_1)$$
$$\leq |A \setminus (X_2 \setminus X_1)| + |A \cap (X_2 \setminus X_1)|$$
$$= |A|.$$

For the proof of the reverse direction we perform induction on $|E|$. Let $t = \max\{|A| : A \in \mathscr{F}_1 \cap \mathscr{F}_2\}$ and consider some minimal element e in the poset P. The contractions $\mathscr{F}_1/e$ and $\mathscr{F}_2/e$ are again supermatroids on $P \setminus e$. A maximal common feasible set in the contracted supermatroids has cardinality at most $t - 1$.

By induction, there exist ideals X_1 and X_2 containing e such that $X_1 \setminus e$ and $X_2 \setminus e$ are complementary ideals in $P \setminus e$ and $(r_1(X_1) - 1) + (r_2(X_2) - 1) - (|X_1 \cap X_2| - 1) \le t - 1$, or equivalently,

(5.5) $$r_1(X_1) + r_2(X_2) - |X_1 \cap X_2| \le t .$$

Then $i_j(X_j) = X_1 \cap X_2$ or $i_j(X_j) = (X_1 \cap X_2) \setminus e$ for $j = 1, 2, F(X_1 \setminus X_2) = E \setminus X_2$ and $F(X_2 \setminus X_1) = E \setminus X_1$. If e is an isthmus of X_1 in $(E, \mathcal{F}_1)$, then $i_2(X_2) \subseteq X_1 \cap X_2 \subseteq i_1(X_1)$ and we can apply Lemma 5.3 to finish the proof.

So we may assume $i_1(X_1) = i_2(X_2) = (X_1 \cap X_2) \setminus e$. Consider now the deletion $\mathcal{F}_1 \setminus F(e)$ and $\mathcal{F}_2 \setminus F(e)$ on $P \setminus F(e)$. The maximum size of a feasible set common to both deletions is at most t. By induction, we obtain ideals Y_1 and Y_2 not containing e with $i_1(Y_1) = Y_1 \cap Y_2 = i_2(Y_2)$, $(E \setminus F(e)) \setminus Y_1 \subseteq F(Y_2 \setminus Y_1)$, $(E \setminus F(e)) \setminus Y_2 \subseteq F(Y_1 \setminus Y_2)$ and

(5.6) $$r_1(Y_1) + r_2(Y_2) - |Y_1 \cap Y_2| \le t .$$

The properties of the four sets obtained by induction imply the following relations:

(5.7) $$[(X_1 \cap X_2) \setminus (Y_1 \cup Y_2)] \setminus e \subseteq [i_1(X_1) \cap D_1(Y_1)] \setminus Y_1 ,$$

(5.8) $$(Y_1 \cap Y_2) \setminus (X_1 \cup X_2) \subseteq [i_2(Y_2) \cap D_2(X_2)] \setminus X_2 .$$

Now these two conclusions and the tightened submodularity condition of Lemma 5.2 applied to the sum of (5.5) and (5.6) yield:

$$r_1(X_1 \cap Y_1) + r_2(X_2 \cup Y_2) - |(X_1 \cap Y_1) \cap (X_2 \cup Y_2)| + r_1(X_1 \cup Y_1)$$
$$+ r_2(X_2 \cap Y_2) - |(X_1 \cup Y_1) \cap (X_2 \cap Y_2)|$$
$$\le r_1(X_1) + r_1(Y_1) - |[(X_1 \cap X_2) \setminus (Y_1 \cup Y_2)] \setminus e| + r_2(X_2) + r_2(Y_2)$$
$$- |(Y_1 \cap Y_2) \setminus (X_1 \cup X_2)| - |(X_1 \cap Y_1) \cap (X_2 \cup Y_2)|$$
$$- |(X_1 \cup Y_1) \cap (X_2 \cap Y_2)|$$
$$= r_1(X_1) + r_2(X_2) + r_1(Y_1) + r_2(Y_2) - |X_1 \cap X_2| - |Y_1 \cap Y_2| + 1$$
$$\le 2t + 1 .$$

Hence either for $A_1 = X_1 \cap Y_1$ and $A_2 = X_2 \cup Y_2$, or for $A_1 = X_1 \cup Y_1$ and $A_2 = X_2 \cap Y_2$ we have

$$r(A_1) + r(A_2) - |A_1 \cap A_2| \le t.$$

Suppose this is the case for $A_1 = X_1 \cap Y_1$, $A_2 = X_2 \cup Y_2$ (the other case being similar). We now claim that A_1 and A_2 satisfy the requirements of Lemma 5.3, and so the lemma yields the pair of complementary ideals in the theorem.

(i) $F(A_2 \setminus A_1) = E \setminus A_1$. Only the inclusion $E \setminus A_1 \subseteq F(A_2 \setminus A_1)$ needs a proof. Now
$$E \setminus A_1 = (E \setminus X_1) \cup (E \setminus Y_1)$$
$$\subseteq F(X_2 \setminus X_1) \cup [F(Y_2 \setminus Y_1) \cup F(e)]$$
$$\subseteq F(A_2 \setminus A_1).$$

(ii) $A_1 \cap A_2 \subseteq i_1(A_1)$. This follows from (5.1) since

$$
\begin{aligned}
A_1 \cap A_2 &= (X_1 \cap Y_1) \cap (X_2 \cup Y_2) \\
&= (X_1 \cap X_2 \cap Y_1) \cup (X_1 \cap Y_1 \cap Y_2) \\
&= [((X_1 \cap X_2) \setminus e) \cap Y_1] \cup (X_1 \cap Y_1 \cap Y_2) \\
&= (i_1(X_1) \cap Y_1) \cup (X_1 \cap i_1(Y_1)) \\
&\subseteq i_1(A_1).
\end{aligned}
$$

(iii) $i_2(A_2) \subseteq A_1 \cap A_2$. We only have to show that $i_2(A_2) \subseteq A_1$. Consider an element $x \in i_2(A_2)$ and distinguish 3 subcases.

(a) If $x \in X_2 \cap Y_2$, then by (5.1), $x \in i_2(X_2) \cap i_2(Y_2) \subseteq X_1 \cap Y_1 = A_1$.

(b) If $x \in Y_2 \setminus X_2$, then again by (5.1), $x \in i_2(Y_2) \subseteq Y_1$. Suppose $x \in E \setminus X_1 = F(X_2 \setminus X_1) \subseteq D_2(X_2)$. But then $x \notin i_2(X_2 \cup Y_2)$. Hence $x \in Y_1 \cap X_1 = A_1$.

(c) If $x \in X_2 \setminus Y_2$ and $x \neq e$, then $x \in A_1$ follows as in (b). If $x = e$, then $e \in X_2$ is an isthmus of $X_2 \cup Y_2$, which implies e is an isthmus of X_2, contradicting our assumption. $\qquad\square$

Chapter XII. Topological Results for Greedoids

Topological approaches to combinatorial problems have proved to be fruitful in several fields of combinatorics. Properties of graphs, posets, matroids and other discrete structures have been investigated by assigning topological spaces (usually simplical complexes) to them and then applying powerful results from algebraic topology. The seeming detour from discrete structures to a continuous framework and back again has provided solutions for several combinatorial problems (see Lovász [1978] and Björner [1987] for surveys).

The aim of this chapter is to formulate results about topology in greedoids, which unify and extend some previously known results for the subclasses of matroids, posets, and branchings.

Section 1 contains a short summary of topological prerequisites, as some of these are not easily accessible. In particular, it reviews the notion of topological connectivity, shellability, and contractibility.

Section 2 then introduces two simplicial complexes (independence systems) associated with a greedoid and gives conditions when they are shellable. These results enable us to partially extend the notion of the Tutte polynomial and Tutte invariants from matroids to greedoids. They allow us to investigate percolation problems in greedoids and to study the complexity of certain oracle algorithms for greedoids.

Section 3 contains results on homotopy properties of various simplicial complexes associated with greedoids. We give a topological extension of Helly's theorem for antimatroids. We extend the result on the connectivity of basis graphs (Chapter VI) to higher dimensional connectivity. These results generalize, unify and simplify results of Lovász [1977] on homotopy of branchings and Maurer [1973] on homotopy in matroids.

1. A Brief Review of Topological Prerequisites

This section is devoted to a short summary of the topological background we will need later on. For this, we will leave greedoids for a while and consider independence systems $(E, \mathscr{F})$ (or briefly $\mathscr{F}$ when $E = \cup \mathscr{F}$ is understood). Note that from the point of view of algebraic topology, $(E, \mathscr{F})$ is a **simplicial complex** and has an associated geometric object as follows:

Let $E = \{e_1, \ldots, e_n\}$ and identify each element e_i with the ith unit vector in $\mathbb{R}^n$. For each $X \in \mathcal{F}$, let $s(X)$ be the **simplex** given by

$$s(X) = \text{conv}\{e_i : e_i \in X\}$$

and take

$$B(\mathcal{F}) = \bigcup \{s(X) : X \in \mathcal{F}\}$$

to be the **geometric realization** of $(E, \mathcal{F})$. We may then think of $B(\mathcal{F})$ as a topological space.

Let $f_i = |\{X \in \mathcal{F} : |X| = i\}|$, $1 \leq i \leq r$, count the number of independent sets with i elements. The vector $(f_0, f_1, \ldots, f_r)$ is called the **f-vector** of $(E, \mathcal{F})$. Then the **Euler characteristic** $\chi(\mathcal{F})$ is given by

$$\chi(\mathcal{F}) = \sum_{i=1}^{r} (-1)^{i-1} f_i .$$

Let $(E, \mathcal{F})$ be an independence system and let $X \in \mathcal{F}$ be a nonempty feasible set. Then $(E, \mathcal{F})$ is a **cone with apex** X if

$$(E \setminus X, \ \mathcal{F}/X) = (E \setminus X, \mathcal{F} \setminus X)$$

(i.e. if $Y \in \mathcal{F}$, then also $X \cup Y \in \mathcal{F}$).

Given a poset $P = (E, \leq)$, the **chain complex** $\mathscr{C}h(P)$ of P is the independence system of chains. With an arbitrary set system $\mathcal{F} = (X_i)_{i \in I}$ of nonempty subsets of some set E we associate the **nerve** $\mathcal{N}(\mathcal{F})$ as the independence system $(I, \mathcal{J})$ given by

$$\mathcal{J} = \{J \subseteq I : \bigcap_{i \in J} X_i \neq \emptyset\} .$$

Let T_1 and T_2 be topological spaces and $f, g : T_1 \to T_2$ continuous mappings. We say that f and g are **homotopic** (in T_2) if there exists a continuous mapping $F : T_1 \times [0, 1] \to T_2$ such that $F(., 0) = f$ and $F(., 1) = g$. We denote this relation by $f \sim g$.

We say that two topological spaces T_1 and T_2 are **homotopy equivalent** if there exist two continuous functions $f : T_1 \to T_2$ and $g : T_2 \to T_1$ such that $f \circ g \sim id_{T_2}$ and $g \circ f \sim id_{T_1}$. A space T is **contractible** if it is homotopy equivalent to a single point. A trivial example of a contractible space is the geometric realization of any cone. Two independence systems are **homotopy equivalent** if their geometric realizations are homotopy equivalent.

The **dimension** of an independence system $(E, \mathcal{F})$ is $\max\{|X| : X \in \mathcal{F}\} - 1$. We define the **homotopy dimension** of an independence system as the least dimension of an independence system to which it is homotopy equivalent. So the homotopy dimension is 0 if and only if every connected component of the independence system is contractible.

A topological space T is **topologically k-connected** if for all $0 \leq d \leq k$ every continuous mapping f of the d-dimensional sphere S^d into T can be extended to a continuous mapping of the $(d + 1)$-dimensional ball B^{d+1} with boundary S^d into T. Equivalently, $f \sim 0$ in T, where 0 is any mapping which maps S^d onto a single point of T. An independence system $(E, \mathcal{F})$ is **topologically k-connected** if its geometric realization is topologically k-connected. It is easy to see that if T is a topologically k-connected space and T' is homotopy equivalent to T, then T' is topologically k-connected.

In the remainder of this section we summarize some topological results without proof. These may be found in, e.g., Alexandroff and Hopf [1972], Spanier [1966] and Björner et al. [1985].

Theorem 1.1. *For any set system $(E, \mathcal{F})$, the nerve $\mathcal{N}(\mathcal{F})$ is homotopy equivalent to the hereditary closure $\mathcal{H}(\mathcal{F})$.* □

A generalization of this Nerve theorem is the following.

Theorem 1.2. *Let $(E, \mathcal{F})$ be an independence system and suppose that $\mathcal{F} = \mathcal{F}_1 \cup \ldots \cup \mathcal{F}_k$ where each $\mathcal{F}_i$ is an independence system.*

(i) *Assume that for every sequence $1 \leq i_1 < \ldots < i_r \leq k$ of indices, $\mathcal{F}_{i_1} \cap \ldots \cap \mathcal{F}_{i_r}$ is either $\{\emptyset\}$ or contractible. Then $\mathcal{F}$ is homotopy equivalent to the nerve $\mathcal{N}(\cup \mathcal{F}_1, \ldots, \cup \mathcal{F}_k)$.*

(ii) *Assume that for every sequence $1 \leq i_1 < \ldots < i_r \leq k$ of indices, $\mathcal{F}_{i_1} \cap \ldots \cap \mathcal{F}_{i_r}$ is either $\{\emptyset\}$ or k-connected. Then the nerve $\mathcal{N}(\cup \mathcal{F}_1, \ldots, \cup \mathcal{F}_k)$ is k-connected if and only if $\mathcal{F}$ is k-connected.* □

A **cross-cut** in a finite lattice is an antichain that intersects every maximal chain. A cross-cut C in a lattice gives rise to the **cross-cut system** $(C, \mathcal{F})$, where $\mathcal{F} = \{X \subseteq C : \wedge X \neq 0 \text{ or } \vee X \neq 1\}$.

Theorem 1.3. *A cross-cut system in a lattice L is homotopy equivalent to the chain complex of $L \setminus \{0, 1\}$.* □

An independence system is called **pure** if all its bases have the same rank. A **shelling** of a pure independence system $(E, \mathcal{F})$ is an ordering $B_1, \ldots, B_n$ of its bases such that for any $1 < i \leq n$ and any $j < i$ there exists a $k < i$ such that

$$B_j \cap B_i \subseteq B_k \cap B_i \text{ and } |B_k \cap B_i| = |B_i| - 1 .$$

We say that $(E, \mathcal{F})$ is **shellable** if it is pure and has a shelling.

Theorem 1.4. *A shellable independence system $(E, \mathcal{F})$ of rank r is topologically $(r - 2)$-connected.* □

Consider a shelling $B_1, \ldots, B_k$ of a pure independence system $(E, \mathcal{F})$ of rank r. For $1 \leq i \leq k$, set

$$R(B_i) = \{x \in B_i : B_i \setminus x \subseteq B_j \text{ for some } j < i\}$$

and

$$h_j = |\{i : |R(B_i)| = j\}| , \ 0 \leq j \leq r .$$

The vector $(h_0, h_1, \ldots, h_r)$ is called the **h-vector** of $(E, \mathscr{F})$.

Theorem 1.5. *Let (f_i) and (h_i) be the f-vector and h-vector of a shellable independence system $(E, \mathscr{F})$ of rank r and set $h(t) = \sum_{i=0}^{r} h_i t^{r-i}$ and $f(t) = \sum_{i=0}^{r} f_i t^{r-i}$. Then*

(i) $h(1 + t) = f(t)$,

(ii) $h_r = f(-1) = (-1)^{r-1} (\chi(\mathscr{F}) - 1)$.   □

As a consequence of this theorem, the h-vector turns out to be independent of the particular shelling order chosen.

Theorem 1.6. *A shellable independence system $\mathscr{F}$ is contractible if and only if $\chi(\mathscr{F}) = 1$.*   □

With every element $e \in E$ we may associate two derived independence systems, the deletion $\mathscr{F} \setminus e = \{X \in \mathscr{F} : e \notin X\}$ of e and the link $\mathscr{F} \triangle e = \{X \in \mathscr{F} : e \notin X, X \cup e \in \mathscr{F}\}$. An independence system $\mathscr{F}$ is **vertex-decomposable** if either $\mathscr{F}$ consists only of the empty set or both $\mathscr{F} \setminus e$ and $\mathscr{F} \triangle e$ are vertex-decomposable for some $e \in \cup \mathscr{F}$.

Theorem 1.7. *Let $\mathscr{F}$ be an independence system and $e \in E$. If $\mathscr{F} \triangle e$ has homotopy dimension at most $d - 1$ and $\mathscr{F} \setminus e$ has homotopy dimension at most d, then $\mathscr{F}$ has homotopy dimension at most d.*   □

Theorem 1.8. *Vertex-decomposable independence systems are shellable.*

Proof. Provan and Billera [1980].   □

A closely related property is the following. We call an independence system F **non-evasive** if either $|\cup \mathscr{F}| = 1$ or, for some $e \in \cup \mathscr{F}$, both $\mathscr{F} \setminus e$ and $\mathscr{F} \triangle e$ are non-evasive. The notion of evasiveness was motivated by the following algorithmic consideration (Kahn, Saks and Sturtevant [1984]).

Let $(E, \mathscr{F})$ be an independence system and let $X \subseteq E$. Suppose that X is not given explicitly but only by an oracle which tells us, for each $e \in E$, whether or not $e \in X$. We want to determine if $X \in \mathscr{F}$. Which independence systems are the "worst" in the sense that for every algorithm that solves this problem there is an instance X such that the algorithm has to make $|E|$ queries to the oracle? It is easy to show that these "worst" independence systems are exactly the evasive ones.

The following lemma relates evasiveness to the more classical topological notion of contractibility:

Lemma 1.9. *Every non-evasive independence system is contractible.*   □

Every non-evasive independence system is vertex-decomposable (but not the converse, as is shown by the matroid $U_{2,3}$). The exact relation is given by the following lemma (Björner, Korte and Lovász [1985]).

Lemma 1.10. *Let $(E, \mathscr{F})$ be a vertex-decomposable independence system. Then $\mathscr{F}$ is non-evasive if and only if $\chi(\mathscr{F}) = 1$.*   □

2. Shellability of Greedoids and the Partial Tutte Polynomial

There are basically two pure independence systems associated with greedoids. For the family $\mathscr{B}$ of bases of a greedoid, denote by $\mathscr{B}^* = \{X : E \setminus X \in \mathscr{B}\}$ the family of basis complements. The **primal system** $(E, \mathscr{H}(\mathscr{B}))$ consists of all subsets of bases whereas the **dual system** $(E, \mathscr{H}(\mathscr{B}^*))$ is the family of subsets of basis complements. Observe that the dual system of a matroid is just the primal system of the dual matroid. Recall that an isthmus e of a matroid is an element which occurs in every basis.

Lemma 2.1. *Let $(E, \mathscr{F})$ be a greedoid and $e \in E$. Then*

(i) $(E \setminus e, \mathscr{H}(\mathscr{B}^*) \setminus e) = (E \setminus e, \mathscr{H}(\mathscr{B}/e)^*)$ *if $e \in \mathscr{F}$,*

(ii) $(E \setminus e, \mathscr{H}(\mathscr{B}^*) \triangle e) = (E \setminus e, \mathscr{H}(\mathscr{B} \setminus e)^*)$ *if e is not an isthmus of $\mathscr{F}$,*

(iii) $(E, \mathscr{H}(\mathscr{B}^*)) = (E \setminus e, \mathscr{H}(\mathscr{B}/e)^*)$ *if $e \in \mathscr{F}$ and e is an isthmus of $\mathscr{F}$.*

Proof. (i) If $A \in \mathscr{H}(\mathscr{B}/e)^*$, then $A \subseteq E \setminus B$ for some $B \in \mathscr{B}$ with $e \in B$. Hence $e \notin A$, i.e. $A \in \mathscr{H}(\mathscr{B}^*) \setminus e$. Conversely, let $A \in \mathscr{H}(\mathscr{B}^*) \setminus e$, that is $e \notin A$ and $A \subseteq E \setminus B$ for some $B \in \mathscr{B}$. If $e \in B$, we are done. Otherwise $e \in \mathscr{F}$ can be augmented from B to some $B' \in \mathscr{B}$. Then $A \subseteq E \setminus B'$ and hence $A \in \mathscr{H}(\mathscr{B}/e)^*$.

(ii) Observe that $A \in \mathscr{H}(\mathscr{B}^*) \triangle e$ is equivalent to $A \subseteq E \setminus (B \cup e)$ for some $B \in \mathscr{B}$, $e \notin B$, which again is equivalent to $A \in \mathscr{H}(\mathscr{B} \setminus e)^*$.

(iii) follows from (i) since $(E, \mathscr{H}(\mathscr{B}^*)) = (E \setminus e, \mathscr{H}(\mathscr{B}^*) \setminus e)$ for any isthmus $e \in \mathscr{F}$.   □

Using this lemma we can now prove

Theorem 2.2. *The dual system of a greedoid is vertex-decomposable and hence shellable.*

Proof. By the above lemma we may assume that any feasible $e \in \mathscr{F}$ is not an isthmus of $\mathscr{F}$. Then for any $e \in \mathscr{F}$, the deletion and the link of the dual system with respect to e are again dual systems of greedoids and hence vertex-decomposable by induction.   □

A particular shelling of the dual system can be obtained by using an arbitrary linear order $\leq$ on the ground set E. This induces a linear order (the lexicographic order) on E^* by setting $x_1 \ldots x_k \leq y_1 \ldots y_l$ if either $x_1 \ldots x_k$ is a prefix of $y_1 \ldots y_l$ or for the least i with $x_i \neq y_i$ we have $x_i \leq y_i$. For each $B \in \mathscr{B}$, let α_B be the lexicographically smallest feasible ordering of B (note that this ordering can easily be constructed by the greedy algorithm).

Lemma 2.3. *Every linear ordering of E induces a shelling of $(E, \mathscr{H}(\mathscr{B}^*))$ by*

$$(E - B_1) < (E - B_2) \text{ if and only if } \alpha_{B_1} < \alpha_{B_2} \text{ for } B_1, B_2 \in \mathscr{B}.$$

Proof. In terms of the primal system $(E, \mathcal{H}(\mathcal{B}))$, we have to show that for any two bases $B_1, B_2 \in \mathcal{B}$ with $\alpha_{B_1} < \alpha_{B_2}$ there exists a $B_3 \in \mathcal{B}$ with $\alpha_{B_3} < \alpha_{B_2}$, $B_3 \cup B_2 \subseteq B_1 \cup B_2$ and $|B_3 \cup B_2| = |B_2| + 1$.

Let $\alpha_{B_1} = \beta x \gamma$ and $\alpha_{B_2} = \beta y \delta$ for some $x \neq y$. Then $x < y$ by the choice of B_1 and B_2. Augment βx from $\beta y \delta$ to obtain a word $\beta x \epsilon$. Since $\beta x \epsilon < \beta y \delta = \alpha_{B_2}$, we must have $x \notin B_2$. Let $B_3 \in \mathcal{B}$ be the set underlying $\beta x \epsilon$. Then

$$\alpha_{B_3} \leq \beta x \epsilon < \beta y \delta = \alpha_{B_2} \text{ and } B_3 \cup B_2 = B_2 \cup x \subseteq B_1 \cup B_2 . \qquad \square$$

The duality for matroids implies that the primal system of a matroid, i.e. the matroid itself, is shellable. Hence also the primal systems of exchange systems and Gauss greedoids (cf. Chapter IX) are shellable. For greedoids in general this is not true. There is, however, one other class whose primal system is shellable.

Theorem 2.4. *The primal system of a balanced interval greedoid is shellable.*

Proof. For given $B_1, B_2 \in \mathcal{B}$, let B_3 be constructed as in the proof of Lemma 2.3. Then $B_3 \cap B_2 = B_3 \setminus x$. Assume $x \in E_j$, the j-th block of the partition of the ground set, and let $B_2 \cap E_j = y$. Then $B_2 \cap B_3 = B_2 \setminus z$. Since $x \in B_1$, we must have $z \notin B_1$. Hence $B_1 \cap B_2 \subseteq B_2 \setminus z = B_2 \cap B_3$. $\qquad \square$

Theorem 2.5. *Let $(E, \mathcal{F})$ be a balanced interval greedoid with respect to the partition $E = E_1 \cup \ldots \cup E_k$. Suppose there exists a subset $A \subseteq \Gamma(\emptyset)$ such that $A \cap B \neq \emptyset$ for all $B \in \mathcal{B}$ and $|A \cap E_i| \leq 1$ for $1 \leq i \leq k$. Then the primal complex of $(E, \mathcal{F})$ is contractible.*

Proof. Let the ordering of E start with the elements of A and consider the construction in the proof of the previous theorem. For $B \in \mathcal{B}$, let $x \in E_j$ be the smallest element in $B \cap A$. Then x is also the first letter in α_B. Suppose $B \setminus x \subseteq B' \in \mathcal{B}$. Then $B' = B \setminus x \cup y$ for some $y \in E_j$, and if $x \neq y$, then by assumption $y \notin A$. Hence $\alpha_B < \alpha_{B'}$.

In terms of the shelling of the primal system, this implies $x \in R(B)$, i.e. $B \neq R(B)$. Thus $h_r = 0$, where r is the rank of $(E, \mathcal{F})$. Then Theorems 1.5 and 1.6 imply that $(E, \mathcal{H}(\mathcal{B}))$ is contractible. $\qquad \square$

For matroids, Tutte [1954] introduced a polynomial which is closely related to enumerative questions as well as to shellability properties. We now extend (at least partially) this concept to greedoids.

Again, fix an ordering of the ground set. For $B \in \mathcal{B}$ and $x \notin B$ call x **externally active in B** if

$$\alpha_B < \alpha_{B \setminus y \cup x} \text{ for all } y \in B \text{ with } B \setminus y \cup x \in \mathcal{B} .$$

If $B \setminus y \cup x \notin B$ for all $y \in B$, then x is by definition externally active. The **external activity** $e(B)$ is the number of externally active elements in B.

Theorem 2.6. *Let $(E, \mathcal{F})$ be a greedoid of rank r and let $(h_0, \ldots, h_{n-r})$ be the h-vector of $(E, \mathcal{H}(\mathcal{B}^*))$. Then*

$$\sum_{B \in \mathcal{B}} t^{e(B)} = \sum_{i=0}^{n-r} h_i t^{n-r-i} .$$

Proof. Consider any ordering of the ground set and the induced shelling of $(E, \mathcal{H}(\mathcal{B}^*))$. Let $x \in R(E \setminus B)$ for some $B \in \mathcal{B}$, i.e.

$$x \in E \setminus B \text{ and } E \setminus (B \cup x) \subseteq E \setminus B'$$

for some $B' \in \mathcal{B}$ such that $\alpha_{B'} < \alpha_B$. Equivalently, $B' = B \setminus y \cup x$ for some $y \in B$. Hence x is not externally active in B.

Since obviously,

$$\{x \in E \setminus B : x \text{ is not externally active in } B\} \subseteq R(E \setminus B) ,$$

we get

$$|R(E \setminus B)| = n - r - e(B)$$

and the desired equality between the two polynomials follows.   □

We define the **greedoid polynomial** $\lambda_{\mathcal{F}}(t)$ by

$$\lambda_{\mathcal{F}}(t) = \sum_{B \in \mathcal{B}} t^{e(B)} .$$

By the above theorem, $\lambda_{\mathcal{F}}$ is independent of the particular ordering of the ground set. For matroids, the Tutte polynomial has two variables. For greedoids, only one variable seems to make sense since the dual complex is shellable but the primal is not in general.

Besides interpreting the polynomial in terms of external activities and in terms of the h-vector, we may also express it in terms of **spanning sets,** i.e. sets containing a basis of the greedoid.

Theorem 2.7. *Let $(E, \mathcal{F})$ be a greedoid of rank r and s_j the number of spanning sets of cardinality j, $r \leq j \leq n$. Then*

$$\lambda_{\mathcal{F}}(1 + t) = \sum_{i=0}^{n-r} s_{r+i} t^i .$$

Proof. By Theorem 1.5, the coefficient of t^i in $\lambda_{\mathcal{F}}(1 + t)$ counts the number of independent sets X in the dual system for which $|X| = n - r - i$. Hence $E \setminus X$ is spanning of size $r + i$.   □

As an immediate consequence we obtain the following results:

Corollary 2.8.

(i) $\lambda_{\mathscr{F}}(0) = (-1)^{n-r-i}(\lambda(\mathscr{H}(\mathscr{B}^*)) - 1)$,
(ii) $\lambda_{\mathscr{F}}(1)$ is the number of bases of $\mathscr{F}$,
(iii) $\lambda_{\mathscr{F}}(2)$ is the number of spanning sets of $\mathscr{F}$. $\square$

Like its matroid analogue, the greedoid polynomial can be computed recursively.

Theorem 2.9. Let $(E, \mathscr{F})$ be a greedoid and $e \in \mathscr{F}$. Then

(i) $\lambda_{\mathscr{F}}(t) = \lambda_{\mathscr{F}/e}(t) + \lambda_{\mathscr{F}\backslash e}(t)$ if e is not an isthmus,
(ii) $\lambda_{\mathscr{F}}(t) = \lambda_{\mathscr{F}/e}(t)$ if e is an isthmus,
(iii) $\lambda_{\mathscr{F}}(t) = \lambda_{\mathscr{F}_1}(t) \cdot \lambda_{\mathscr{F}_2}(t)$ if $\mathscr{F}$ is the direct sum of $(E_1, \mathscr{F}_1)$ and $(E_2, \mathscr{F}_2)$.

Proof. (i) Let the ordering of E start with e and denote by $e_1(B)$ and $e_2(B)$ the external activity with respect to $G \backslash e$ and G/e respectively. Since e is feasible and not an isthmus, we get

$$\lambda_{\mathscr{F}}(t) = \sum_{e \in B} t^{e(B)} + \sum_{e \notin B} t^{e(B)}$$

$$= \sum_{B' \in \mathscr{B}\backslash e} t^{e_1(B')} + \sum_{B' \in \mathscr{B}/e} t^{e_2(B')}$$

$$= \lambda_{\mathscr{F}\backslash e}(t) + \lambda_{\mathscr{F}/e}(t) .$$

(ii) As before, with the observation that

$$\sum_{e \notin B} t^{e(B)} = 0 .$$

(iii) Let s_j^1 and s_j^2 be the number of j-element spanning sets in $(E_1, \mathscr{F}_1)$ and $(E_2, \mathscr{F}_2)$. A set $A \subseteq E_1 \cup E_2$ is spanning in $\mathscr{F}_1 + \mathscr{F}_2$ if and only if $A = A_1 \cup A_2$, where A_i is spanning in $\mathscr{F}_i$, $i = 1, 2$. Hence

$$s_j = \sum_{i=r_1}^{|E_1|} s_i^1 s_{j-i}^2 \quad \text{for} \quad r_1 + r_2 \leq j \leq |E_1 \cup E_2| .$$

Thus

$$\lambda_{\mathscr{F}}(1 + t) = \lambda_{\mathscr{F}_1}(1 + t) \cdot \lambda_{\mathscr{F}_2}(1 + t) .$$ $\square$

The above three rules allow us to decompose the greedoid recursively into smaller greedoids and, at the same time, update the greedoid polynomial. The irreducible greedoids with respect to rules (i) and (ii) are those greedoids where all elements are loops. These can then be decomposed into the direct sum of 1-element greedoids of rank 0 by applying rule (iii).

Example. Let $(E, \mathscr{F})$ be the directed branching greedoid of the directed graph in Figure 17. Then $\lambda_{\mathscr{F}}(t) = t^4 + 2t^3 + t^2$.

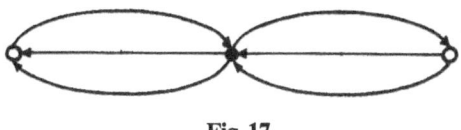

Fig. 17

As with the Tutte polynomial for matroids, we may also introduce a version of the Tutte invariants for greedoids. To this end, let f assign a complex value to every greedoid $(E, \mathscr{F})$. The function f is a **greedoid invariant** if it satisfies the following axioms:

(2.1.1) $f(\mathscr{F}_1) = f(\mathscr{F}_2)$ if $\mathscr{F}_1$ is isomorphic to $\mathscr{F}_2$,
(2.1.2) $f(\mathscr{F}) = f(\mathscr{F} \setminus e) + f(\mathscr{F}/e)$ if $e \in \mathscr{F}$ is not an isthmus,
(2.1.3) $f(\mathscr{F}) = f(\mathscr{F}/e)$ if $e \in \mathscr{F}$ is an isthmus,
(2.1.4) $f(\mathscr{F}) = f(\mathscr{F}_1) \cdot f(\mathscr{F}_2)$ if $\mathscr{F} = \mathscr{F}_1 + \mathscr{F}_2$,
(2.1.5) $f(\mathscr{F}) = 1$ if $E = \{e\}$ and $\mathscr{F} = \{\emptyset, e\}$.

Since a greedoid invariant has the same recursive behaviour as the greedoid polynomial and since we can substitute for t in the greedoid polynomial so that they agree on irreducibles, we have the following:

Theorem 2.10. *Every greedoid invariant is an evaluation of the greedoid polynomial.*
□

As mentioned before, shellable independence systems came up in two fields of application: first in percolation theory and network reliability analysis (cf. Oxley and Welsh [1979], Ball and Provan [1982]) and secondly in the investigation of certain oracle algorithms (cf. Kahn, Saks and Sturtevant [1984]).

As an illustration of the first application, suppose that in a rooted directed graph (a communication network) the links fail with a certain probability p and operate with probability $1 - p$. What is the probability that a message from the root reaches every node ?

Translated into our framework, suppose that the elements $e \in E$ of a greedoid $(E, \mathscr{F})$ are colored red with probability p and blue with probability $1 - p$. What is the probability $\pi_{\mathscr{F}}(p)$ that the set of blue elements is spanning?

Theorem 2.11. *Let $(E, \mathscr{F})$ be a greedoid of rank r. Then*

$$\pi_{\mathscr{F}}(p) = (1 - p)^r p^{n-r} \lambda_{\mathscr{F}}(p^{-1}) .$$

Proof. Set

$$f(\mathscr{F}) = (1 - p)^{-r} p^{r-n} \pi_{\mathscr{F}}(p)$$

and observe that if $|E| = 1$ and $\mathscr{F} = \{\emptyset\}$, then $f(\mathscr{F}) = p^{-1}$. So by Theorem 2.10, the result follows if f is a greedoid invariant.

Assume $e \in \mathscr{F}$ is not an isthmus. Then the probability that the blue elements are spanning is the sum of the corresponding probabilities in $\mathscr{F}/e$ if e is blue and in $\mathscr{F} \setminus e$ if e is red. Hence

$$\pi_{\mathscr{F}}(p) = (1-p)\pi_{\mathscr{F}/e}(p) + p\pi_{\mathscr{F}\setminus e}(p) ,$$

or equivalently

$$f(\mathscr{F}) = f(\mathscr{F}/e) + f(\mathscr{F} \setminus e) .$$

This shows that f satisfies (2.1.2), and the other conditions of (2.1) are verified similarly. Hence f is a greedoid invariant. $\square$

For the second application, we consider the following problem: given a greedoid $(E,\mathscr{F})$ and a fixed but unknown subset $X \subseteq E$, decide whether X is spanning by asking questions of the form: "Is $e \in X$?". The greedoid is called **non-evasive** if there exists an algorithm which requires less than $|E|$ questions in the worst-case. For example, a free matroid is trivially evasive.

The following result characterizes non-evasive greedoids.

Theorem 2.12. *For a greedoid $(E,\mathscr{F})$ the following statements are equivalent:*

(i) $(E,\mathscr{F})$ *is non-evasive.*
(ii)   *The dual system $(E, \mathscr{H}(\mathscr{B}^*))$ is contractible.*
(iii) $\lambda_{\mathscr{F}}(0) = 0$.

Proof. A subset $X \subseteq E$ is spanning if and only if $E \setminus X \in \mathscr{H}(\mathscr{B}^*)$. Hence $(E,\mathscr{F})$ is non-evasive if and only if $(E, \mathscr{H}(\mathscr{B}^*))$ is non-evasive. Since by Theorem 2.2 the dual system is vertex-decomposable, Theorem 1.10 implies that $(E, \mathscr{H}(\mathscr{B}^*))$ is non-evasive if and only if $\chi(\mathscr{H}(\mathscr{B}^*)) = 1$. By Theorem 1.6, this is equivalent to (ii) and by Corollary 2.8, it is equivalent to (iii). $\square$

For an antimatroid $(E,\mathscr{F})$ we have considered one further natural independence system, namely the family $\mathscr{H}$ of the free convex sets. We call $(E,\mathscr{H})$ the **free convex set system** of $(E,\mathscr{F})$.

Theorem 2.13. *The free convex set system of an antimatroid is non-evasive (and hence contractible).*

Proof. This is clear if $|E| = 1$. So let $|E| > 1$, let e be a feasible singleton and assume e is convex. Then it is straightforward to check that $(E \setminus e, \mathscr{H} \setminus e)$ consists of the free convex sets of $(E \setminus e, \mathscr{F}/e)$ and $(E \setminus e, \mathscr{H} \Delta e)$ consists of the free convex sets of $(E \setminus e, \mathscr{F} \setminus e)$. Hence $(E, \mathscr{H})$ is non-evasive by Lemma 1.9. $\square$

This theorem implies that there is a way to decide with fewer than $|E|$ questions whether a "hidden" set $X \subseteq E$ is free and convex; we leave the formulation of such an algorithm to the reader.

Also note that if the antimatroid has Helly number 2, then this theorem implies that its nonempty free convex sets are the edges and nodes of a tree.

3. Homotopy Properties of Greedoids

In this section we link the combinatorial connectivity of a greedoid to the topological connectivity of certain associated simplicial complexes. Let $\mathcal{C}\!h(\mathcal{F}')$ and $\mathcal{C}\!h(\mathcal{D}')$ be the chain complexes of the nonempty feasible and dominating sets, respectively.

Recall from Lemma VI.2.1 that if $(E, \mathcal{F})$ is an interval greedoid, $\mathcal{M}_\emptyset = \{X \subseteq \Gamma(\emptyset) : X \in \mathcal{F}\}$ is a matroid.

Theorem 3.1. *Let $(E, \mathcal{F})$ be an interval greedoid. Then the chain complex $\mathcal{C}\!h(\mathcal{F}')$ is homotopy equivalent to $\mathcal{M}_\emptyset$.*

Proof. By Theorem VI.1.4, the feasible sets together with E form a lattice. If $E \notin \mathcal{F}$, then formally add a new maximal element to get a lattice L. The atoms of L are precisely the feasible singletons $e \in \Gamma(\emptyset)$. Clearly the set of atoms is a cross-cut. By Lemma VI.2.3, its induced independence system equals $\mathcal{M}_\emptyset$. On the other hand it is homotopy equivalent to the chain complex of the nonempty feasible sets by Theorem 1.3. □

Corollary 3.2. *Let $(E, \mathcal{F})$ be a k-connected interval greedoid with $k \geq 2$. Then the chain complex $\mathcal{C}\!h(\mathcal{F}')$ is topologically $(k-2)$-connected.*

Proof. By the above theorem the matroid $\mathcal{M}_\emptyset$ has rank k. Since $\mathcal{M}_\emptyset$ is shellable, $\mathcal{M}_\emptyset$ and hence $\mathcal{C}\!h(\mathcal{F}')$ is topologically $(k-2)$-connected by Theorem 1.4. □

The result for dominating sets is similar. Note that if the greedoid is an antimatroid, then dominating sets are just complements of free convex sets. In this case the chain complex $\mathcal{C}\!h(\mathcal{D}')$ is a cone and hence contractible. But even if the set E is omitted, the chain complex of dominating sets will remain contractible by Theorem 2.13.

Theorem 3.3. *Let $(E, \mathcal{F})$ be a k-connected interval greedoid with $k \geq 2$. Then the chain complex $\mathcal{C}\!h(\mathcal{D}')$ is topologically $(k-2)$-connected.*

Proof. We apply induction on $|E|$ and, for $x \in \Gamma(\emptyset)$, set

$$\mathcal{D}'(x) = \{A \in \mathcal{D}' : x \in A\} .$$

We want to show that the set systems $\mathcal{D}'(x)$, $x \in \Gamma(\emptyset)$, satisfy the conditions of Theorem 1.2.

For this purpose let $\{A_1, \ldots, A_j\} \in \mathcal{C}\!h(\mathcal{D}')$ be a chain of dominating sets, i.e. $A_1 \subset A_2 \subset \ldots \subset A_j$. Then since $A_1 \in \mathcal{F}$, it contains a feasible element $x \in \Gamma(\emptyset)$. Hence $A_i \in \mathcal{D}'(x)$ for all $1 \leq i \leq j$ and

$$\mathcal{C}\!h(\mathcal{D}') \subseteq \bigcup \{2^{\mathcal{D}'(x)} : x \in \Gamma(\emptyset)\} .$$

Secondly, let $X = \{x_1, \ldots, x_j\} \subseteq \Gamma(\emptyset)$ be arbitrary and assume that

$$\mathscr{D}'(X) = \mathscr{D}'(x_1) \cap \ldots \cap \mathscr{D}'(x_j) = \{A \in \mathscr{D}' : X \subseteq A\} \neq \emptyset .$$

We claim that

$$\mathscr{C}h(\mathscr{D}') \cap 2^{\mathscr{D}'(X)} = \mathscr{C}h(\mathscr{D}'(X))$$

is $(k-2)$-connected.

Consider $A \in \mathscr{D}'(X)$, then in particular $A \in \mathscr{F}$. By Lemmas VI.2.1 and VI.2.3, $X \subseteq A \cap \Gamma(\emptyset)$ is feasible. More precisely, this shows

(3.1) $\mathscr{D}'(x_1) \cap \ldots \cap \mathscr{D}'(x_j) \neq \emptyset$ if and only if $\{x_1, \ldots, x_j\} \in \mathscr{M}_\emptyset$.

Since connectivity is preserved under contraction, $(E \setminus X, \mathscr{F}/X)$ is also k-connected. Obviously the ordered set $\mathscr{D}'(X)$ is isomorphic to the dominating sets in the contraction $(E \setminus X, \mathscr{F}/X)$. So the claim follows by induction if $X \notin \mathscr{D}$. But if $X \in \mathscr{D}$, then $\mathscr{C}h(\mathscr{D}'_X)$ is a cone with apex X. Hence $\mathscr{C}h(\mathscr{D}'_X)$ is contractible and thus topologically k-connected for every k.

In view of Theorem 1.2, it suffices to prove that the nerve $\mathscr{N}(\{\mathscr{D}'_x : x \in \Gamma(\emptyset)\})$ is topologically $(k-2)$-connected. By (3.1) this independence system is isomorphic to $\mathscr{M}_\emptyset$. The k-connectivity implies that $\mathscr{M}_\emptyset$ has rank at least k, so it is topologically $(k-2)$-connected.   □

Theorem 3.4 *Every independence system that is the free convex set system of an antimatroid $(E, \mathscr{F})$ with Helly number h, has homotopy dimension at most $h-2$. (Note that by definition, this free set complex has dimension $h-1$.)*

Proof. Assume that $h \geq 3$; for $h = 2$ the proof follows with slight modifications. Let e be a feasible singleton in $(E, \mathscr{F})$. Then, as we saw earlier, $\mathscr{H} \triangle e$ is the free convex set complex of $\mathscr{F} \setminus e$ and $\mathscr{H} \setminus e$ is the free convex set complex of $\mathscr{F}/e$. Hence by induction we may assume that $\mathscr{H} \triangle e$ has homotopy dimension at most $h-3$ and $\mathscr{H} \setminus e$ has homotopy dimension at most $h-2$. Then by Theorem 1.7, $\mathscr{H}$ has homotopy dimension at most $h-2$.   □

Theorem 3.5 *Let $A_1, \ldots, A_m$ be convex sets in an antimatroid $(E, \mathscr{F})$. Then the nerve $\mathscr{N}(\{A_1, \ldots, A_m\})$ is homotopy equivalent to a subsystem of the free convex set system of $(E, \mathscr{F})$. In particular, the homotopy dimension of $\mathscr{N}(\{A_1, \ldots, A_m\})$ is at most the Helly number of $(E, \mathscr{F})$ minus 2.*

Proof. Let $\mathscr{H}_i$ be the collection of free convex sets contained in A_i. Then by Theorem 2.15, each $\mathscr{H}_i$ is contractible and so is $\mathscr{H}_i \cap \ldots \cap \mathscr{H}_{i_r}$ provided that $A_{i_1} \cap \ldots \cap A_{i_r} \neq \emptyset$. Hence the Nerve theorem applies and we get that $\mathscr{N}(\{A_1, \ldots, A_m\})$ is homotopy equivalent to $\mathscr{H}_1 \cup \ldots \cup \mathscr{H}_m$.   □

This theorem may be viewed as a topological generalization of Helly's theorem. In fact, if there were $h+1$ convex sets $A_0, \ldots, A_h$ such that $\bigcap_i A_i = \emptyset$ but $\bigcap_{i \neq i_0} A_i \neq \emptyset$ for each $0 \leq i_0 \leq h$, then $\mathscr{N}(\{A_0, \ldots, A_h\})$ would be the boundary of an h-dimensional simplex, contradicting the theorem.

To give another application of this result, recall from Chapter III that if a poset P has order dimension d, then it can be represented by convex sets

$\{A_i : i \in P\}$ in an antimatroid with Helly number (in fact, Erdös-Szekeres number) at most d. Assume that $P = L \setminus \{0, 1\}$ is a truncated lattice. Then the nerve $\mathcal{N}(\{A_i : i \in P\})$ is homotopy equivalent to $\mathscr{C}\!h(P)$ (this follows by applying the Nerve theorem to the family of subcomplexes $\{\mathscr{C}_i : i \in P\}$, where $\mathscr{C}_i$ consists of all chains $A_{i_1} \subset \ldots \subset A_{i_k}$ such that $A_{i_k} \subseteq A_i$). So it follows that the homotopy dimension of $\mathscr{C}\!h(P)$ is at most $d - 2$. Thus we have proved the following result of Kahn and Ziegler [1987]:

Theorem 3.6 *Let P be a truncated lattice with order dimension $d \geq 2$. Then the homotopy dimension of $\mathscr{C}\!h(P)$ is at most $d - 2$.*   □

As a special case, consider the truncated lattice of faces of a d-dimensional convex polytope. The chain complex of this is homotopy equivalent to the boundary of the polytope, i.e. to the $(d-1)$-dimensional sphere (this follows easily from the Nerve theorem). Hence this chain complex has homotopy dimension at least $d - 1$. Thus the lattice of faces has order dimension at least $d + 1$.

References

Alexandroff, P., and Hopf, H. [1972]: Topologie. Springer, Berlin Heidelberg New York 1972

Anstee, R.P., and Farber M. [1984]: Characterization of totally balanced matrices. J. Alg. **5** (1984) 215–230

Ball, M.O., and Provan, J.S. [1982]: Bounds on the reliability polynomial for shellable independence systems. SIAM J. Alg. Discr. Methods **3** (1982) 166–181

Bell, D.E. [1977]: A theorem concerning the integer lattice. Stud. Appl. Math. **56** (1977) 187–188

Birkhoff, G. [1935]: Abstract linear dependence in lattices. Amer. J. Math. **57** (1935) 800–804

Birkhoff, G. [1967]: Lattice theory, 3rd edn. Amer. Math. Soc. Colloq. Publ. 25, Providence, USA

Björner, A. [1983]: On matroids, groups and exchange languages. In: L. Lovász and A. Recski (eds.) Matroid Theory and its Applications. Conference Proceedings, Szeged, September 1982. Colloquia Mathematica Societatis János Bolyai, Vol. 40, North-Holland, Amsterdam Oxford New York 1985, pp. 25–60

Björner, A. [1985]: Combinatorics and topology. Notices Amer. Math. Soc. **23** (1985) 339–345

Björner, A., Lovász, L., and Shor, P. [1988]: Chip-firing games on graphs. Preprint

Björner, A., Korte, B., and Lovász, L. [1985]: Homotopy properties of greedoids. Advances in Applied Mathematics **6** (1985) 447–494

Björner, A., and Ziegler, G.M. [1987]: Introduction to greedoids. Working Paper, August 1987

Bland, R.G. [1977]: A combinatorial abstraction of linear programming. J. Comb. Theory B **23** (1977) 33–57

Bollobás B. [1978]: Extremal graph theory. Academic Press, New York London 1978

Borůvka, O. [1926]: O jistém problému minimálnim (with extended abstract in German). Acta Societatis Scientiarium Naturalium Moravicae. Tomus III, Fasc. 3, Signatura: F**23** (1926) 37-58

Boyd, E. A., and Faigle, U. [1990]: An algorithmic characterization of antimatroids. Discr. Appl. Math. **28** (1990) 197–205

Brylawski, Th. H., and Dieter E. [1986]: Exchange systems. Preprint, University of North Carolina, Chapel Hill, USA

Chvátal, V., and Klincsek, G. [1980]: Finding largest convex subsets. Congressus numeratium **29** (1980) 453–460

Crapo, H. [1965]: Single-element extensions of matroids. J. Research Nat. Bureau of Standards **69B** (1965) 55-65

Crapo, H. [1982]: Selectors. A theory of formal languages, semimodular lattices and shelling processes. Adv. Math. **54** (1984) 233–277

Dietrich, B.L. [1987]: A circuit set characterization of antimatroids. J. Comb. Theory B **43** (1987) 314–321

Dietrich, B.L. [1989]: Matroids and antimatroids – a survey. Discrete Mathematics **78** (1989) 223–237

Dijkstra, E.W. [1959]: A note on two problems in connexion with graphs. Numerische Mathematik **1** (1959) 269-271

Dilworth, R.P. [1940]: Lattices with unique irreducible decompositions. Ann. Math. **41** (1940) 771–777

Dilworth, R.P. [1950]: A decomposition theorem for partially ordered sets. Ann. Math. **51** (1950) 161–166

Ding, L.Y., and Yue M.-Y. [1987]: On a generalization of the Rado-Hall theorem to greedoids. Asia-Pacific Journal of Operational Research **4** (1987) 28–38

Doignon, J.P. [1973]: Convexity in cristallographic lattices. J. Geom. **3** (1973) 71–85

Duffus, D., and Rival, I. [1976]: Crowns in dismantlable partially ordered sets. In: Colloquia Mathematica Societatis János Bolyai, 18. Combinatorics. Keszthely, 1976, pp. 271–292

Dunstan, F.D.J., Ingleton, A.W., and Welsh, D.J.A [1972]: Supermatroids. In: D.J.A. Welsh and D.R. Woodall (eds.) Combinatorics. The Institute of Mathematics and its Applications, 1972

Edelman, P.H. [1980]: Meet-distributive lattices and the anti-exchange closure. Algebra Universalis **10** (1980) 290-299

Edelman, P.H. [1986]: Abstract convexity and meet–distributive lattices. Contemporary Mathematics **57** (1986) 127–150

Edelman, P.H., and Jamison, R.E. [1985]: The theory of convex geometries. Geometriae dedicata **19** (1985) 247–270

Edelman, P.H., and Saks, M.E. [1988]: A combinatorial representation and convex dimension of convex geometries. Order **5** (1988) 23–32

Edmonds, J. [1965]: Paths, trees and flowers. Can. J. Math. **17** (1965) 449-467

Edmonds, J. [1967a]: Systems of distinct representatives and linear algebra. J. Research Nat. Bureau of Standards **71B** (1967) 241-245

Edmonds, J. [1967b]: Optimum branchings. J. Research Nat. Bureau of Standards **71B** (1967) 233-240

Edmonds, J. [1970]: Submodular functions, matroids and certain polyhedra. In: R. Guy, H. Hanani, N. Sauer, J. Schonheim (eds.) Combinatorial Structures and their Applications. Gordon and Breach (1970), pp. 69-87

Edmonds, J. [1971]: Matroids and the greedy algorithm. Mathematical Programming **1** (1971) 127–136

Edmonds, J. [1973]: Edge disjoint branchings. In: R. Rustin (ed.) Combinatorial Algorithms. Academic Press, New York 1973, pp. 91–96

Edmonds, J., and Fulkerson, D.R. [1970]: Bottleneck extrema. J. Comb. Theory **8** (1970) 299–306

Edmonds, J., and Giles, R. [1977]: A min-max relation for submodular functions on graphs. In: P.L. Hammer et al. (eds.) Ann. Discr. Math. **1** (1977) 185–204

Erdös, P., and Szekeres, G. [1935]: A combinatorial problem in geometry. Comp. Math. **2** (1935) 463–470

Faigle, U. [1979]: The greedy algorithm for partially ordered sets. Discr. Math. **28** (1979) 153–159

Faigle, U. [1980]: Geometries on partially ordered sets. J. Comb. Theory B **28** (1980) 26–51

Faigle, U. [1985]: Submodular combinatorial structures. Habilitation Thesis, University of Bonn, 1985

Faigle, U., Gierz, G., and Schrader, R. [1985]: Algorithmic approaches to setup minimization. SIAM J. Comput. **14** (1985) 954–965

Faigle, U., Goecke, O., and Schrader, R. [1986]: Church-Rosser decomposition in combinatorial structures. Preprint, Institut für Operations Research, Universität Bonn

Faigle, U., and Schrader, R. [1984]: Comparability graphs and order invariants. In: U. Pape (ed.) Graph theoretic concepts in computer science. Trauner, Linz, 1984, pp. 136–145

Faigle, U., and Schrader, R. [1987]: Setup optimization problems with matroid structure. Order 4 (1987) 43–54

Farber, M. [1983]: Characterizations of strongly chordal graphs. Discr. Math. 43 (1983) 173–189

Farber, M., and Jamison, R.E. [1986]: Convexity in graphs and hypergraphs. SIAM J. Alg. Discr. Methods 7 (1986) 433–444

Folkman, J., and Lawrence, J. [1978]: Oriented Matroids. J. Comb. Theory B 25 (1978) 199–236

Fujishige, S. [1991]: Submodular functions and optimization. Ann. Discrete Math. 47 (1991)

Fulkerson, D.R., and Gross, O.A. [1965]: Incidence matrices and interval graphs. Pacific J. Math. 15 (1965) 835–855

Gavril, F. [1972]: Algorithms for minimum coloring, maximum clique, minimum covering by cliques and maximum independent sets of a chordal graph. SIAM J. Comp. 1 (1972) 180–187

Gelfand, I.M., and Serganova, V.V. [1987]: Combinatorial geometry and strata of tori on homogeneous compact manifolds. Usp. Mat. Nauk 42 (1987) 107–134

Goecke, O. [1986]: Eliminationsprozesse in der kombinatorischen Optimierung – ein Beitrag zur Greedoid-Theorie. PhD Dissertation, Universität Bonn

Goecke, O. [1988]: A greedy algorithm for hereditary set systems and a generalization of the Rado-Edmonds characterization of matroids. Discr. Appl. Math. 20 (1988) 39–49

Goecke, O., Korte, B., and Lovász, L. [1989]: Examples and algorithmic properties of greedoids. In: Simeone, B. (ed.) Combinatorial optimization. Como 1986. Lecture Notes in Mathematics, vol. 1403. Springer, Berlin Heidelberg New York 1989, pp. 113–161

Goecke, O., and Schrader, R. [1990]: Minor characterization of undirected branching greedoids – a short proof. Discr. Math. 82 (1990) 93–99

Golumbic, M.C. [1980]: Algorithmic graph theory and perfect graphs. Academic Press, New York London San Francisco 1980

Golumbic, M.C., and Goss, M.F. [1978]: Perfect elimination and chordal bipartite graphs. J. Graph Theory 2 (1978) 155–163

Golumbic, M.C., and Jamison, R.E. [1985]: The edge intersection graphs of paths in a tree. J. Comb. Theory B 38 (1985) 8–23

Grätzer, G. [1978]: General lattice theory. Birkhäuser, Basel Stuttgart 1978

Graham, R.L., and Hell, P. [1985]: On the history of the spanning tree problem. Ann. History of Comp. 7 (1985) 43–57

Graham, R.L., Simonovits, M., and Sós, V.T. [1980]: A note on the intersection properties of subsets of integers. J. Comb. Theory A 28 (1980) 106–110

Greene, C., and Kleitman, D.J. [1976]: Strong versions of Sperner's theorem. J. Comb. Theory A 20 (1976) 80–88

Greene, C., and Markowsky, G. [1974]: A combinatorial test for local distributivity. IBM Technical Report No. RC 5129, 1974

Gröflin, H. [1987]: On switching paths polyhedra. Combinatorica 7 (1987) 193–204

Gröflin, H., and Liebling, Th. [1981]: Connected and alternating vectors: polyhedra and algorithms. Mathematical Programming 20 (1981) 233–244

Grötschel, M., Lovász, L., and Schrijver, A. [1981]: The ellipsoid method and its consequences in combinatorial optimization. Combinatorica 1 (1981) 169–197

Grötschel, M., Lovász, L., and Schrijver, A. [1988]: Geometric algorithms and combinatorial optimization. Springer, Berlin Heidelberg Tokyo New York 1988

Hall, P. [1935]: On representations of subsets. J. London Math. Soc. **10** (1935) 26–30

Hexel, E. [1988]: On shelling structures – support and sum. J. Inf. Process. Cybern. EIK **24** (1988) 6, pp. 293–305

Hoffman, A.J. [1974]: A generalization of max flow–min cut. Math. Programming **6** (1974) 352–359

Hoffman, A.J. [1978]: Some greedy ideas. IBM Research Report RC 7279 1978

Hoffman, A.J. [1979]: Binding constraints and Helly numbers. Ann. New York Acad. Sci. **319** (1979) 284–288

Huet, G. [1980]: Confluent reductions: Abstract properties and applications to term rewriting systems. J. Assoc. Comp. Mach. **27** (1980) 797–821

Jamison, R.E. [1970]: A development of axiomatic convexity. Technical Report 48, Clemson University Math., pp. 15–20

Jamison, R.E. [1981]: Partition numbers for trees and ordered sets. Pacific J. Math. **96** (1981) 115–140

Jamison, R.E. [1982]: A perspective on abstract convexity: Classifying alignments by varieties. in: D.C. Kay and M. Breen (eds.) Convexity and related combinatorial geometry. Marcel Dekker, New York 1982, pp. 113–150

Kahn, J., Saks, M., and Sturtevant, D. [1984]: A topological approach to evasiveness. Combinatorica **4** (1984) 297–306

Kahn, J., and Ziegler G.M. [1987]: Personal communication

Kay, D.C., and Womble, E.W. [1971]: Axiomatic convexity theory and relationships between the Caratheodory, Helly and Radon numbers. Pac. J. Math. **38** (1971) 471–485

Korte, B., and Lovász, L. [1981]: Mathematical structures underlying greedy algorithms. In: F. Gécseg (ed.) Fundamentals of Computation Theory. Lecture Notes in Computer Sciences vol. 117. Springer, Berlin Heidelberg New York 1981, pp. 205–209

Korte, B., and Lovász, L. [1983]: Structural properties of greedoids. Combinatorica **3** (1983) 359–374

Korte, B., and Lovász, L. [1984a]: Greedoids, a structural framework for the greedy algorithm. In: W.R. Pulleyblank (ed.) Progress in Combinatorial Optimization. Proceedings of the Silver Jubilee Conference on Combinatorics, Waterloo, June 1982, pp. 221–243. Academic Press, London New York San Francisco 1984

Korte, B., and Lovász, L. [1984b]: Shelling structures, convexity and a happy end. In: B. Bolobás (ed.) Graph theory and combinatorics, Proceedings of the Cambridge Combinatiorial Conference in Honour of Paul Erdös. Academic Press, London 1984, pp. 219–232

Korte, B., and Lovász, L. [1984c]: Greedoids and linear objective functions. SIAM Journ. Algebr. Discr. Methods **5** (1984) 229–238

Korte, B., and Lovász. L. [1985a]: A note on selectors and greedoids. Europ. J. Comb. **6** (1985) 59–67

Korte, B., and Lovász, L. [1985b]: Posets, matroids and greedoids. In: L. Lovász and A. Recski (eds.) Matroid theory and its applications. Conference Proceedings, Szeged, September 1982. Colloquia Mathematica Societatis János Bolyai, vol. 40, North-Holland, Amsterdam Oxford New York 1985, pp. 239–265

Korte, B., and Lovász, L. [1985c]: Polymatroid greedoids. J. Comb. Theory B **38** (1985), pp. 41–72

Korte, B., and Lovász, L. [1985d]: Relations between subclasses of greedoids. Zeitschrift für Operations Research, Series A **29** (1985) 249–267

Korte, B., and Lovász, L. [1985e]: Basis graphs of greedoids and two-connectivity. In: R.W. Cottle (ed.) Essays in Honor of George B. Dantzig, Part I. Mathematical Programming Study, no. 24, North-Holland, Amsterdam 1985, pp. 158–165

Korte, B., and Lovász, L. [1986a]: On submodularity in greedoids and a counterexample. Sezione di Matematica Applicata, Dipartimento di Matematica, Università di Pisa 128 (1986) 1–13

Korte, B., and Lovász, L. [1986b]: Non–inter 'al greedoids and the transposition property. Discr. Math. 59 (1986) 297–314

Korte, B., and Lovász, L. [1986c]: Homomorphisms and Ramsey properties of antimatroids. Discr. Appl. Math. 15 (1986) 283–290

Korte, B., and Lovász, L. [1988]: Intersections of matroids and antimatroids. Discr. Math. 73 (1988/89) 143–157

Korte, B. and Lovász, L. [1989]: Polyhedral results for antimatroids. In: Ann. New York Acad. Sci. 555 Combinatorial Mathematics: Proceedings of the Third International Conference on Combinatorial Mathematics, New York 1989, pp. 283–295

Korte, B., and Schrader, R. [1981]: On the existence of fast approximation schemes. In: O. Mangasarian, R. Meyer and S. Robinson (eds.) Nonlinear Programming 4 Academic Press. 1981, pp. 415–437

Kruskal, J.B. [1956]: On the shortest spanning subtree of a graph and the traveling salesman problem. Proc. American Math. Society 7 (1956) 48–50

Las Vergnas, M. [1975]: Matroides Orientables. C.R. Acad. Sci. Paris 280 (1975) 61–64

Lawler, E.L. [1973]: Optimal sequencing of a single machine subject to precedence contraints. Management Science 19 (1973) 544–546

Lehel, J. [1982]: Tree structures, representation and covering problems in hypergraphs. Thesis (in Hungarian)

Lehman, A. [1979]: On the width-length inequality. Mathematical Programming 16 (1979) 245–259

Levi, F.W. [1951]: On Helly's theorem and the axioms of convexity. J. Indian Math. Soc. 15 (1951) 65–76

Lovász, L. [1977]: A homology theory for spanning trees of a graph. Acta Mathematica Academiae Scientiarum Hungaricae 30, 3–4 (1977) 241–251

Lovász, L. [1979a]: Topological and algebraic methods in graph theory. In: J.A. Bondy and N.P.R. Murty (eds.) Graph theory and related topics. Academic Press, New York San Francisco London 1979, pp. 1–14

Lovász, L. [1979b]: Combinatorial problems and exercises. North-Holland, Amsterdam New York Oxford 1979

Lovász, L. [1983]: Submodular functions and convexity. In: A. Bachem, M. Grötschel, B. Korte (eds.) Mathematical Programming. The State of the Art. Springer, Berlin Heidelberg New York Tokyo 1983, pp. 235–257

MacLane, S. [1936]: Some interpretations of abstract linear dependence in terms of projective geometry. Amer. J. Math. 58 (1936) 236–240

Maurer, S.B. [1973]: Matroid basis graphs I. J. Comb. Theory B 14 (1973) 216–240

Monge, G. [1781]: Déblai et remblai. Mémoires de l'Academie des Sciences 1781

Monjardet, B. [1985]: A use for frequently rediscovering a concept. Order 1 (1985) 415–416

Newman, M.H.A. [1942]: On theories with a combinatorial definition of "equivalence". Ann. Math. 43 (1942) 223–243

Oxley, J.G., and Welsh, D.J.A. [1979]: Tutte polynomial and percolation. in: J.A. Bondy and U.P.R. Murty (eds.) Graph theory and related topics. Academic Press, New York San Francisco London 1979, pp. 329–339

Perfect, H. [1966]: Symmetrized form of P. Hall's theorem on distinct representatives. Quart. J. Math. (Oxford) 17 (1966) 303–306

Provan, J.S., and Billera, L.J. [1980]: Decompositions of simplicial complexes related to diameters of convex polyhedra. Math. Oper. Res. **5** (1980) 576–594

Rado, R. [1942]: A theorem on independence relations. Quart. J. Math. (Oxford) **13** (1942) 83–89

Rado, R. [1957]: Note on independence functions. Proc. London Math. Soc. **7** (1957) 300–320

Randow, R. von [1975]: Introduction to the theory of matroids. Springer, Berlin Heidelberg New York 1975

Recski, A. [1989]: Matroid theory and its applications. Springer, Berlin Heidelberg New York 1989

Scarf, H. [1977]: An observation of the structure of production sets with indivisibilities. Proc. Math. Acad. Sci. **74** (1977) 3637–3641

Schmidt, W. [1985a]: Strukturelle Aspekte in der kombinatorischen Optimierung: Greedoide auf Graphen. PhD Dissertation, Universität Bonn 1985

Schmidt, W. [1985b]: Greedoids and searches in directed graphs. Preprint, Institut für Operations Research, Universität Bonn

Schmidt, W. [1985c]: A min-max theorem for greedoids. Preprint, Institut für Operations Research, Universität Bonn

Schmidt, W. [1988]: A characterization of undirected branching greedoids. J. Comb. Theory B **45** (1988) 160–182

Schrijver, A. [1983]: Min–max results in combinatorial optimization. In: Mathematical Programming – The state of the art (A. Bachem, M. Grötschel, B. Korte, eds.) Springer, Berlin Heidelberg New York Tokyo 1983, pp. 439–500

Schrijver, A. [1986]: Theory of linear and integer programming. Wiley, Chichester, 1986

Spanier, E.H. [1966]: Algebraic topology. McGraw-Hill, New York 1966

Tardos, E. [1990]: An intersection theorem for supermatroids. J. Comb. Theory B **50** (1990) 150–159

Tutte, W.T. [1954]: A contribution to the theory of chromatic polynomials. Can. J. Math. **6** (1954) 80–91

Tutte, W.T. [1959]: Matroids and graphs. Trans. Amer. Math. Soc. **90** (1959) 527–552

Waerden, B.L. van der [1937]: Moderne Algebra, 2nd edn. Springer, Berlin 1937

Welsh, D.J.A. [1976]: Matroid theory. Academic Press, London New York San Francisco 1976

White, N. [1986]Theory of matroids. Cambridge University Press, Cambridge 1986

Whitney, H. [1935]: On the abstract properties of linear dependence. Amer. J. Math. **57** (1935) 509–533

Ziegler, G.M. [1988]: Branchings in rooted graphs and the diameter of greedoids. Combinatorica **8** (1988) 217–234

Notation Index

Author Index

Subject Index

Algorithms and Combinatorics

Editors: R. L. Graham, B. Korte, L. Lovász

Combinatorial mathematics has substantially influenced recent trends and developments in the theory of algorithms and its applications. Conversely, research on algorithms and their complexity has established new perspectives in discrete mathematics. This new series is devoted to the mathematics of these rapidly growing fields with special emphasis on their mutual interactions.

The series will cover areas in pure and applied mathematics as well as computer science, including: combinatorial and discrete optimization, polyhedral combinatorics, graph theory and its algorithmic aspects, network flows, matroids and their applications, algorithms in number theory, group theory etc., coding theory, algorithmic complexity of combinatorial problems, and combinatorial methods in computer science and related areas.

The main body of this series will be monographs ranging in level from first-year graduate up to advanced state-of-the-art research. The books will be conventionally type-set and bound in hard covers. In new and rapidly growing areas, collections of carefully edited monographic articles are also appropriate for this series. A subseries with the subtitle "Study and Research Texts" will be published in softcover and camera ready form. This will be mainly on outlet for seminar and lecture notes, drafts of textbooks with essential novelty in their presentation, and preliminary drafts of monographs. Refereed proceedings of meetings devoted to special topics may also be considered for this subseries. The main goals of this subseries are very rapid publication and the wide dissemination of new ideas.

Prospective readers of the series ALGORITHMS AND COMBINATORICS include scientists and graduate students working in discrete mathematics, operations research and computer science.

Volume 1

K. H. Borgwardt

The Simplex Method

A Probabilistic Analysis

1987. XI, 268 pp. 42 figs. in 115 sep. illustrations.
Softcover DM 85,–
ISBN 3-540-17096-0

Springer-Verlag
Berlin
Heidelberg
New York
London
Paris
Tokyo
Hong Kong
Barcelona
Budapest

Algorithms and Combinatorics

Editors: R. L. Graham, B. Korte, L. Lovász

Springer-Verlag
Berlin
Heidelberg
New York
London
Paris
Tokyo
Hong Kong
Barcelona
Budapest